Testing during Concrete Construction

Other RILEM Proceedings available from Chapman and Hall

Testing during Concrete Construction

Proceedings of the International Workshop held by RILEM
(The International Union of Testing and Research Laboratories
for Materials and Structures) and organized by the Institut für
Massivbau, Technical University of Darmstadt, Germany.

Mainz
March 5–7, 1990

EDITED BY

H. W. Reinhardt

CHAPMAN AND HALL

LONDON · NEW YORK · TOKYO · MELBOURNE · MADRAS

MOUNT PLEASANT LIBRARY
TEL. 051 207 3581 Ext. 3701.

UK   Chapman and Hall, 2–6 Boundary Row, London SE1 8HN

USA   Van Nostrand Reinhold, 115 5th Avenue, New York NY10003

JAPAN   Chapman and Hall Japan, Thomson Publishing Japan, Hirakawacho
              Nemoto Building, 7F, 1-7-11 Hirakawa-cho, Chiyoda-ku, Tokyo 102

AUSTRALIA   Chapman and Hall Australia, Thomas Nelson Australia, 102 Dodds
              Street, South Melbourne, Victoria 3205

INDIA   Chapman and Hall India, R. Seshadri, 32 Second Main Road, CIT
              East, Madras 600 035

First edition 1991

© 1991 RILEM

Printed in Great Britain
at the University Press, Cambridge

ISBN 0 412 39270 4   0 442 313 896 (USA)

All rights reserved. No part of this publication may be reproduced or
transmitted, in any form or by any means, electronic, mechanical,
photocopying, recording or otherwise, or stored in any retrieval
system of any nature, without the written permission of the copyright
holder and the publisher, application for which shall be made to the
publisher.
The publisher makes no representation, express or implied, with
regard to the accuracy of the information contained in this book and
cannot accept any legal responsibility or liability for any errors or
omissions that may be made.

British Library Cataloguing in Publication Data
Available

Library of Congress Cataloging-in-Publication Data
Available

Contents

Preface

This volume is the result of a three day workshop at Mainz in Germany from 5–7 March 1990 organized under the auspices of RILEM, the International Union of Testing and Research Laboratories for Materials and Structures. Specialists from various fields were invited to discuss the subject of testing during concrete construction. This was an interdisciplinary task since it involves testing methods, materials knowledge, building construction technology, statistics and quality assurance schemes.

I was very glad that most of the invited persons had agreed to prepare a written contribution in advance of the workshop, to present their papers, to serve as a chairman, and, what was very important, to discuss the material with great enthusiasm. The participants came from contractors, material producers, research laboratories, universities and building authorities. This means that there were representatives of all parties present which are involved in the building process and in testing. I would like to thank all contributors for their very valuable effort.

Special thanks are due to the members of the Scientific Committee, Professor H.K. Hilsdorf, University of Karlsruhe, Dr C.D. Pomeroy, British Cement Association, Wexham Springs, and Professor G. Schickert, Federal Institute for Materials Research and Testing, Berlin, for their useful advice during the preparations for the workshop.

The Deutsche Forschungsgemeinschaft (German Research Community DFG) supported the workshop very generously and the Deutscher Akademischer Austauschdienst (German Academic Exchange Service DAAD) facilitated the stay of our colleagues from East European countries. This encouragement and support is gratefully acknowledged.

Without the assistance of several persons the workshop would not have taken place and this volume would not have been produced. Especially the tireless efforts of Mrs B. Leinberger with the typing and layout of the papers and the efficient organization by Mr T. Fehlhaber are higherly appreciated.

Finally, thanks are due to Mr J.N. Clarke, Senior Editor in the E. & F.N. Spon group of Chapman and Hall, for sympathetic co-operation in editing the book.

H.W. Reinhardt
Darmstadt, March 1990

PART ONE
INTRODUCTION

1 RELEVANCE OF TESTING DURING CONSTRUCTION

H.W. REINHARDT
Darmstadt University of Technology, Darmstadt, Germany

Abstract
After a short consideration of quality control, those fields are discussed which need testing during concrete construction. They are related to concrete, reinforcing and prestressing elements, and concrete repair. It is emphasized that there is a special demand for non- destructive methods which allow quick storage and processing of data.

1 Introduction

The design of a structure is based on the assumption of certain properties of the materials to be used. The main property is strength which should lead to a sufficient capacity to carry dead, imposed, and wind loads. The actual strength of a material is a variable quantity which depends on variations in mix constituents, placing and compaction, hardening temperature, curing, and on the testing procedure; the bearing capacity of a structural component varies as function of material strength, placing of bars in a reinforced section, and geometry of the cross-section.

It is common practice to measure compressive strength on concrete cubes or cylinders which are produced during the construction process and stored according to the standard. If the 28-days values satisfy the requirements of the prescribed strength, compliance with the standard is assumed. Of course, there are preceding tests on the constituents of concrete: the cement is controlled and has to meet the specification of the cement standard, aggregates have to comply with specifications, admixtures and additives have to be approved and have to meet the requirements. The same is true for reinforcing and prestressing steel, for chairs, ducts, and other parts which are used in the structure. However, these tests are performed outside the building process, the properties are recorded by certificates.

Properties of fresh concrete which are sometimes explicitely specified and measured are cement content, water/ cement ratio, workability, air content, temperature, and density. Some of these properties are directly related to properties of hardened concrete - like entrained air content to freeze-thaw resistance - or indirectly related

- like workability to appropriate placing and compaction and thus to strength and durability.

Continuous concrete production allows continuous measurements of strength at various concrete ages. Suppliers of ready-mixed concrete may assess continually testing. Using previously established relations between 28-day and 24-hour accelerated strength, it is possible to control the estimated strength at an early age. This procedure is very useful in steering the building process into a quality conscious direction and to avoid faults and errors at an early stage.

All of these measures mentioned so far aim at the compliance with standard specifications and at the reduction of variability. This procedure is inherent to quality control and is absolutely necessary for quality products. Most of the quality control measures are concerned with strength. However, there are more properties which determine the overall quality of a structure and which are not covered by the usual procedure.

2 Needs for testing

2.1 Fresh concrete
In case of designed mixes, the design specifies the performance, and the concrete producer determines the actual mix proportions. The producer may also use admixtures and additives in order to improve some properties of the concrete such as workability of fresh concrete and impermeability of hardened concrete. He may also use another tpye of cement than originally specified. In general, there are many questions with respect to the proper composition of concrete which may arise and which can be solved only by testing. Producer, contractor, and owners can discuss matters objectively if they have objective data from tests. Consumer and producer have to take some risk as the quality is concerned, but there should be agreement between the parties as the material used is concerned.

2.2 From fresh to hardened concrete
In the transition between fresh concrete and hardened concrete there are many aspects which govern the construction progress, the cost and the quality of a concrete structure. From setting measurements, conclusions can be drawn with respect to the type of cement, water/cement ratio, and expected hardening. These measurements are also sensitive to temperature changes during the construction of a large project. A non-destructive testing method, which is accurate, easy to handle and continuous in measuring, would be very helpful.

Concrete pavements, pretensioned hollow-core slabs, pipes, paving stones, and other structures or products rely upon the stability and strength of the green concrete. Appropriate compaction is necessary for an accurate geometry and for sufficient strength in the hardened state. Since the density is in close correlation with strength

- for a given concrete mix - any variation in density should be monitored directly after placing of concrete. If too large deviations from the specified property occur, the production and/or placing and compaction procedure can be adjusted.

Strength development depends strongly on temperature. In a thick walled element, temperature rises due to heat of hydration, and temperature is higher in the centre than at the edges. This makes that strength develops faster in the centre than at the edges. By monitoring of the temperature and accompanying testing of strength on separate specimens, stripping times or partial post-tensioning times can be determined. On the other hand, temperatures may become too high and artificial cooling is required. Interactive monitoring, control and construction works will lead to faster and cost saving building. All testing methods supporting this interaction are most valuable.

2.3 Young concrete

Let us call concrete young until the end of curing which is one to seven days. The main properties of the concrete skin have developed, which means also that the resistance against weathering, abrasion, erosion, and caviation is determined. The same is true with respect to permeability and thus to the ingress of carbon dioxide, water, oxygen, and various ions, which means that the quality with respect to durability is determined. Efforts should be made in order to measure the relevant properties of young concrete. In case that the concrete does not comply with the specifications measures would be taken to resume curing and to improve the concrete quality with respect to durability.

Sometimes there are reasons to assume deficient compaction of concrete or the presence of large voids or delamination between concrete batches. These deficiencies are not visible on the surface, or the structure may be even situated in soil or water. Testing methods and devices should be able to trace these deficient parts of the structure. Of course, measurements which can be performed on young concrete can also be performed at larger age. However, it should be emphasized that testing should take place as early as possible in order to supply the tools for correction.

2.4 Reinforcement

The right size, quality and number of bars and their correct placing are essential for the bearing capacity of a reinforced concrete cross-section. All these properties should be checked during construction if not regularly, then randomly. Besides bearing capacity, durability depends mainly on the thickness of the cover. Therefore an economic method with easy performance should be available which is able to distinguish between bar diameter and - simultaneously - concrete cover. If such a device were used on the building site, the durability question would not be as urgent as it is now.

2.5 Post-tensioning

The time of application of tensioning forces depends on the actual strength of concrete which should be determined on the structure. The amount of post-tensioning is a relation of construction method, concrete strength and static system of the structure. The designer prescribes the force and the contractor realizes the force by measuring the hydraulic pressure in the jack. However, there are friction losses which may be accounted for in an erraneous way. The direct measurement of forces would be to help to ascertain the correctness of the applied forces.

With respect to corrosion protection of the prestressing steel, the ducts should be completely filled by grout. The grouting procedure is easy if there are straight ducts and air vents. However, if there are high and low reference points, the grout may settle and large voids may be created. A second grouting procedure can help to avoid this problem. However, it would be more reliable if uncomplete grouting could be detected and located by a non-destructive method.

2.6 Concrete repair

Good workmanship during concrete repair is the main guarantee for a durable repair. Quality of repair work can be checked visually, but should also be tested by standard methods and reliable devices. In this respect, there are a few aspects: adhesion of repair layers, moisture of the substrate concrete, thickness of coatings and films on a mineral surface. The testing methods should be non-destructive, because any destruction will impair the durability of the structure after repair.

3 Type of testing

Several times, non-destructive testing has been mentioned. In the author's opinion, non-destructive testing should always be preferred, because it has several advantages compared to destructive testing: the structure is not affected, many methods use electric devices leading to signals which can be amplified, stored and processed automatically, the results can be visualized on screens and plotted as tables and/or graphs. The progress in the micro-electronic field is enormous which allows quick interpretation of results (see for instance Lew, 1988, and Schickert & Schnitger, 1985). Furthermore, testing on site can be combined with simultaneous analysis leading to interactive assessment of quality of a structure.

Besides electric devices, there should also be mechanical devices which are easy to handle such that a person could use them without advanced education in testing. It is no secret that already the presence of a controlling device can improve quality.

On the other hand, more persons involved in the construction procedure should be educated in testing on the site. The quality of a product increases with the extent of testing. But, since testing costs money, there is an optimum of cost and revenue. At the moment, the building industry suffers from a bad image as quality is concerned. Testing is not an aim as such but should supply the basis for correct decisions during construction and for objective facts in discussions between the inherent parties of a building project.

4 Goals of the workshop

The goal of the workshop is at least fourfold: first, it should be tried to gather as much knowledge as possible in the various testing fields which are relevant for the construction process. It is not intended to write a state-of- the-art report, but the main methods, strategies, results and experiences should come up; second, the contributions and discussions should provide a catalogue of testing needs and should clearly state the lack of knowledge; third, research needs should be specified, i.e. those fields of lacking knowledge should be marked where further development seems to promise great success with respect to economy, durability, and safety of structures; fourth, RILEM would like to receive some suggestions with regard to new international activities which could serve the goal of increasing the knowledge and improve the testing methods and devices for testing during concrete construction. RILEM as an international organization was active in testing during her life and will help to coordinate international efforts in this field.

5 Conclusions

Testing during concrete construction will

- improve the quality of structures since it allows corrections of errors and faults during an early stage,

- provide objective data for decisions which will make the construction process more economic and faster,

- provide data for objective negociations between designer, producer, contractor, sub-contractor, and owner,

- lift up the image of the building industry as a quality conscious industrial branch.

Therefore, testing methods and devices should be further developed and disseminated among the relevant parties in order to proceed in the right direction. Non-destructive testing should be promoted.

References

[1] Lew, H.S. (ed.) (1988) Nondestructive testing, ACI SP-112, Detroit

[2] Schickert, G., Schnitger, H. (eds.) (1985) Zerstörungsfreie Prüfung im Bauwesen. DGZFP, Berlin

2 JUDGEMENT SYSTEM FOR TESTING REPORTS DURING CONSTRUCTION

R. SPRINGENSCHMID and R. BREITENBÜCHER
Institute of Building Materials, Technical University
of Munich, Germany

Abstract
Before concrete constructions can be build, a lot of technological tests are required and a number of such tests have to be done during concrete production. This is also true for the aggregates, cement and other constituents of the concretes. The results of these tests have to be studied critically. This judgement, however, has often to be done by engineers who usually are confrontated with problems in construction or management only. For this purpose a judgement system with check-lists was developed.

1 Introduction

In the case of building construction contracts, the award should go to the company which offers the most economical bid. For this purpose, not only the price itself, but also the quality of the work to be done has to be considered. This, however, is difficult. Therefore, the contract is usually awarded to the lowest bidder. The tenders are often calculated not by engineers, but by purchasing clerks who are well informed on low prices but not on technological and construction details. In order to be successful against competitive bidders, they often calculate the bids on (or below) the lowest price limit. Thus contractors have to look for cheap subcontractors because they are not able to do the work for the price offered. All the materials such as readymixed concrete, asphalt, bitumen or cement, and aggregates are bought from the supplier who offers the lowest price. For some construction materials such as cement, the engineer on the site can usually rely on the quality of the material and the certificates based on random tests. Experience has shown, however, that there are other products where it can be of great advantage for the purchaser

- to ask for all specified quality test reports

- to study carefully all test reports

- to compare the materials delivered with the result of quality tests taken from the reports, based on visual examination and if necessary on additional tests.

This work has to be done by the site supervision engineer and is of increased importance if bids are low and competition among suppliers and construction companies is hard.

Usually the contractor's site engineer is anxious to produce high quality and takes care that all quality tests which are required by the site superviser are fulfilled. However, he has to take into account that his competitive contractors may rely on the chance that the tests are not fully carried out. The aim of the contractor often is to survive the period of guarantee - mostly 5 years - without any fault. He has to consider that his competitor will not spend any money on quality beyond the period of guarantee.

As soon as the concrete has hardened, no improvement of the quality is possible. In some cases protective layers or additional structural members can compensate for low quality. This, however, is time consuming and costs a lot of money. Therefore the only tune for control of the concrete mix is before the concrete leaves the mixing plant.

2 System of Quality Control in the FRG

In the Federal Republic of Germany a quality control system in the construction industry is established, which is based on two points: the manufacturer's own in-plant control (internal control) and an externally monitored control by an independent organisation or institute (external control). In the internal control the quality of materials is tested continuously or in small intervals during production, and the results are registered. The external controller visits the plant only twice a year, i.e. this is only a spot check and faults during production must be found and eliminated by the manufacturer himself. In the external control the results of the internal control are checked, the instruments and equipment for the internal control are inspected, and separate samples are taken for testing in an independent testing institute. When all requirements are met, the products and the delivery note are signed as "externally controlled according to standard no. ...". If some faults are ascertained during the external control, a second control (repetition) will be performed within about four weeks. If the faults are not eliminated in the meantime, the products are not allowed to be signed as "externally controlled".

3 Quality Control of Concrete from the Practical Point of View

For the concrete production and placement itself the volume of internal and external control is specified in the German Standard DIN 1084. Within the scope of the internal control a qualification test must be carried out a few weeks before concreting starts. Here it has to be checked whether it is possible to reliably obtain the required properties with the proposed ingredients and mix-composition. During construction work and daily concrete production, the producer must control the procured ingredients (mostly by visual inspection) and check the function of the technical equipment (material batcher, mixer etc.). Furthermore, he has to check in a quality control whether the required properties (e.g. compressive strength, air content etc.) are as specified.

Twice a year the external controller checks the concrete production and placement. This can only be done when the concrete is produced or placed on the job site. Therefore the external controller often asks at the job-site whether concreting will take place within the next few days. This indirect announcement, however, is not in accord with the real intention of a non-announced spot check by the external control.

Since these internal and external controls are only minimum controls, the purchaser often also takes samples for an additional control in the case of important structures.

4 Judgement of Test and Control Reports

Only on very large sites, the quality of the concrete and its ingredients is judged by competent experts, who are taken into consultation by the purchaser. Normally this judgement is performed by non-specialized engineers, who are primarily engaged in a lot of other tasks at the same time, e.g. management and settlement of account. Often only the actual situation of expenses and final dates during the construction period are considered.

It is ordered by law that the structural analysis and constructive details must be controlled by an independent engineer with particular experience before the construction can be built. On the other hand, there is no requirement for a judgement or a control of the qualification of the concrete and other materials by an independent engineer before beginning concreting. Rather there is a free competition within a lot of laboratories, which often are connected with ready-mix concrete plants or big contractors.

In the past, faults and damages in structures owing to mistakes in statics occured only rarely. Mostly they were caused by non-suitable or poor materials and/or faults in the execution. To prevent such damages in the future, it seems necessary to install also an independent expert for concrete-/material technology. His task would be to check the control test results as specified by DIN or other standards. For jobs where concrete is exposed to unusual or severe influences, another task for the expert in materials technology would be of even more importance. He has to check whether the specified standard tests and acceptance criteria are sufficient to secure the expected longtime behaviour of the structure. If necessary, additional tests or modified criteria have to be specified. This can be done only by competent materials engineers with great experience.

This report, however, is limited to the standard tests as specified for usual structures. It offers a system to facilitate the control of test results.

5 Problems and Sources of Errors with Test Reports

For the judgement of concrete qualification and quality, usually only the test reports of the concrete itself (e.g. compressive strength, water penetration depth etc.) are submitted and controlled. Frequently the external control reports for the different ingredients (aggregates, cements, additives etc.) are not placed at disposal in time or are not available. In this case injunctions or remarks to the quality or the concrete ingredients are lost or cannot be taken into account. But also when these reports are present, they must be read (and understood) - often they only are registered and put away without consideration of the content. This problem is more and more intensified when the concrete is not produced on the job site but in a ready-mix concrete plant of a supplier. The way of external control reports (R) is very long:

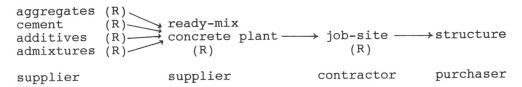

What should be prevented by a careful judgement?

In the last few years about 120 different concretes for big tunnel and bridge constructions were judged by the authors whereof only a relative small number could be accepted from the beginning. The following faults were observed frequently:

- missing or obsolete test reports of the external control for the ingredients, ready-mix concrete plant and the job site

- external control reports for other aggregate size ranges than used, partly even reports for other plants than used

- missing or invalid licences for additives or admixtures

- use of aggregates or cements during construction other than tested in the qualification test (sometimes only because they could be obtained cheaper from another supplier)

- qualification tests were carried out very (too) late - only a few days before starting concreting - i.e. it was impossible - if necessary - to change the mix composition or ingredients and to repeat the qualification tests.

- required properties (e.g. water penetration depth) were not tested

- special properties are tested only partly: e.g. for concrete with high freeze-resistance only the air content was controlled while the freeze-resistance of the aggregates was not proved.

- during the construction period the external control reports were not available, mostly in those cases, when they included negative remarks.

To enable better judgement of the concrete quality for special structures, check-lists were developed for the concrete ingredients, the concrete qualification test and the external control reports (see Table 1 and 2). By this method, a good continuous survey on required and satisfied proofs is obtained.

Table 1

UNIV.-PROF. Dr. RUPERT SPRINGENSCHMID
ORDINARIUS FÜR BAUSTOFFKUNDE UND WERKSTOFFPRÜFUNG
TECHNISCHE UNIVERSITÄT MÜNCHEN

Judgement of Concrete Ingredients for Qualification Tests

Concrete plant: _____

Object: _____

Member:

1. Aggregates

Size range	Type	Plant	Date of external control	Requirements DIN 4226 satisfied	High Freezing Resistence

2. Cement

Type	Strength Class	Cement plant	Date of external control	Requirements DIN 1164 satisfied	

3. Additives

Type	Designation	Plant	Expiry date of licence	Requirements satisfied	

4. Admixtures

Type	Designation	Plant	Expiry date of licence	Requiremnts satisfied	

Test of efficiency of the air-entraining agent combined with the simultaneous use of_____ % Superplasticizer

Date of the test:_____Requirements satisfied:_____

5. Judgement

The proposed concrete ingredients can be used for the above mentioned members.

The following proofs have to be completed:_____

_____

_____

Number of Judgement_____

Munich,

Univ.-Prof. Dr. R. Springenschmid

Table 2a

UNIV.-PROF. Dr. RUPERT SPRINGENSCHMID
ORDINARIUS FÜR BAUSTOFFKUNDE UND WERKSTOFFPRÜFUNG
TECHNISCHE UNIVERSITÄT MÜNCHEN

Judgement of Qualification-Tests

Concrete plant:_____

Object:_____

Member:

Range of application: T_{FC} above 18 °C (Summer)/ T_{FC} below 18 °C (Winter)

Type-No:_____

Concrete Strength class:_____

1.Qualification of the ingredients

Requirement satisfied:_____Judgement-No._____ Date of Judgement:_____

2. Mix-Composition

2.1 Aggregates

Size range	Type	Plant	Grading analysis of the size range	Granulometry percentage (Vol.-%)

Grading analysis of the total aggregates: Requirements satisfied:_____

2.2 Cement

Type	Strength class	Cement plant	Cement content (kg/m^3)	

2.3 Additives

Type	Designation	Plant	Content (kg/m^3)	

2.4 Admixtures

Type	Designation	Plant	Content (% of Cem.w.)	

Table 2b

2.5 Water   Water-content (kg/m^3):_____

2.6 w/c-ratio   Water-cement-ratio:_____ w/c-ratio including puzzolane:_____

3. Results of fresh concrete tests and early strength tests

3.1 Fresh concrete

Density (kg/dm^3):_____Consistency ("Ausbreitmaß", cm):_____Air-content (Vol.%):_____

3.2 Temperatures and early strength

Temperature of the fresh concrete	(°C)							
Storage-Temperature	(°C)							
Compressive Strength after 12h [1]	(N/mm^2)							

4. Results of Tests on hardened concrete

Age	days	1	7	28			
Compressive Strength[1]	(N/mm^2)						
Flexural-/-tensile splitting strength	(N/mm^2)	–					
E-Modulus	(N/mm^2)	–					
Water-penetration depth	(cm)						

[1] with respect to 200 mm-Cubes

5. External control

5.1 Ready-mixed concrete plant

External control according to DIN 1084 T.3: Date:_____Requirements satisfied:_____

5.2 Job-site

External control according to DIN 1084 T.1: Date:_____Requirements satisfied:_____

6. Judgement

The exposed qualification-test satisfies the requirements for the above mentioned members:

The following proofs have to be completed:_____
_____
_____
_____

Number of Judgement:_____

Munich,

Univ.-Prof. Dr. R. Springenschmid

PART TWO

CONSTITUENTS IN THE MIX

3 ANALYSIS OF FRESH CONCRETE – A SURVEY

H. GRUBE
Forschungsinstitut der Zementindustrie, Düsseldorf, Germany

Abstract
This paper describes the objective of the analysis of fresh concrete and specifies a lot of testing procedures. Many of them are well suited as rapid tests for quality assurance. But no particular test nor any combination is able to varify all important components of a single concrete batch by type and mass in a short time and with a sufficient accuracy. Some laboratory test methods e.g. X-ray Fluorescence or X-ray diffraction may give further information but not for on-line verification and engagement during the mixing process.

1 Introduction

Concrete consists of three main components: cement, aggregate and water. It may contain additions and admixtures. Harmful ingredients influencing e.g. strength, durability, corrosion protection of the reinforcement should not be included. The new European standard for concrete pr ENV 206 [1] gives definitions for the components mentioned above. Beyond that there are standards being prepared for cement, aggregates, mixing water, additions (fly-ash) and admixtures. Each of the components is influencing more or less the properties of the fresh and hardened concrete.

During concrete production the components are weighed and mixed according to the recorded mixing instruction. Mixing should reach homogenity. It is no problem in modern plants to hold the weight of the components within ± 1 % of the nominal value. So the scatter must not exeed the limits given by pr ENV 206 for all masses with ± 3 % (± 5 %) for admixtures) or the nominal value when the plant is operated by skilled personnel. Quality assurance (QA) requires control of the single components of the concrete when being delivered and continuous check of the weighing and mixing units. This is the first essential to fulfil the qualifications. Testing the hardened concrete is only an additional check to detect lacks in the QA-system and to avoid rough faults in stability of a building.

The second essential is to check properties of the fresh concrete. The test may be in the interest of the producer or the contractor or the owner of the building.

2 Analysis of the fresh concrete

Reliable results of the following characteristic properties of fresh concrete may be expected by analysing the fresh concrete:

a) water cement ratio (w/c, ω)

b) water content W (kg/m^3)

c) cement content C (kg/m^3) and type of cement

d) aggregate content G (kg/m^3), type and grain size distribution

e) content (kg/m^3) and type of additions

f) content (mass per kg of cement) and type of admixtures

g) air content (Vol.-%) and pore size distribution

h) bulk density (kg/cm^3)

i) workability (consistency)

k) harmful ingredients.

There is a lot of literature giving a review on different testing methods e.g. [2,3,4,5,6]. As it is very difficult to analyse a suspension, the dispersed material of which may be introduced by different components, it seems to be expedient making difference between two problems:

- tests for systematical quality assurance. Every test result (e.g. water content, workability, content of fines) lying between settled limits cuts down the risk of accepting or placing a wrong concrete mix. For this objective the test methods may be assessed to be satisfactory, which are summed up in the Tables 1 to 4.

- total analysis of the fresh concrete without any preinformation. The subsequent determination of all components by quality and quantity seems not to be possible when looking for a certain accuracy and for the short time between production and placing of the concrete. Without any doubt it would be advantageous to have an analyser working by push-bottom and producing reliable results within about 10 minutes. But looking at it practically, this type of analysis seems to be dispensable excluding the instance of deception.

So it may be normal to have some preinformation about the mix design, some single components etc.

There are some tests needing no preinformation for getting results of high accuracy when sampling is perfect (e.g. according to RILEM CPC 1 [7]. There are: water

20

content (b) and bulk density (h), aggregate content, type and grain size distribution for diameters > 0.25 mm (d), total content of fines < 0.25 mm (c) + (e), total air content (g) and consistency (i).

Looking at this we can realize that efforts are to be concentrated on the analysis of the fine solids < 0.25 mm and on the materials in aquous solution.

The following Tables 1 to 4 contain approved test methods for fresh concrete characterizing special properties. Some of the tests are standardized, others are published in the technical literature. The order from a) to k) is the same as mentioned at the beginning of this chapter. If tests are to be conducted later, the concrete sample can be conserved by deep freezing [8].

Whenever callibration tests in the tables are required, the reader may realize that preinformation is needed. Unsatisfying results can also be created by uncorrect callibration.

3 Conclusions

The following conclusions can be drawn:

- Fresh concrete analysis is not so independent, complete and rapid that it could be used for adjusting production devices on-line.

- Besides fresh concrete analysis the QA should be supported by continuous control of the concrete components and the production devices.

- Each of the test methods mentioned in the tables relates to another property of the fresh concrete. For purpose of quality assurance it is necessary to choose test methods which give the best information at minimum effort and cost.

- The most important values are cement and water content. This being checked, the planned consistency and the expected appearance of the concrete are approving that main possible defects are not likely.

- The investigations show that cement and water content cannot be determined in one single test with the wanted [3] accuracy ($\Delta C \leq 10$ kg/m^3; $\Delta W \leq 5$ kg/m^3; $\Delta \omega \leq 0.04$) in a time of less than 20 minutes.

Table 1.

(a) Testing the water cement ratio (w/c, ω)		(Look also at separate determination of cement content and water content)	
Name of the test	Objective and short description of the test	Literature and accuracy	Remarks
Chromate test a.1)	Chemical test. Direct determination of w/c-ratio by concentration of chromate-ions in the free water of the concrete. The chromate has been inserted by the cement and is tested in 2÷5 ml of the liquid.	Lit. /8/ accuracy of w/c ≈ ± 5 % ($\Delta\omega \approx \pm 0,03$) of the nominal value; testing time (tt.) ≈15 min.	- callibration required - chromate content of cement may not be constant - time after mixing and temperature are influencing the result.
Vibrating tester a.2)	Physical test. Indirect test of w/c-ratio by determining the viscosity of the paste. The vibration of two disks is damped by the concrete.	Lit. /9/ accuracy of w/c ≈ ± 7 % ($\Delta\omega \approx \pm 0,04$); tt.: ≈ 1 minute	- callibration required - exchange in content of fines (cement/fly-ash) will not be noticed - stiffening will be measured as a lower w/c - concrete with high-range water-reducing admixtures cannot be assessed.
Thaulow-test a.3)	Physical test. Indirect determination of w/c-ratio from the bulk density (air-less) of the concrete. Exact weighing is necessary.	Lit. /10, 10a, 11, 12/ coefficient of variation for w/c ≈ ± 4 % ($\Delta\omega \approx \pm 0,03$ (to 0,05)); tt.: ≈ 10 to 15 minutes.	- mixing ratio C/G and the density of cement ρ_c and aggregate ρ_a must be known
Pyknometer-test a.4)	Physical test, similar to a.3)	Lit. /13, 14/ look at a.3)	look at a.3)

Table 1. continued

(b) Testing the water content

Name of the test	Objective and short description of the test	Literature and accuracy	Remarks
Hot Plate (Drying) b.1)	Physical test. Drying of concrete by rapid heating on e.g. gas oven. Water content determined by loss of weight.	Lit. /5, 15, 16/ c.o.v. for W $\approx$ 3% (to 5%) tt.: $\approx$20 minutes, cooling included	- water absorbed by the aggregate must be determined separately. - no further preinformation required if water is the only volatile material.
Microwave (Drying) b.2)	Physical test, similar to b.1), heating by microwave oven	Lit. /5, 6, 7/ $\Delta W \approx \pm$ 6 kg/m^3 tt.: $\approx$ 25÷60 minutes.	- similar to b.1)
Kelly-Vail-Method b.3)	Chemical test. Mixing the concrete sample with a chloride solution of known concentration the solution will be diluted. New concentration is measured by titration. It is proportional to the free water content of the concrete.	Lit. /5, 6, 17, 18, 19/ $\Delta W \approx \pm$ 7 dm^3/m^3 tt.: $\approx$ 15 minutes	- Parallel test for getting the initial Cl-content in the free water of the concrete is required.
AM-test b.4)	Physical test. Concrete sample is doused with spirit and ignited. Procedure is repeated until mixing water is evaporated. Loss of water is determined by loss of weight.	Lit. /20/ $\Delta W \approx$ 7 dm^3/m^3 tt.: $\approx$ 30 to 40 minutes	- similar to b.1) - degree of drying not defined as in b.1) - the handling of spirit near to open fire may be dangerous.

Table 1. continued

CM-method b.5)	Physical test. Addition of carbide to the concrete sample in a pressure containment. Increase of pressure is proportional to the content of free water.	Lit. /21/ $\Delta W \approx 8$ dm³/m³ as a mean of 4 tests tt.: ≈ 10 min/test	- testing of free water - the mass of the concrete sample is very poor (200 g), so it cannot be representative in one test.
Neutron-moisture gauge b.6)	Physical test. Moderating of neutrons when striking H-atoms of the water in the fresh concrete sample. Counting of moderated neutrons.	Lit. /22, 23, 24, 25/ $\Delta W \approx 5 \div 7$ dm³/m³ tt.: ≈ 30 sec.	- calibration required - water absorbed in the aggregate must be determined separately - protection of personnel.

Table 2.

(c) Testing the cement content (and type of cement)			
Name of the test	Objective and short description of the test	Literature and accuracy	Remarks
Wash-out test c.1)	Physical test. Passing the fresh concrete by adding water through a set of screens, the finest of which is (0,15 mm) 0,25 mm. The cement content is calculated by the fresh weight of concrete minus - aggregate > 0,25 mm (dry) - water (see tests b)) - fine aggregate and additions ≤ 0,25 mm (to be tested separately).	Lit. /15, 16/ cement content $\Delta C \approx 15$ to 20 kg/m³ tt.: ≈ 30 minutes	- content of fine aggregates (< 0,15 mm) < 0,25 mm and additions must be known or determined separately - when fines are collected they can be analysed with respect to fly-ash, blast furnace slag, sand (quarzite or calcite) or klinker; look at c.6) and c.9) to c.11) or /16a/.

Table 2. continued

Rapid Analysis Machine (RAM) c.2)	Physical test. Same principal as c.1). Mechanical washing and partition of the concrete sample. Smallest mesh of the sieve is 0,15 mm.	Lit. /3, 5, 26/ $\Delta C \approx 15$ to 20 kg/m³ tt.: ≈ 10 minutes	as for c.1)
Electrical Conductivity c.3)	Physical test. The amount of ions will accumulate in the mixing water of the fresh concrete by the cement. Adding a surplus of water to the concrete the electrical conductivity of the solution can be a measure for the cement content.	Lit. /27/ $\Delta C \approx 5$ to 10 % of nominal weight tt.: ≈ 40 minutes	- extensive callibration tests are required - time after mixing and temperature are of great influence - test may not be specific for different charges of cement.
Flotation c.4)	Physical test. Specific fines (e.g. the cement) in a suspension show a hydrophobe behaviour when getting in contact with special chemicals (so-called collectors). The hydrophobe particles will swim up by taking up air bubbles wich are entrained into the suspension. They can be collected, separated, weighed and analysed.	Lit. /3, 4, 28/ $\Delta C \approx \pm 5$ % of the nominal weight tt.: $\approx 10 \pm 15$ minutes	- Extensive callibration is required - Separation of cement and additions is difficult.
Centrifuge c.5)	Physical test. Separation of cement and aggregate in a suspension of a heavy liquid, the density of which lies between cement ($S_c \approx 3,1$) and aggregate ($S_a \approx 2,6$)	Lit. /5, 18, 18a/ $\Delta C \approx \pm 20$ kg/m³ tt.: ≈ 1 h to 2 h	- Callibration tests are required - Laboratory test with extreme safeguards to protect personnel and environment.

Table 2. continued

Density test c.6)	Physical test. The solids of a concrete sample ≤ 0,15 mm are washed out and collected. The suspension of the fines in a constant volume of water has a significant density which can be measured by an araeometer.	Lit. /29/ $\Delta C \approx \pm 3$ % of nominal weight tt.: ≈ 20 minutes	- Callibration test is required. - Analysation of different fines is possible but difficult.
Heat of reaction c.7)	Chemical test. Mixing the concrete sample with a special amount of HCl the reaction leads to temperature rise. In an insulated containment the temperature difference can be determined as a measure of cement content.	Lit. /30, 31/ $\Delta C \approx \pm 10$ % of nominal weight tt.: ≈ 10 minutes	- Callibration tests are required - Reactive aggregates are (e.g. lime stone) influencing the result.
Concrete Quality Monitor (CQM) and Kelly/Vail c.8)	Chemical test. Dissolution of the washed-out fines in acid and determination of Ca-content by titration or by other Ca-analysers.	Lit. /5, 17, 18/ $\Delta C \approx 15$ to 20 kg/m^3 tt.: ≈1 hour	- Callibration tests are required - Reactive aggregates are influencing the result
Nuclear cement content gauge c.9)	Physical test. Absorption of X-rays from a radioactive source by Calcium is measured.	Lit. /5, 32/ $\Delta C \approx 15$ to 20 % of nominal weight tt.: >1 hour	- Callibration tests are required. - Laboratory test with extensive safeguards and preparation work.

Table 3.

Name of the test	Objective and description of the test	Literature and accuracy	Remarks
X-ray fluorescenz detection c.10)	Physical test. Quantitative analysis of different elements in a special sample containing washed-out fines of the concrete.	Lit. /5, 33/ $\Delta C \approx 10$ to 20 % of nominal weight tt.: several hours	- Calibration tests are required
Microscopical analysis c.11)	Physical test. Washed-out fines < 0,15 mm are inactivated, dried and classified in groups of different grain size. Within the groups a qualitative and partly quantitative analysis is possible.	Lit. /34, 35, 36/ $t.t > 1$ hour	- skilled personnel required

(d) Testing the aggregate content (and type and grain size distribution)

Name of the test	Objective and description of the test	Literature and accuracy	Remarks
Washout test, sieve analysis d.1)	Physical test. Collecting the aggregate (>0,15 mm) > 0,25 mm of test c.1). Making a complete sieve analysis. Drying and weighing of the different grain sizes.	Lit. /15/ High accuracy according to representative sample tt.: $\approx$ 20 minutes	- No preinformation needed - Assessment of grain size distribution and mineralogical composition

(e) Content and type of mineral additions

e.1) Procedures according to c.1), c.2), c.4), c.6), c.10), c.11) may be suited alone or in combination.

Table 3. continued

(f) Content and type of admixtures

IR-spectroscopy f.1)	Physical test. Extraction of mixing water from the concrete. Analysis in the IR-spectrometer.	Lit. /37/	- Callibration tests are required

(g) Testing the air content and the air pore size distribution

Name of the test	Objective and short description of the test	Literature and accuracy	Remarks
Pressure adjustment test g.1)	Physical test. Decrease in pressure of a particular air volume which acts upon a particular volume of compacted fresh concrete is a measure for total air content in the concrete.	Lit. /38, 39/ $\Delta V \approx \pm 0{,}5$ Vol.-% absolute tt.: ≈ 5 to 8 minutes	- No preinformation - Only for concrete with dense aggregates
"Roll-a-meter" g.2)	Physical test. A compacted volume of concrete is confined in a containment of particular volume, supplied with a burette on top. The remaining air volume and the burette are filled with water to a destined level. When the concrete is suspended in the water by "rolling" the container, the reading on the burette shows the air content of the concrete.	Lit. /40/ $\Delta V \approx \pm 0{,}5$ Vol.-% absolute tt.: ≈ 10 minutes	- No preinformation

Table 3. continued

Air-meter g.3)	Physical test. A sample of mortar taken from the concrete is being suspended in water and glycerin. The escaping air-bubbles are captured under water. Their buoyancy depending on time gives a measure of the pore size distribution and the total air content (principle of Stokes).	Lit. /41/ accuracy is still being tested. tt.: ≈ 30 minutes	- Mortar content of concrete sample must be known.

Table 4.

Name of the test	Objective and short description of the test	Literature and accuracy	Remarks
Vibration test g.4)	Physical test. Measurement of air content and pore size distribution with the apparatus according to g.2). Pore size distribution is determined without suspending the concrete but only by additional vibration in intervals and air volume readings on the burette.	Lit. /42/	- Results depend on the viscosity of the paste (w/c, cement) - For high and mean viscosity the procedure gives reliable indication for a stable system of small air pores.

(h) Testing the bulk density of the concrete

Bulk density h.1)	Physical test. Filling a container of known volume with concrete up to the rim during vibration. The bulk density is the quotient weight/volume.	Lit. /16, 38, 43/ $\Delta\rho \approx \pm 1\%$ of the nominal value tt.: ≈ 5 to 8 minutes.	- Concrete sample must be representative - Determination of air content in the same test according to g.1) is recommended.

Table 4. continued

(i) Testing the workability (consistency) of the concrete

Name of the test	Objective and short description of the test	Literature and accuracy	Remarks
Consistency i.1)	Physical test. Measuring the deformation of the fresh concrete effected by a force or an impact (slump test, VEBE-test, compaction test, spreading table).	Lit. /44, 45, 46/ tt.: ≈ 5 minutes	- No preinformation

(k) Testing harmful ingredients

| Harmless-ness-test k.1) | Physical or chemical test. Investigation with respect to materials in the con-crete which may impair the strength or durability of the concrete (e.g. wood or peat) or the corrosion pro-tection of the reinforce-ment. The procedure depends on the suspected ingredients (e.g. c.1), d.1) → peat or wood), (e.g. b.3) → Chlorides), (e.g. f.1) → organic compounds) | - | |

References

[1] pr ENV (Feb. 1989) Concrete-Performance, Production, Placing and Compliance Criteria

[2] Walz, K.: Prüfung der Zusammensetzung des Frischbetons (Frischbetonanalyse). beton 27 (1977), H. 7, S. 282/287, H. 8, S. 313/317 und H. 9, S. 347/352

[3] Nägele, E., und H.K. Hilsdorf: Die Frischbetonanalyse auf der Baustelle. beton 30 (1980) H. 4, S. 133/138; ebenso Betontechnische Berichte 1980/81, Beton-Verlag GmbH, Düsseldorf 1982, S. 33/49

[4] Nägele, E., und H.K. Hilsdorf: Die Bestimmung des Wasserzementwertes auf der Baustelle. DAfStb-Heft 349 Verlag W. Ernst & Sohn, Berlin 1984

[5] Tom, J.G., und A.D. Magoun: Evaluation of procedures used to measure cement and water content in fresh concrete. National Cooperative Highway Research Program Report 284, TRSB, June 1986

[6] Head, W.J., H.M. Phillippi, P.A. Howdyshell und D. Lawrence: Evaluation of selected procedures for the rapid analysis of fresh concrete. American Society for Testing and Materials 5 (1983) No. 2, S. 88/102

[7] Tentative Recommendations: No. 1 - Sampling fresh concrete in the field

[8] Clear, C.A.: Delayed analysis of fresh concrete for cement and water content by freezing. Magazine of Concrete Research 40 (1988) No. 145, S. 227/233

[9] Carlsen, R., und I. Gukild: Eine direkte Methode zur Bestimmung des W:Z-Verhältnisses in Frischbeton. Zement-Kalk-Gips 24 (1971) H. 6, S. 268/273

[10] Ohgishi, S., I. Tanahashi, H. Ono und K. Nizutani: The applications of test method to measure constituent materials of fresh concrete for the quality control of concrete. Transactions of the Japan Concrete Institute 6 (1984), S. 75/82

[11] Thaulow, S.: Field testing of concrete. Norsk Cementforening, Oslo 1952

[12] Wellenstein, R.: Einfluß der Meßfehler charakterisierender Stoffwerte im Frischbeton auf den Wasserzementwert. Betonstein-Zeitung 35 (1969) H. 7, S. 442/446

[13] Blaut, H.: Erfahrungen bei der Messung des Wasser-Zement-Wertes nach dem Thaulow-Verfahren. Betonwerk + Fertigteil-Technik 38 (1972) H. 8, S. 602/608

[14] Naik, T.R., und B.W. Ramme: Determination of the Water-Cement-Ratio of Concrete by the Buoyancy Principle. ACI Materials Journal 86 (1989) No. 1, S. 3/9

[15] Foth, J.: Baustoff-Untersuchungen mit dem Pyknometer. Betonstein-Zeitung 35 (1969) H. 11, S. 667/671

[16] Lackner, R., und D. Smolen: Die Bestimmung des Wasserzementwertes an Frischbeton mit dem Luftpyknometer. Betonstein-Zeitung 37 (1971) H. 4, S. 215/217

[17] DIN 52 171 E (1989) Bestimmung der Zusammensetzung von Frischbeton

[18] Hentrich, J.: TGL 33432/01 - Ein neuer Standard zur Mischungsanalyse des Frischbetons. Betontechnik 6 (1986) H. 2, S. 47/49

[19] Wierig, H.-J., und H. Winkler: Zur quantitativen Bestimmung der Hauptbestandteile von Zementen. Zement-Kalk-Gips 37 (1984) H. 6, S. 308/310

[20] Kelly, R.T., und J.W. Vail: Rapid analysis of fresh concrete. Concrete 2 (1968), Nr. 4, S. 140/146 und Nr. 5, S. 206/210

[21] Williamson, R.: Methods for determining the water and cement content of fresh concrete. Materials and Structures 18 (1986) No. 108, S. 269/278

[22] Hime, W.G., und R.A. Willis: A method for the determination of the cement content of plastic concrete. ASTM Bulletin YNr 209, Okt. 1955, S. 37/43

[23] Rousselin, R.J., R. Baroux und Y. Charonnat: Application des traceurs pour l'étude du malaxage des matériaux de construction. Bull. liaison labo. P. et Ch. (1985) No. 140, S. 113/118

[24] Nischer, P.: Die Bestimmung des Wassergehaltes im Frischbeton mit der Spiritusmethode. Zement und Beton 22 (1977) H. 4, S. 145/146

[25] Nischer, P.: Das Karbidverfahren zur raschen Bestimmung des Wassergehalts von Betonzuschlag und Beton. Beton- und Stahlbetonbau 81 (1986) H. 7, S. 183/184

[26] Foth, J.: Radiometrische Wassergehaltsbestimmung an Frischbeton während seiner Herstellung. Wissenschaftliche Zeitschrift der Hochschule für Architektur und Bauwesen, Weimar, 17 (1970) H. 4, S. 367/368

[27] Rosinski, F.: Die Betonherstellung beim Donaukraftwerk Ottensheim-Wilkering, Zement und Beton 1972 Nr. 61/62, S. 22/26

[28] Forschungsgesellschaft für das Straßenwesen; Arbeitsgruppe Untergrund-Unterbau: Merkblatt über die Anwendung radiometrischer Verfahren zur Bestimmung der Dichte und des Wassergehaltes von Böden. Köln 1975

[29] Koch, E.: Das Feuchtemeßverfahren mit Neutronen bei der Herstellung von Beton. Informationsheft des Büro EUROSOTOP Nr. 55; Strahlungs- und Isotopenanwendung im Bauwesen, Band II, Brüssel 1972, S. 607/620

[30] Forrester, J.A., P.F. Black und T.P. Lees: An apparatus for the rapid analysis of fresh concrete to determine its cement content. Technical Report 42.490 Cement and Concr. Ass. London, April 1974

[31] Chadda, L.R.: The rapid testing of cement content in concrete and mortar. Indian Concrete Journal 29 (1955) Nr. 8, S. 258/260

[32] Nägele, E., und U. Schneider: Frischbetonanalyse durch Flotation, Möglichkeiten und Grenzen. beton 39 (1989) H. 6, S. 252/257

[33] Murdock, L.J.: The determination of the proportions of concrete. Cement and Lime Manufacture 21 (1948) Nr. 5, S. 91/96

[34] Farkas, E., F. Tamas und F. Wittman: Thermometrische Bestimmung des Zementgehaltes von Transportbeton. Silikattechnik 29 (1978) H. 7, S. 195/197

[35] Mullick, A.K., S.C. Maiti und R.C. Wason: Monitoring quality of fresh concrete in structures. National Council for Cement and Building Materials, Technology Digest, Vol. 2 (1988) No. 9, New Delhi

[36] Mitchell, T.M.: Nuclear gage for measuring the cement content of plastic concrete. Public Roads 38 (1975) Nr. 4, S. 134/139

[37] Merkblatt über die Präparationsverfahren für die Röntgenfluoreszenzanalyse von Stoffen der Zementindustrie (Fassung Mai 1978). Zement-Kalk-Gips 31 (1978), H. 11, S. 558/564

[38] Merkblatt zur Bestimmung des Hüttensandgehaltes von Eisenportland- und Hochofenzementen nach DIN 1164 (Fassung Feb. 1975). Zement-Kalk-Gips 28 (1975), H. 5, S. 214/216

[39] Mikroskopie des Zementklinkers - Bilderatlas. Verein Deutscher Zementwerke e.V., Beton-Verlag GmbH, Düsseldorf, 1965

[40] Biggs, L.D., und J.J. Bruns: Transmitted and reflected visible light microscopy of two bituminous flyashes. Mat. Res. Soc. Sym. Proc. 43 (1985), S. 21/29

[41] Diehm, P., und K. Krehl: Identifizierung und Klassifizierung von Betonzusatz-mitteln mit der Infrarot-Analyse und anderen chemischen und physikalischen Untersuchungsmethoden. Betonwerk + Fertigteil-Technik 41 (1975), H. 6, S. 299/302 und H. 7, S. 341/345

[42] DIN 1048 Teil 1, Prüfverfahren für Beton; Frischbeton

[43] ISO 4848: Concrete - Determination of air content of freshly mixed concrete - pressure method

[44] ASTM C 173-78: Standard Test Method for Air Content of Freshly Mixed Concrete by the Volumetric Method

[45] DBT Luftporenmesser, Serie 2/1. Dansk Beton Technik A/S, Hellerup

[46] Johansen, R.I.: Characterisation of the air-void system in concrete by a vibration test. Int. Symp. on Admixtures for Mortar and Concrete, Brussels 30. Aug. - 1. Sept. 1967

[47] ISO 6276 Concrete, compacted fresh - Determination of density

[48] ISO 4109, ISO 4110, ISO 4111, ISO/DP 9812, Fresh Concrete - Determination of consistency

[49] Grube, H., und J. Krell: Prüftechnische Einflüsse auf die Bestimmung des Aus-breitmaßes von Beton. beton 34 (1984) H. 10, S. 409/414

[50] Krell, J., R. Stratmann-Albert und E. Bielak: Verbesserung des Ausbreitversu-ches zur genaueren Konsistenzbestimmung. beton 39 (1989) H. 5, S. 211/215

4 MEASUREMENT OF CONSTITUENTS

H. HUBER
Tauernkraftwerke AG, Materialversuchsanstalt, Strass, Austria

Abstract
1.4 hm^3 of concrete were required for Zillergründl arch dam, 186 m high. All measures to obtain a uniform concrete without cracks were taken, such as fabrication of a homogeneous cement with fly ash, preparation of aggregates from deposits using an automatic sand grading method and special electronic control and supervision of concrete mix. These measures have proved entirely successful from the points of view of technology and economy and allowed to produce a concrete of suitable quality.

1 Introduction

Production of mass concrete has to take account of specific technological and economic conditions and therefore it has always been an incentive to dam constructors to obtain a concrete quality as uniform as possible. Only a uniform concrete quality allows to meet the various and often contrasting requirements (strength, heat generation, durability, machines workability). To obtain a homogeneous concrete, optimal components and their exact batching and blending at the mixing plant must be assured.

Zillergründl arch dam (Fig. 1) is an example to demonstrate the possibilities available today of producing homogeneous concrete. Zillergründl arch dam is the main structure of the Ziller hydroelectric development which is a high-head pumped storage scheme storage constructed between 1980 and 1987 in the Tyrol (Austria). The 186 m high arch dam is 42 m at its base and has a concrete volume of 1.4 hm^3.

2 Mixing plant controlled by microprozessor

It was in 1970 that the mixing plants of large construction sites in Austria were equipped for the first time with neutron monitoring probes to measure the natural moisture content of the aggregates in combination with an automatic weighing control. This system was developed further and is today a perfect microprocessor

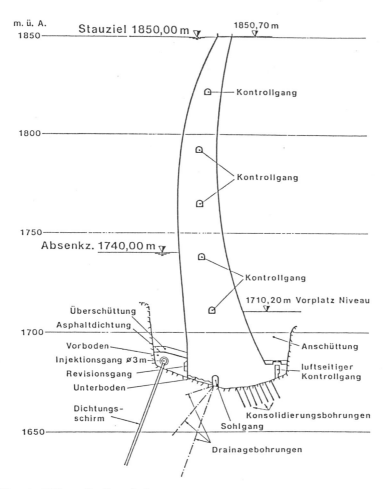

Fig. 1. Zillergründl arch dam

controlled weighing, batching and process control system. The advantages afforded by this electronic control are not only found in the easy handling of the input keyboard for all functional data and the display of the main data at the monitor, but especially in the functions allowing a uniform batching of concrete components and a permanent production monitoring (Table 1).

The decisive quality control of concrete is done by the electronic plant through permanent comparison of actual and desired values for any concrete component. Whenever a chosable and permissible deviation which had been determined to range between 2 and 10 kg/m^3 depending upon the component in question, is exceeded, all weighing containers are blocked and the relevant error component is signalized. This component can be corrected or has to be eliminated as mismix together with the whole

Table 1. Functions of the microprocessor controlled mixing plant

1	Operation keyboard	Function and numerical keyboard for input of weighing, batching, mixing and function data
2	Memory	Component list, operational data, stored data for recording
3	Moisture measurement by means of neutron measuring probes	Determination of the natural moisture content of the aggregates to take account of the amount of water required, water/cement ratio
4	Electronic weighing units	Exact weighing with pressure pickups ($\pm$ 0.5 %)
5	Batching monitoring, optimisation of batching vibrating troughs	Perfect accuracy of batching of the concrete components ($\pm$ 0.5 %)
6	Variance control	Adjustable and checkable maximum deviation of the actual value from the desired value
7	Consistency measured	Consistency value obtained via mixer power
8	Indicating monitor	Clear display of all weighing and batching data at the screen
9	Recording	Record of concrete components used, delivery note, batching record, deviation record for actual values shift record with statistics
10	Safe emplacement of the different concretes	Signal converter storage equipment from mixer to emplacement location

quantity originally weighed in. Such elimination was necessary only rarely due to the integrated sophisticated batching system.

The variance control together with the recording of all measured data via a matrix fast printer (Table 2) constitues the most efficient production monitoring for concrete and replaces such to a large extent the conventional quality test in the laboratory. As a result of all measures taken to homogenise concrete production for Zillergründl dam concrete was obtained during all the three years of concreting which was in good accordance with the desired cement content, the deviation being only 0.5 kg/m^3 and

Table 2. Record of Concrete Constituents

```
                                    ZEIT 10.49  DATUM  01.09.83

CHARGENPROTOKOLL

CHARGEN NR.: 0345

VERWENDET FUER BLOCK: 00/09   00/07   00/00

MISCHER   REZ-NR   MENGE   REST-M.   KONTROLLWERT
  4        2/00     1.0     0.0      46891109470450


DOS-T  KOMP   SOLL-W    IST-W    IST-TOL   IST-ABW   FEUCHTE
  8     01    131 KG   130 KG    5 KG      1 KG      00.0 %
  7     01    765 KG   757 KG    10 KG     8 KG      00.0 % #
 10     01    1.91 l   1.84 l    0.10 l    0.07 l    00.0 %
  1     01    2511 KG  2508 KG   20 KG     3 KG      05.8 %
  2     01    1470 KG  1476 KG   20 KG     6 KG      03.3 %
  3     01    1829 KG  1848 KG   40 KG     19 KG     01.4 %
  4     01    193 KG   184 KG    40 KG     9 KG      01.8 %
  5     01    2302 KG  2296 KG   40 KG     6 KG      00.0 %
  6     01    1430 KG  1426 KG   40 KG     4 KG      00.0 %
  9     01    115 KG   116 KG    5 KG      1 KG      00.0 %
```

Table 3. Results from quality test of core concrete for Zillergründl dam

	Compressive strength N/mm^2		
	Mean value	Standard deviation	10 % fractile
After 28 days	19.3	2.7	15.8
90 days	26.5	2.8	22.9
180 days	29.7	2.8	26.1
	Required minimum value		24.0
365 days	32.6	3.1	28.0
	Required minimum value		27.0
Batching of cementing material (2/3 cement, 1/3 fly ash)	170 kg/m^3		
Water/cement ratio	0.65		

which due to a high degree of uniform consistency was easy to work and emplace and for which all properties of mass concrete specified were achieved without involving any difficulties (Table 3). The standard deviation amounting to 2.8 N/mm^2 (180 days) recorded during the compression test to which the mainly emplaced "core concrete" was exposed is already close to the deviation of testing so that further improvement of the results would call for improved test methods.

The uniform concrete components and the electronically controlled concrete production allowed to fulfill the high specifications for the concrete of Zillergründl dam with low amounts of cementing material.

References

[1] Huber, H. (1984) Massnahmen bei der Herstellung von massigen Betonbauteilen, Österr. Betontag Graz

[2] Huber, H. (1984) Gleichmässige Betonherstellung, **Zement und Beton**, 1

[3] Huber, H. (1988) Measures to improve the Quality of Mass concrete, ICOLD San Francisco

[4] Widmann, R. (1977) Massenbetonprobleme beim Bau der Bogenmauer Kölnbrein, Deutscher Betontag, Berlin

[5] Widmann, R. (1984) Grundlagen für den Entwurf der Bogenmauer Zillergründl, **Wasserwirtschaft**, 3

[6] Widmann, R. (1985) How to Avoid Therman Cracking of Mass Concrete, ICOLD, Q. 57-R. 15

5 FLY ASH DETERMINATION IN FRESH CONCRETE USING FLOTATION

E. NÄGELE and U. SCHNEIDER
Gesamthochschule Kassel, Germany

Abstract
Three different methods are described to determine the fly ash content of a fresh concrete mix by means of flotation. Two methods are pure flotation tests using different collectors and pH-values. The third method requires an additional determination of the density of the flotation product.

1 Introduction

The determination of the fly ash content of fresh concrete is of major importance in modern concrete analysis. This is due to the fact that the use of fly ash for concrete-making is so widespread today that the amount of fly ash added to the concrete is considered to be a decisive parameter for concrete quality. On the other hand it is astonishing to note that no method yet exists, which allows a rapid and reliable determination of the fly ash content in fresh concrete.

Generally, two different types of fly ash additions are in use in most countries, the direct addition of fly ash to the concrete on site while making the concrete and the use of fly ash premixed with cement. It would be of great practical importance to have a method capable of distinguishing between these two types of fly ash admixtures.

The flotation method allows a rapid and reliable determination of the cement content and the binder content respectively, if the fresh concrete contains fine admixtures. Many efforts have been made to modify the flotation method in such a way that a selective flotation of cement and fly ash will become possible.

Thus, this paper summarizes the state-of-the-art concerning the determination of the fly ash content of fresh concrete by flotation.

2 Fundamentals of cement flotation

Flotation is a separation method for fine powder materials. The powder is suspended in water and the material to be separated from the others is made water-repellant by the use of special reagents, called collectors. If air is blown into the suspension,

40

the hydrophobic material adsorbs to the bubbles and floats to the surface where it is removed mechanically.

Thus the basic step of any flotation is the selective adsorption of the collectors to the material to be separated. Such collectors have been developed for cement [1,2,3], allowing for a cement content determination by flotating the cement out of the fresh concrete.

In this process, the materials in fresh concrete can be subdivided into three groups, cement, aggregate and fine admixtures.

The collectors used thus far have been designed for cement content determination only, i.e. they do not distinguish between cement, puzzolanic or other hydraulic material. With these collectors only the sum of cement and fine admixtures, the binder content can be determined.

The results of such an analysis are independent from the age of the mix and the type of cement used, i.e. no information concerning w/c-ratio or type of cement is needed to perform a reliable analysis, except if the concrete contains fine admixtures.

3 Further separation of the binder

The cement content determination by flotation is a method also applicable on site. But if one wishes to separate the binder in more detail, new collectors and methods have to be developed.

At present, a fly ash determination can only be performed in a laboratory. Further, for optimal adjustment of the collectors and the other reagents more information must be available concerning the materials used for making the concrete to be analysed. If samples of the cement, the fly ash and the aggregate are available, it is possible to use the flotation method for a fly ash determination in fresh concrete. As the method is a separation method rather than an analytical procedure it is not possible to use the flotation as a test whether the concrete contains fly ash or not.

Fine admixtures in fresh concrete can be subdivided with this respect into three groups [4]:

a) Latent hydraulic materials like blast furnace slag. Standard collectors float these materials like cement. However, collectors are known that permit a separation of Portland cement clinker (and gypsum) from the blast furnace slag of a blast furnace slag cement. This permits also a flotation analysis of fresh concrete which contains blast furnace slag as an aggregate.

b) Puzzolanic materials like fly ash or trass. Standard collectors float these materials also like cement. Surprisingly, the flotation of fly ash is more difficult than that of blast furnace slag, although the physico-chemical differences between fly ash and cement are much greater than those between blast furnace slag and cement.

c) Fine other admixtures like powdered quartz or limestone and silica-fume respectively. These fine admixtures also float with the standard collectors. For limestone powder, additional reagents are known, which prevent its flotation. Hence for a flotation analysis of fresh concrete containing fine admixtures, the amount of quartzitic or fume admixtures must be known, whilst the amount of limestone powder can be determined.

4 Experimental

The flotation experiments described in this paper have been performed with a commercially available laboratory flotation machine, manufacured by KHD, type MN 934. The performance of a flotation test is described in detail elsewhere [1,2,4].

Sample size was 100 g of binder and, if used, 200 g of quartzitic aggregate 0-2 mm throughout the tests. A German Portland cement PZ 35 F, a German blast furnace slag cement HOZ 35 L and two German fly ashes named I and II have been used. The fly ashes correspond to the respective German standards for fly ashes for concrete and have been selected for their x-ray-diffraction patterns, because the qualitiy of the separation by flotation was checked by x-ray diffraction analysis according to [1].

Thus, for an optimal performance of the x-ray diffraction analysis fly ash I was used only with the Portland cement and fly ash II was used only with the blast furnace slag cement. However, this does not affect the general validity of the results, as all collectors in the test perform in the same manner with all of the about 20 fly ashes tested in the recent years.

Two series of experiments were performed: One series was aimed at finding suitable collectors for the separation of fly ash and cement.

Preliminary experiments showed, that alkaline activation can seriously affect the performance of a collector like in ordinary cement flotation [1]. Thus, these experiments were performed at pH-values of 10 and 12.5. The other series dealt with fly ash determination from a single flotation experiment and a subsequent determination of the density of the froth product [4].

5 Results and discussion

5.1 Collectors for fly ash determination

Some typical results obtained with the best collectors are shown in Table 1. The column named "%FAS" shows the percentage of fly ash in the froth removed, "AFA" gives the amount of fly ash floated in relation to the amount originally present in the sample, and "AZ" indicates the amount of cement floated in relation to the amount of cement originally present in the sample.

We see from Table 1 that it is not possible to achieve a full separation of cement

Table 1: Results of flotation tests (in Wt.-%)

Collector No.	Type	pH-value	Portland Cement % FAS	AFA	AZ	Blast Furnace Slag Cement % FAS	AFA	AZ
1	Sulfate	12	51	100	98,6	50,5	100	95
2	Succinamide	12	56	88	70	71	99	41
3	Phosphate	10	76	96,8	30	69	70	30
4	Carboxylate	12	66	96,5	49	62	68	41
5	Ammonium-derivate	12	63,5	75	43	66	92	47
6	Mixture	12	67	72	75	68	91	70
7	No.1 + Modifier	12	53	71	44	73	86	42
8	Amine	12	64,5	100	57	61,5	99	62
9	Sulfonate	12	52	60	55	73	100	38
10	Sulfonate	12	65	48	29	76	77	24
11	Phosphate	10	70	100	44	78	56	16
12	Ether-sulfate	10	29	24	58	54	29	25
13	Ether-phosphate	10	77	100	38	76	100	32

and fly ash by a single flotation step. Furthermore it is known that it is also not possible to float a floated material again, thus preventing a multi-step process.

Several chemical types of collectors yield a fairly good separation of cement and fly ash. On the other hand, there exist large differences inside the single groups of collectors. Thus, the ability of a collector to separate fly ash and cement is an individual property of the respective collector and at our present state of knowlegde does not correspond to its chemical or structural composition. However, in general, those collectors with good results with Portland cement produce not so good results with blast furnace slag cement and vice versa. Thus, it is reasonable to use different collectors for the two types of cement.

Two procedures can be derived from Table 1 for a fly ash determination by flotation. Both methods are complicated compared to the cement content determination by flotation and both are not suited for site application but are pure laboratory methods which require a well equipped laboratory in connection with a skilled operator for the flotation machine.

Both procedures work in principle without any pre-information, however a knowledge of the type of cement used in the concrete is preferable for procedure one in order to select the best collectors available. Both procedures cannot be applied with fresh concrete containing fine siliceous admixtures.

Procedure 1:

1. Flotation of the fresh concrete with standard reagents.
   Result: Sum of fly ash, cement and fine admixture if present.

2. Flotation with a suppressor for the fine calcareous admixtures
   Result: Sum of fly ash and cement

   The difference of test 1 and 2 thus gives the content of fine calcareous aggregate of the fresh concrete.

3. Flotation with collector No. 3 if Portland cement is used and collector No. 2 or 13 if blast furnace slag cement is used. If the type of cement is not known, collectors No. 8 or No. 13 are recommended.

   Result: All fly ash and a constant fraction of the cement. The fractions of the cement floated with the respective collectors are given in Table 2.

Table 2. Fractions of cement floated with fly ash collectors

Collector No.	Type of cement	Fraction of cement floated Wt.-%
3	Portland cement	30
2	Blast furnace slag cement	41
13	"	32
13	Portland cement	38
8	"	57
8	Blast furnace slag cement	62

44

From tests No. 1 and 3 the fly ash content may then be calculated according to the following equations, where Z and FA denote the amounts of cement and fly ash respectively (in g or kg), m is the amount floated in test No. i and x denotes the fraction of cement floated according to Table 2.

$$m_1 = Z + FA \tag{1}$$

$$m_3 = FA + xZ/100 \tag{2}$$

(1) in (2) yields

$$m_1 - m_3 = Z - xZ/100 \tag{3}$$

$$Z = \frac{m_1 - m_3}{\left(1 - \frac{x}{100}\right)} \tag{4}$$

and from (1):

$$FA = m_1 - Z \tag{5}$$

Procedure 2:

1. Like in procedure 1

2. Like in procedure 1

3. Flotation with collector 5 at pH = 12, 11 and lower.
   The pH-value has to be kept constant by continuous automatic titration with hydrochloric acid.

This collector has a special pH-dependence of the amount and composition of the material floated. At pH = 12 only little material floats. Its composition corresponds to that of the original sample. This amount has to be subtracted from the yield of the subsequent tests. With decreasing pH-value the amount floated increases rapidly, the material floated consisting mostly of fly ash. At pH-values below 7 also aggregate floats. If all of the binder floats at pH = 10 no subsequent tests are necessary, else the pH-value has to be reduced by one and the flotation repeated until all of the binder but no aggregate is removed. Usually this occurs between pH = 10 and 7. The pH must not be lower than 7 because then considerable amounts of aggregate float too, thus preventing a fly ash determination. It should be noted as a remark, that at pH = 3 this collector is able to float aggregate particles of up to 4 mm diameter!

The fly ash content is then determined as follows:

Take the flotation test at that pH-value that is 2 higher than that pH required to float all of the binder. This material consists of all of the fly ash and of about 20 % of the cement. The calculation of the fly ash content is performed as given above, setting x=20 and taking the corrected flotation yield as m_3.

5.2 Fly ash determination from a single flotation test
This method requires just a single flotation test using standard reagents. It also requires samples of the fly ash used for concrete-making but no samples of the cement.

The flotation product is dried at 105°C and homogenized subsequently. Then the density of the powder and of the reference sample of the fly ash is determined using a pycnometer. From the amount of binder floated and the determined densities of the floated material and of the fly ash, the fly ash content of the fresh concrete is calculated according to the following equations, where m denotes the amount of material floated (in g or kg) and ρ_i is the density of the component i.

$$m = Z + FA \tag{6}$$

$$\frac{m}{\rho_m} = \frac{FA}{\rho_{FA}} + \frac{Z}{\rho_Z} \tag{7}$$

$$FA = m\{\frac{\rho_Z^{-1} - \rho_m^{-1}}{\rho_Z^{-1} - \rho_{FA}^{-1}}\} \tag{8}$$

Table 3 [4] shows some typical results obtained by this method.

Table 3. Fly ash determination by flotation and subsequent determination of the density of the froth product
Fly Ash 1: $\rho = 2,88$ g/cm^3
Fly Ash 2: $\rho = 2,63$ g/cm^3

| Flotationsansatz (g) | | | | Gehalt an FA | Ausbeute | Gehalt an FA |
PZ	HOZ	FA1	FA2	Einwaage %	(g)	bestimmt durch Flotation (%)
100	-	100	-	50	199,0 - 1	50,2 ± 0,4
-	100	-	100	50	197,3 - 1,5	49,1 ± 0,6
100	-	-	100	50	$200 \pm_2^0$	50,5 ± 0,5
-	100	100	-	50	199 ± 1,5	49,6 ± 0,6
100	-	50	-	33,3	149± 1	32,8 ± 0,8
-	100	-	50	33,3	$147 \pm_1^3$	33,4 ± 0,6
100	-	25	-	20	123 ± 1	18,6 ± 1,4
-	100	-	25	20	122 ± 2	19,0 ± 2
100	-	10	-	9,1	110 ± 2	9 ± 2
-	100	-	10	9,1	108 ±	8,6 ± 2,4

6 Summary

The determination of the fly ash content of fresh concrete is of major importance in modern concrete analysis because the use of fly ash for concrete-making is so widespread today, that the amount of fly ash added to the concrete is considered to be a decisive parameter for its quality.

The collectors used thus far in cement flotation have been designed for cement content determination only. Thus, they do not distinguish between cement, puzzolanic or other hydraulic material. With these collectors only the sum of cement and fine admixtures, the binder content can be floated away from the aggregates.

Two series of experiments were performed in this work: One series was aimed at finding suitable collectors for the separation of fly ash and cement. The other series dealt with fly ash determination from a single flotation experiment and a subsequent determination of the density of the froth product.

Several chemical types of collectors yield a fairly good separation of cement and fly ash. However, those with good results with Portland cement produce not so good results with blast furnace slag cement and vice versa. Thus, it is reasonable to use different collectors for the two types of cement.

Three procedures have been elaborated for a fly ash determination by flotation. All methods are complicated compared to the cement content determination by flotation and they are not suited for site application but are pure laboratory methods which require a well equipped laboratory in connection with a skilled operator for the flotation machine.

Two procedures are pure flotation tests which do not require any pre-information on the concrete to be analysed but require several flotations using different collectors and pH-values respectively and the results obtained therefrom must be regarded rather as estimations than as analytical results.

The third method requires just a conventional flotation followed by a determination of the density of the floated material. If reference samples of the fly ash are available, a precise and reliable determination of the fly ash content of fresh concrete is possible.

Considering the fact that no other method presently exists (except for counting the spheres under the microscope) to determine the fly ash content of fresh concrete, the proposed flotation methods indicates new possibilities in this field.

References

[1] Nägele, E. (1981) Application of the flotation method to analytical problems of concrete technology. Diss. Universität Karlsruhe, FRG,

[2] Hilsdorf, H.K., Nägele, E. (1980) A new method for cement content determination of fresh concrete. Cem. Concr. Res. 10, 23-34

[3] Hilsdorf, H.K., Nägele, E. (1980) Die Frischbetonanalyse auf der Baustelle. Beton, 30, 133

[4] Nägele, E., Schneider, U. (1989) Frischbetonanalyse durch Flotation. Beton 39, 252

6 DIRECT MEASUREMENT OF CHLORIDE IN FRESH CONCRETE DELIVERED TO SITE

F. TOMOSAWA
Department of Architecture, Faculty of Engineering,
The University of Tokyo, Japan

Abstract
Circumstances of increase of chloride content in concrete in 1970's in Japan were reviewed. As the regulations on chloride content in fine aggregate had not functioned well, regulations on the total chloride content of fresh concrete was proposed. Target performance of direct measurement method of chloride ion in fresh concrete was settled by Ministry of Construction in 1984, and several types of the methods based on different principles have been developed in 2 to 3 years and evaluated by the evaluation committee entrusted by the Ministry. Some of the developed methods and their performance are introduced. Legal systems for the regulation of the total chloride content in ready-mixed concrete have been established and the content of choride ion in concrete have been decreased to a satisfactory level.

Introduction

It was in mid 1960's that sea sand was first used as fine aggregate for concrete in Japan, and the use increased so rapidly that in 5 years about 40% of the demand of fine aggregate was supplied from sea sand. As the sea sand is produced and used mainly in south-west half of Japan, it means that about 70 to 80% of fine aggregate for concrete comes from the bottom of sea in those districts of Japan. This tendency of the use of sand resources has not changed up to today, and approximately 40 million cubic meters of sea sand are dredged and used yearly (See Fig. 1).

The problem of using sea sand as an ingredient of concrete is, of course, the introduction of chloride ion into concrete which may cause corrosion of steel reinforcement in the concrete. The investigation of chloride content in the concrete of existing structures carried out by the Ministry of Construction in 1979 showed that quite a lot of the concrete structures are

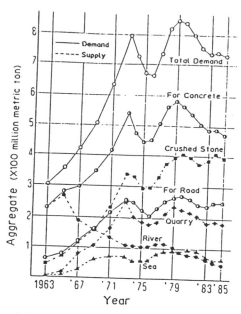

Fig. 1 Demand and supply of aggregate

contaminated by the chloride ion from sea sand and some are susceptible to heavy corrosion of the reinforcement. The establishment of an effective controlling system of chloride content in concrete brought about through sea sand was a matter of urgency at that time. The direct measurement methods of chloride ion in fresh concrete were required and developed in this circumstance.

NEED OF DIRECT MEASUREMENT OF CHLORIDE IN FRESH CONCRETE

The chloride content had been long regulated to be below 0.01% of weight of sand as NaCl since 1953 in Japan Architectural Standard Specification for Reinforced Concrete Structure, published from Architectural Institute of Japan. The overwhelming increase of the use of sea sand, however, had not allowed to follow this stipulation satisfactorily in actual construction activities, and a large quantities of concrete containing a dangerous amount of chloride had been manufactured and placed in the structures.

The regulation value of chloride content of sand 0.01% had been amended to 0.02% in 1975 and a new limit of 0.1% had been allowed if some effective countermeasures against a risk of corrosion of steel in concrete were taken. The allowable limit of 0.02% were again increased to 0.04% of sand in 1977. But these regulations were not respected as well.

The reason that the quality control on the chloride content of sand had not functioned well was considered to lie in the diversification of the responsability to the chloride content of sand. The sand dredgers were not well aware of the stipulations on the chloride content of their products and simply sold those to the concrete manufacturers, the concrete manufacturers considered that they were not responsible for washing out the chloride from the products that they bought from the sand deliverer, whereas the construction engineers had no measures to test and check the chloride content of sand which were already mixed up in concrete and were not interested in the sourse of ingredients of concrete supplied conveniently from readymixed concrete plant through concrete deriverers.

These are the backgrounds which required the introduction of the regulation of the total quantity of chloride in fresh concrete delivered to site. The total quantity of chloride shall be tested and controlled at the acceptance of concrete at site by the responsible construction engineers and shall be reported to the administrative authority of the district. This controlling system is only possible when an easy and rapid test method of quantitative chloride content of fresh concrete is available.

This concept and preliminary exprimental study was first presented by the author and others in 1980. The following conclusions are stated in this study with a large amount of experimental data.
1) No corrosion is recognized if NaCl/sand is less than 0.04% and water/cement ratio is less than 0.65. Corrosion is permissible if NaCl/sand is limited below 0.1% and water /cement ratio is less than 0.55. When NaCl/sand is 0.2%, corrosion rate becomes significant so as to require intense countermeasures such as the employment of lower water/cement ratio and an anti-corrosive agent.
2) Allowable chloride content shall be expressed in terms of a quantity of chloride in a unit volume of fresh concrete in place of chloride percentage of fine aggregate in order to facilitate the acceptance test for the responsible construction engineers.
3) Chloride content of fresh concrete can be directly determined with a satisfying accuracy for acceptance test by "Quantab" method or by an ion-selective detector newly developed for the direct measurement of chloride in fresh concrete (Fig. 2).

DEVELOPMENT OF THE TECHNIQUE PROMOTED BY MINISTRY OF CONSTRUCTION

In 1984 The Ministry of Construction has decided to promote the development of the testing apparatus for the direct measurement of chloride content in fresh concrete to establish a regulation system of an allowable total quantity of chloride in concrete delivered to site. A committee for evaluation of testing

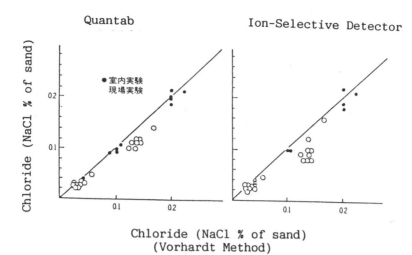

Fig. 2 Direct Measurement of Chloride in Fresh Concrete
(in terms of NaCl % of sand)
"Quantab" and Ion-Selective Detector

apparatus of chloride in fresh concrete was settled and the
performance requirements of developed apparatus were determined
and announced. The committee received the applicant developer
companies and tested and evaluated their developed methods on the
bases of the performance requirements. The performance
requirements were as follows.

1) Possible detection range of the concentration of chloride
   ion in fresh concrete should be from 100g to 1000g per cubic
   meter, which corresponds to the range of 0.05 to 0.5% of
   chloride ion concentration in the water solution.
2) The accuracy should be within +_10% of the control values for
   the range of 0.1 to 0.5% chloride ion concentration when
   measured directly in fresh concrete, whereas the control
   values are the chloride concentration of the distilled
   water from test concretes determined by chemical analysis.
3) The apparatus should be easy to operate, should have a short
   response time and can be applied in any ordinary job site
   environment.
4) The apparatus should be sufficiently durable and reasonably
   economic for use in construction job site or in readymixed
   concrete plant.

The possibility of detection and the accuracy were tested by
diferrent two public materials test centers using 24 diferent
concrete mixes which contained various amount of chloride ion.
Tests were proceeded according to the operating manuals prepared
by the developers of each detector. Then the detectors were
actually employed at the selected readymixed concrete plant for
about one month and evaluated from the view point of
reproduceability, durability and convenience of handling.

VARIOUS DIRECT MEASUREMENT METHODS DEVELOPED AND EVALUATED

Seventeen detectors are already evaluated in 1986 and 1987 to satisfy the proposed performance requirements mentioned above. They are classified into four groups by the detection principles employed. Eleven of them are the ion-selective detectors, some being equipped with digital LED display and printer, three of them are test strip (improved "Quantab") or test tube based on the "Mohr" method, two are based on the voltammetric sensor which measures oxidation and reduction current on a silver electrode, equipped with microcomputer to control and display the measurement procedure and result, and the last one is based on the coulometry by silver ion.

Some examples of these detectors are shown in Figures 3. to 6. with the performance test results.

LEGAL REQUIREMENT ON TOTAL QUANTITY OF CHLORIDE IN FRESH CONCRETE

In 1986, after the first series of evaluation of the detectors for direct measurement of chloride ion in fresh concrete, the Ministry of Construction stipulated allowable total quantities of chloride in fresh concrete to be placed in a reinforced concrete structure by its Technical Standard Notification. The stipulation says as follows.

1) Total quantity of chloride ion in fresh concrete for reinforced concrete and post-tensioned prestressed concrete for public work structures shall be equal to or less than 0.60kg for 1 cubic meter of concrete.
2) It shall be equal to or less than 0.30kg for 1 cubic meter of concrete for pre-tensioned prestressed concrete, grouting materials for prestressed concrete and concrete cured by autoclave.
3) Total quantity of chloride ion in fresh concrete to be used for structural part of building shall be equal to or less than 0.30kg for 1 cubic meter of concrete.
4) If the total quantity of chloride ion becomes more than 0.30kg and 0.60kg for 1 cubic meter of concrete for the building structure, the following countermeasures shall be taken:
   -water cement ratio shall be below 0.55
   -slump of concrete shall be below 18cm and water reducing agent shall be used
   -anti-corrosive agent shall be used
   -Cover thickness of bottom reinforcement shall be over 3cm
5) The total quantity of chloride ion shall be inspected at the acceptance by the construction engineers by means of evaluated detectors of chloride ion in fresh concrete.

This stipulation has been actually in execution since April 1987.

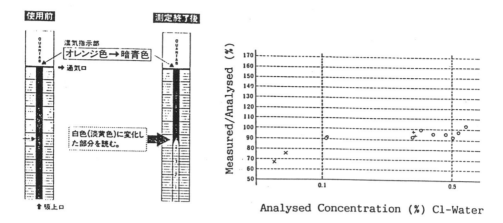

Instrument

Accuracy Test Result

Fig. 3 "Quantab" Method (Onoda Co.Ltd.)

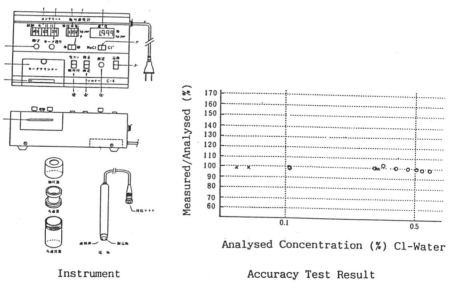

Instrument

Accuracy Test Result

Fig. 4 Voltammetric Sensor System (Yoshikawa Industry Co.Ltd.)

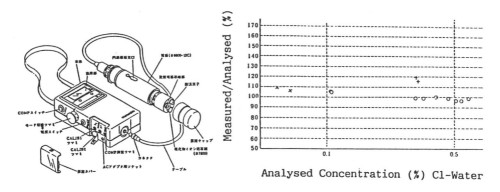

Measured/Analysed (%)

Analysed Concentration (%) Cl-Water

Instrument   Accuracy Test Result

Fig. 5   Ion-Selective Detector (Horiba Industry Co.Ltd.)

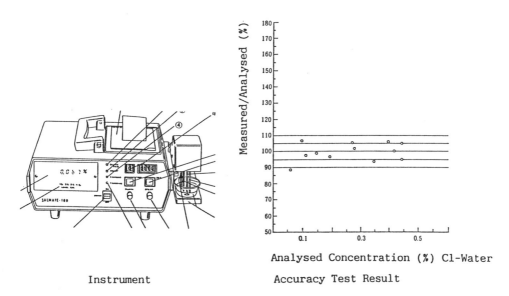

Measured/Analysed (%)

Analysed Concentration (%) Cl-Water

Instrument   Accuracy Test Result

Fig. 6   Coulometry Chloride Ion Detector (Asahi Life Science Co.)

EFFECT ON ACTUAL CHLORIDE CONTENT IN READY MIXED CONCRETE

Japanese Industrial Standard (JIS) 5308 "Ready Mixed Concrete" has been also revised according to this stipulation October in 1986. All the ready mixed concrete is now checked of its chloride content before derivery in the plant by the manufacturers and also checked at the acceptance if needed by the purchasers. The content of chloride in ready mixed concrete has been drastically decreased since the execution of the regulation of total chloride quantity in fresh concrete associated with the direct measurement system of chloride. Figure 7 shows the result of investigation on chloride content in ready mixed concrete. This investigation has been done in 221 ready mixed concrete plant in all over Japan for November 1986 to February 1987 and 2080 concrete samples in total have been measured just before derivery to site. The chloride content has been less than 0.1kg per cubic meter of concrete in more than 80% of the samples.

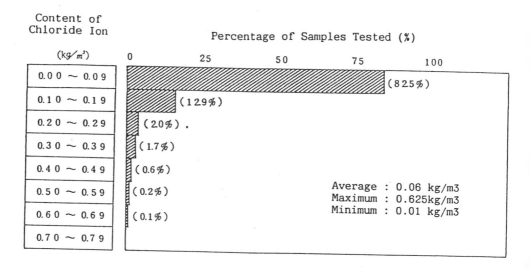

Fig. 7   Chloride Ion Content in Ready Mixed Concrete
         After the Regulation of Total Quantity of Chloride
         (Nov. 1986-Feb. 1987)

CONCLUSION

The regulation system on total quantity of chloride content in fresh concrete and the direct measurement methods of chloride ion in fresh concrete developed for its execution have been stated. The system has been on execution since 1987 and the instruments for the direct measurement of chloride in fresh concrete are now in use in almost all the ready mixed concrete plant and construction site and have made a great contribution to the diminution of chloride in fresh concrete to be used in the structure.

REFERENCE

1) Aggregate Statistics : Ministry of International Trade and Industr
2) F. Tomosawa, I. Fukushi : The state-of-the-art report on the use of sea sand, Concrete Journal, V.16, No.9, Sept. 1978
3) K.Kishitani, F. Tomosawa, I Fukushi : Study for establishing a technical standard for the use of sea sand as aggregate for concrete, Conference of Architectural Institute of Japan, Oct. 1980
4) BRE Information : Simplified method for the detection and determination of chloride in hardened concrete, July 1977
5) A. Yamada et al : A voltammetric detection system for measuring the concentration of chloride ions in fresh concrete, Bull. Chem. Soc. Jpn., 59, 2785-2788(1986)
6) F. Tomosawa : Regulation system of total quantity of chloride in fresh concrete, Cement and Concrete, No.477, Cement Assoc. Jpn. Nov. 1986
7) Regulation of total quantity of chloride in fresh concrete and countermeasures to alkali aggregate reaction : Building Center of Japan, 1986
8) Evaluation reports on the instrument for direct measurement of chloride in fresh concrete : Japanese Institute of Construction and Engineering, 1986 and 1987

7 TESTING METHODS OF CEMENT AND WATER CONTENT IN FRESH CONCRETES AND ESTIMATION OF 28 DAY STRENGTH AT EARLY AGE – STATE OF THE ART IN JAPAN

Y. KASAI
Nihon University, Narashino, Japan

Abstract
The rapid and the early age test methods estimating both the 28 days strength and the mix parameters of fresh concretes and their state-of-the-art in Japan are briefly reviewed.
Twenty three test methods for the unit amount of cement, the unit amount of water and the water-cement ratio to estimate indirectly the 28 days strength are given, while ten direct methods based on accelerated curing are presented as well.
These thirty three methods are arranged in terms of measuring principles and their characteristics.
A comparative study of each test method is given with respect to the hardness, the accuracy, the time required and the expense for the equipments.

1 INTRODUCTION

A demand for predicting the 28 days strength rapidly at an early age is increasing in conjunction with the noticeable increase in recent construction works.

Inadequate fresh concretes have been occasionally produced after the first Oil Crisis (1973~1976) in Japan, thereby a committee on the rapid test method for concrete was organized by The Architectural Institute of Japan (AIJ) in 1975, and a symposium on rapid test method for early age prediction of concrete quality was held by JCI (Japan Concrete Institute) in 1979, with 27 papers, to meet the urgent demand arising from the industry.

As a result, JCI issued a draft "Tentative standard test methods for the early age prediction of concrete quality", while AIJ Committee published " A collection of the rapid, early age test methods predicting the concrete quality" both in 1985.

Experimental studies on the test method predicting the mix proportion of fresh concrete and strength have been followed and reported successively.

The present paper deals with the test method of unit amount of cement, unit amount of water and water-cement ratio in fresh concretes as well as the accelerated strength tests in high temperature cured concretes and mortars to estimate the 28 days strength, and present a brief review on their significance, their principles and assumptions and their characteristics. Each test method is arranged for comparative evaluation in such manner as described in AIJ publication.

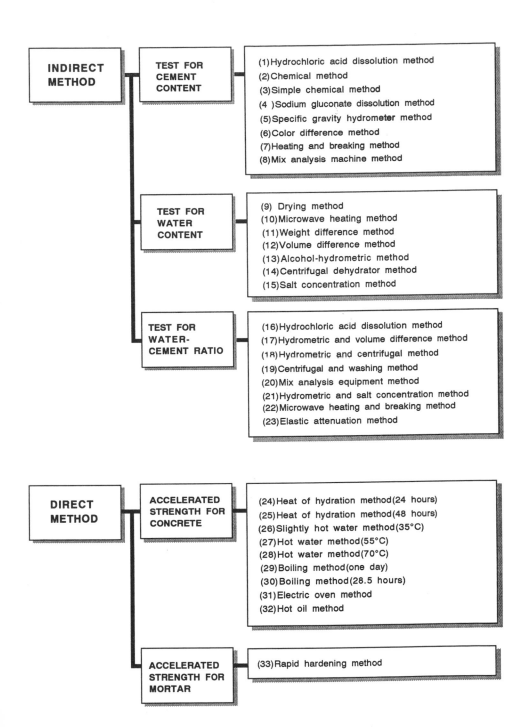

Fig.1 The Rapid And Early Age Test Method Estimating Concrete Strength

2. CLASSIFICATION OF TEST METHODS

The rapid, early age test method predicting the strength of concretes can be classified as shown in Figure 1, though they are based on different test principles.

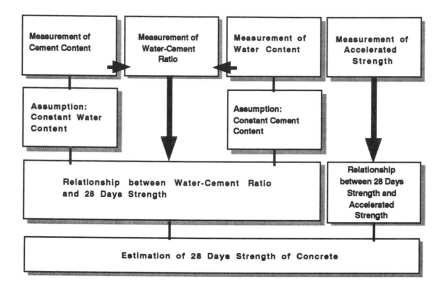

Fig.2 Early Age Test Methods For Concrete And Their Relevance To The Estimation Of 28 Days Strength

The indirect and direct method are included: the former comprises test methods for unit amount of cement, unit amount of water and water-cement ratio, while the latter consists of the accelerated strength tests including hot water curing, high temperature curing in a sealed, restricted mold and accelerated curing method for wet screening mortar and so on.

3. CHARACTERISTICS AND INTERRELATIONS OF EACH TEST METHODS

An interrelation between early the age test methods and the estimated 28 days strength is shown in Figure 2. The characteristics and conditions of applicability of each test methods are discussed hereafter.

1) Prediction method by testing unit amount of cement

This test confirms both the designed compressive strength at 28 days and unit amount of cement assuring the standard concrete quality based on the measurement of unit amount of cement in a fresh concrete.

To estimate the 28 days compressive strength from this test method, the unit amount of water must be kept constant during the production of concrete. It is generally agreed that the water-cement ratio of a concrete is uniform and the designed compressive strength at 28 days is then confirmed, provided that the unit amount of cement and aggregates are accurately weighed and its slump is kept constant.

However, compared with the method based on the water-cement ratio discussed later, the strength predicted by this method shows considerable scatter, since an assumption of the constant unit amount of water is introduced.

2)Prediction method by testing unit amount of water

This test method confirms the designed 28 days strength by measuring the unit amount of water in a fresh concrete. It has an another significance that the upper limit of the unit amount of water specified as 185 kg/m^3 by the new JASS 5 (Japanese Architectural Standard Specification: Reinforced Concrete Work) can be checked at the same time. It is necessary for this method that the concrete must be produced under the constant value of unit amount of cement.

It is generally accepted that the unit amount of water is unvaried if the slump of a fresh concrete is kept constant, however the variation in weighing and size distribution of aggregates as well as fine-coarse aggregate ratio may cause substantial change in unit amount of water even though the apparent change in slump is small. In the presence of superplasticizer, slump is controlled entirely by it so that the direct measurement of unit amount of water is of great significance.

The error in predicting the 28 days strength by this method is relatively large, since the assumption of constant unit amount of cement is introduced.

3)Prediction method by testing water-cement ratio

The strength of hardened concrete is entirely dependant upon the water-cement ratio of the fresh concrete, which can be determined by direct test or calculation from the measured value of unit amount of water and unit amount of cement in a fresh concrete. The 28 days strength can be predictable by referring the relationship between water-cement ratio and concrete strength.

The specimens for the tests of unit amount of water and of unit amount of cement must be sampled from the same fresh concrete to make this method possible. The formula predicting the 28 days strength by the water-cement ratio normally employed at ready mixed concrete plants in Japan has relatively small error.

4)Accelerated curing Method

An outline of this test method is as follows. The high temperature curing to develop strength in a short term is applied to concrete or wet-screening mortar specimens. This accelerated strength value is compared with 28 days strength of concrete of the same source to derive a formula, which is then employed to calculate 28 days strength from the results of the accelerated strength test.

Among factors influencing the accuracy of this method, the formula to assess the 28 days strength out of the accelerated strength test is most substantial since it may vary according to the sort of aggregates (coarse aggregate in particular) and of cement. The preliminary test must be accompanied in each case to predict precisely.

Some skills for the quick preparation of the specimen may be required when a set accelerating agent together with the high temperature curing is applied to the wet-screening mortar. In this case, the prediction formula has similar problems as in concrete.

5)Comparison of these 4 Methods

Above four methods based on **a**) unit amount of cement, **b**) unit amount of water, **c**) water-cement ratio and **d**) accelerated curing are compared as follows.

(1) The accelerated curing method takes longer time than others. It takes 2 to 3 hours in the case of the sealed restricted mold method with high temperature curing , 24 hours in the case of 55°C and 48 hours in the case of 70°C hot water curing, while it takes 2 hours or so in the case of using wet -screened accelerated mortar.

(2) When the formula predicting the 28 days strength is appropriately determined, the reliability of prediction using accelerated curing method **d**) is superior than others. The method using water-cement ratio **c**) is superior than **a**) and **b**), which are rather considered to be a quality assurance test than the strength prediction. In the simultaneous application of **a**) and **b**), the resulting water-cement ratio can clarify the relationship with the strength .

4 EVALUATION, CHARACTERISTICS AND PRINCIPLES OF THE   TEST METHODS

Principles and characteristics of the test methods mainly listed in AIJ publication (A) are rearranged in tables 1 to 4 with some modifications.

A comparison of these test methods is also shown in table 5, which is meant to be a brief introduction and is nevertheless of relativity and still insufficient. Some test method originally developed in Japan will be described in detail.

1)Testing the unit amount of cement

A. The hydrochloric acid dissolution method (**1**)---A method based on the heat of solution of the cement dissolved by Hydrochloric acid. This method can determine the unit amount of cement in wet-screened mortar. The heat of dissolution by hydrochloric acid is proportional to the unit amount of cement as shown in Fig. 3 and 4 by Kanda et al. in /1/1971.

Table 1. Principles and Characteristics of methods testing the cement content in fresh concrete

TEST METHOD	PRINCIPLE	CHARACTERISTICS	REFERENCES
(1) Hydrochloric acid dissolution method	Measuring the heat of dissolution when hydrochloric acid is applied to mortar sieved from concrete.	Impossible when lime components are present in aggregates too. Heat is measured in an adiabatic flask.	Ref./1/ contains 9 refs. Proposed by K. KANDA.
(2) Chemical method	Amount of NaOH titrating the cement solution with HCl is inversely proportional to the cement content in the mortar sample sieved from concretes.	Preliminary test to determine a calibration curve is necessary. Impossible when a lime component in the aggregate is present. It takes ca. 30 min within a range of ±2% using a simple apparatus.	Ref./3/ contains 2 refs. Proposed by Y. KASAI and I. MATSUI.
(3) Simple chemical method	The same as above but simplified to check the cement content by its appearance.	Special sampler is necessary to take mortars from the concrete. A comparison with the designed cement content can be seen by color change taking ca.15 min with accuracy of 5%	Ref./3/ Proposed by M. MORI and S. YASU.
(4) Sodium gluconate dissolution method	Sodium gluconate has no effect on $CaCO_3$ but can dissolve cement only, and the resulting compound is chelatometrically detected.	Test can be executed even though the concrete specimen contains shells or lime stone as aggregates.	Ref./5/ Proposed by K. KATO
(5) Specific gravity hydrometer method	The specific gravity of suspension made of mortar and an additional water is proportional to the cement content of the original mortar sieved from fresh concretes.	Preliminary test to determine a calibration curve is necessary. Impossible when fine powder additives such as fly ash is present. It takes 15 min within a range of ±5% using a simple apparatus.	Ref./2/ contains 18 refs. Recent research was made by K. SHIINA.
(6) Color difference method	Fresh concrete specimen with additional water is colored by Calcein, and color difference is compared with the calibration curve of cement content.	Test is simple but needs an expensive aparatus.	Ref./5/ Proposed by M. NAKAGAWA, I. MIURA and T. KEMI.
(7) Heating and breaking method	Mortar specimen from fresh concretes is heated by microwave oven and broken by a hammer. The quantity passing 88μm sieve is to be cement component.	Test method and apparatus is simple.	Ref. /6/ contains 4 refs. Proposed by S. TUSNODA and T. AKASHI.
(8) Mix analysis machine method	8-1. RAM method Cement component in a part of the slurry of washed concrete is sedimented and the cement content is calculated. 8-2. Mix analysis method Washing and sedimenting cement component and then drying.	Cement content can be measured directly. The coefficients necessary for the calculation must be previously determined. An automated machine is commercially available. It takes ca.10 min with accuracy of ±2%. A prototype has been designed and manufactured.	Ref. /8/ contains 2 refs. The machine is made in U.K. Ref. /9/ contains 1 ref. Proposed by I. TANAHASHI

Table 2. Principles and Characteristics of methods
testing the water content in fresh concrete

TEST METHOD	PRINCIPLE	CHARACTERISTICS	REFERENCES
(9) Drying method	A mortar specimen sieved from fresh concretes is heated to a constant weight and water content is calculated.	Testing apparatus is simple and operated easily. A correction related to the drying of water in aggregates is necessary. It takes ca. 30 min with accuracy of ±2%.	Ref./11/ contains 3 refs. Proposed by Y. KASAI and I. MATSUI.
(10) Microwave heating method	A mortar specimen sieved from fresh concretes is heated by microwave and the weight loss is measured.	Test is very easy and takes 4 ~ 5 min.	Ref. /6/, /7/
(11) Weight difference method	The difference in weight between in air and in water is calculated for mortar specimen sieved from fresh concretes, Water content is then determined by subtracting the under water weight of cement and aggregate.	The amount of cement and aggregates and their specific gravity in the fresh concrete must be previously known.	Ref. /1/
(12) Volume difference method	The volume of degassed mortar sieved from fresh concretes is measured and water content is calculated by subtracting the volume of cement and aggregate.	The amount of cement and aggregates and their specific gravity in the fresh concrete must be previously known. The error is considerable.	Ref./2/
(13) Alcohol-hydro-metric method	When ethyl alcohol is added to concrete, concentration of ethyl alcohol decreases in proportional with the water content of the concrete. If the specific gravity of the ethyl alcohol is measured, water content in the fresh concrete can be estimated.	The water content can be directly measured by simple apparatus. Effects of water in aggregates, dissolution of cement in ethyl alcohol and temperature must be previously considered. It takes ca. 20 min with accuracy of ±5%.	Ref./13/ Proposed by H. NOUMACHI and M. SENOUE
(14) Centrifugal dehydrator method	A water content can be measured from weight loss of concrete or wet-screened mortar dehydrated centrifugally.	Testing procedure is simple, but a calibration curve is needed as the complete dehydration can not be possible.	Ref./17/ Proposed by Kumagaigumi Construction Co.
(15) Salt concentration method	A NaCl solution with known concentration is added to concrete. Water content of the concrete can be determined by measuring the change in salt concentration by the sodium concentration meter.	The salt concentration of small amount of sample from the leachate has been measured by small detector.	Ref. /14/ Changes in concentration of salt in concrete has been measured for long time. A simple measuring device is employed by A. SIMIZU

Table 3. Principles and Characteristics of methods testing the
water-cement ratio in fresh concrete

TEST METHOD	PRINCIPLE	CHARACTERISTICS	REFERENCES
(16) Hydrochloric acid dissolution method	The heat of dissolution of mortar, sieved from concrete, with hydrochloric acid is proportional to the cement content. The water content of the mortar is determined by weighing in air and in water.	The calibration curve of the relationship between cement content in mortar and the temperature increase is included in a commercially available equipment. It should not be applied when aggregates contain lime components. It takes ca. 30 min with accuracy of ca. ±2%	Ref. /1/ contains 9 refs. Proposed by M. KANDA
(17) Hydrometric and volume difference method	The cement content measured by hydrometry and the water content measured by weight difference in air and in water are used for calculation of water-cement ratio of mortar.	The time necessary for execute two kind of measurements is ca. 30 min with accuracy of ca.±5%.	Ref. /2/ contains 18 refs. Recent study was made by K. SHIINA
(18) Hydrometric and centrifugal method	The cement content of mortar sieved from concrete is measured by hydrometry, and the water content is calculated from the aggregate content separated centrifugally.	The time necessary for execute two kind of measurements is ca. 30 min.	Ref. /15/,/16/ Proposed by K. SHIINA and M. TOYODA
(19) Centrifugal and washing method	The water content is measured by centrifugal dehydrator, and the cement content is calculated from the weight loss of the mortar or concrete washed with a special filter cloth.	Effects of surface moisture and clay components in cement and aggregate must be determined by preliminary experiments. The centrifugal dehydrator is commercially available. it takes ca. 30 min with accuracy of ±3%.	Ref. /17/ Proposed by KUMAGAIGUMI Construction CO.
(20) Mix analysis equipment method	The cement content is determined by washing the concrete by machine, and the water content is calculated by subtracting the amount of cement and fine aggregate from the quantity of the specimen.	An automatic analyzer is necessary. It takes ca. 30 min.	Ref. /8/ Proposed by I. TANAHASHI

Continued to next page

Table 3. continued

TEST METHOD	PRINCIPLE	CHARACTERISTICS	REFERENCES
(21) Hydrometric and salt concentration method	The water content is determined by the change in the salt concentration and then the hydrometric method is applied to estimate the cement content	The same concrete specimen is subjected to the two tests. A prototype equipment is developed.	
(22) Microwave heating and breaking method	The water content is determined by microwave heating method. The mortar specimen is then crushed and sieved with 88μm sieve, and the cement content is measured	The same mortar specimen is subjected to the two tests. The equipment is simple and easily available. The accuracy is within ±1%.	Ref. /6/ Proposed by S. TSUNODA and T. AKASHI
(23) Elastic attenuation method	The attenuation of elastic vibration, introduced by a coherent sine wave generator in concrete, is determined in accordance with the water-cement ratio.	The water-cement ratio can be detectable without effects of slump and coarse aggregates. Equipments are commercially available.	Ref. /19/ Proposed by M. KANDA, Y. SUZUKI, S. ISHIBASHI, M. HAYASHI

Table 4. Principles and Characteristics of the accelerated strength test methods for fresh concrete

TEST METHOD	PRINCIPLE	CHARACTERISTICS	REFERENCES
(24) Heat of hydration method (24 hours)	The molded concrete is cured in a thermally insulated container made of foam polystyrene for 24 hours and subjected to the compressive strength test.	Easy to be executed without any heater. The accuracy of prediction of 28 days strength is insufficient since the rate of acceleration is small.	Ref./20/ contains 9 refs.
(25) Heat of hydration method (48 hours)	The molded concrete is cured in a thermally insulated container made of foam polystyrene for 48 hours and subjected to the compressive strength test.	The accuracy is better than the 24 hours' method	Ref. /21/
(26) Slightly hot water method (35°C)	The concrete specimen is put into water of 35°C just after molding, cured for 23.5 hours demolded and subjected to strength test.	No special equipments except for hot water bath is needed. The accuracy for prediction is poor because of the short curing period.	Ref. /21/

Table 4. Continued

TEST METHOD	PRINCIPLE	CHARACTERISTICS	REFERENCES
(27) Hot water method (55°C)	After demolding and capping, concrete specimen is cured in the hot water bath of 55°C for 20.5 hours, and subjected to the compressive strength test. The 28 days strength is estimated by the formula.	The hot water bath to keep the temp. at 55 °C is needed. A calibration curve must be determined previously. It takes 24 hours with accuracy of ca.10%.	Ref. /22/ contains 13 refs. Proposed by T. SOHIRODA.
(28) Hot water method (70°C)	After 18±6 hours from the molding, concrete specimen is cured in the hot water bath at 70°C for 24 ±1 hours and subjected to the strength test after cooling for 1 hour and capped by sulphur or gypsum. The 28 days strength is estimated by the formula.	This method is supplementally listed in "A collection of the rapid, early age test method predicting the concrete quality" issued by AIJ.	Ref. /23/ Proposed by Research Institute of Constructions and Public Works.
(29) Boiling method (one day)	The molded concrete with sealed cap is put into the boiled water for 3~4 hours and tested for strength at an age of 4.5 ~5 hours.	Attention is paid for a scald. The accuracy is poor because of the short period of curing but a result can be obtained quickly.	Ref. /24/ contains 29 refs. Recent study was made by T. SOSHIRODA
(30) Boiling method (28.5 hours)	The molded concrete with sealed cap is cured at 21°C for 5 hours and then put into the boiled water for 23.5 hours and tested for strength after cooling.	More accurate than method (29).	Ref. /21/
(31) Electric oven method	Concrete specimen is formed in the completely sealed, restricted mold and cured in an electric oven for 2 ~ 3 hours. After cooling within the mold, specimen is demolded and subjected to strength test.	It takes 3 ~ 4 hours. Attentions must be paid for the temperature control and the position of specimens in the oven, and to a scald.	Ref. /25/ contains 2 refs. Proposed by M. OZAWA
(32) Hot oil method	Similar to the above method (31) except for the curing temperature of 150 ~ 180 °C in an hot oil.	It takes 1.5 ~ 3 hours. Attentions for the hot oil.	Ref. /26/ Proposed by R. YOSHIKANE
(33) Rapid hardening method	Mixed with a set accelerating agent, mortar specimen sieved from concrete is molded and cured in high temp. curing container for 50 min and tested for strength.	A steel form capable of molding 3 cubes with $\sqrt{10}x\sqrt{10}x5$ (cm) dimension is used. A high temp. curing container keeping temp. at 65°C ~ 80°C±2°C and R.H. over 95% is needed.	Ref. /27/ contains 3 refs. Proposed by S. IKEDA.

67

Table 5. Evaluation of the Rapid, Early Age Test Methods

Test Method	Expense	Time Required	Hardness	Accuracy	Specimen	Specifications
Test of unit amount of cement						
(1) Hydrochloric acid dissolution method	○	◎	○	◎	Mortar	JCI-1, AIJ test
(2) Chemical method	◎	◎	◎	◎	Mortar	JCI-10, AIJ test
(3) Simple chemical method	◎	◎	◎	○	Mortar	JCI-11, AIJ test
(4) Sodium gluconate dissolution method	◎	◎	◎	◎	Mortar	
(5) Specific gravity hydrometer method	◎	◎	◎	○	Both	JCI-13, AIJ test
(6) Color difference method	△	◎	◎	○	Concrete	
(7) Heating and breaking method	○	○	◎	○	Mortar	
(8) Mix analysis machine method	△	○	◎	◎	Concrete	
Test of unit amount of water						
(9) Drying method	◎	○	◎	○	Mortar	AIJ test
(10) Microwave heating method	△	◎	○	○	Mortar	
(11) Weight difference method	○	○	○	◎	Mortar	
(12) Volume difference method	◎	○	○	△	Both	
(13) Alcohol-hydrometric method	◎	◎	◎	○	Concrete	
(14) Centrifugal dehydrator method	△	○	○	○		
(15) Salt concentration method	◎	◎	◎	◎	Concrete	
Test of water-cement ratio						
(16) Hydrochloric acid dissolution method	○	◎	○	◎	Mortar	JCI-1, AIJ test
(17) Hydrometric and volume difference method	◎	◎	◎	○	Concrete	JCI-13, AIJ test
(18) Hydrometric and centrifugal method	△	○	○	○	Mortar	
(19) Centrifugal and washing method	△	○	○	○	Both	JCI-6,7, AIJ test
(20) Mix analysis equipment method	△	○	○	◎	Concrete	
(21) Hydrometric and salt concentration method	◎	○	◎	○	Concrete	
(22) Microwave heating and breaking method	○	○	○	○	Mortar	
(23) Elastic attenuation method	△	◎	◎	?	Concrete	
Test of accelerated strength						
(24) Heat of hydration method (24 hours)	◎	○	○	△	Concrete	
(25) Heat of hydration method (48 hours)	◎	△	○	○	Concrete	ASTM C-684
(26) Slightly hot water method (35°C)	○	○	○	△	Concrete	ASTM C-684
(27) Hot water method (55°C)	○	○	○	○	Concrete	JCI-14 AIJ test
(28) Hot water method (70°C)	○	△	○	○	Concrete	AIJ-suppl.
(29) Boiling method (one day)	△	◎	△	△	Concrete	
(30) Boiling method (28.5 hours)	△	◎	△	○	Concrete	ASTM C-684
(31) Electric oven method	△	◎	○	◎	Concrete	
(32) Hot oil method	△	◎	○	◎	Concrete	
(33) Rapid hardening method	△	◎	○	○	Mortar	JCI

NOTE
◎ cheap short easy excellent
○ middle middle middle middle
△ expensive long difficult inferior

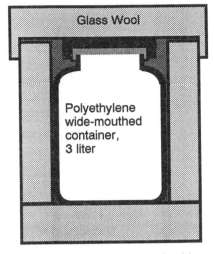

Cap

Glass Wool

Polyethylene
wide-mouthed
container,
3 liter

Cylindrical Can of Galvanized Iron
Sheet, 20ø-30cm

Fig.3 Insulated bottle for the hydrochlolic
acid dissoluton method /1/

Maximum Temperature (°C)

$C=440kg/m^3$,W/C=0.42

$C=230kg/m^3$,W/C=0.80

Shaking Time (second)

Fig.4 Shaking time of dissoluving bottle
vs. maximum temperature /1/

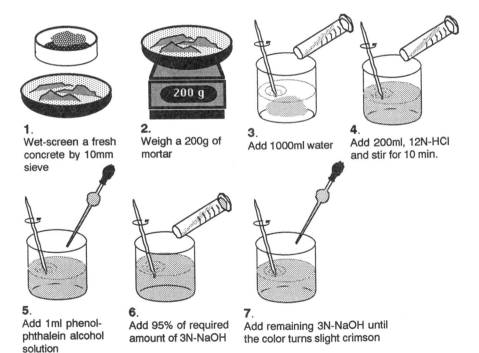

1.
Wet-screen a fresh
concrete by 10mm
sieve

2.
Weigh a 200g of
mortar

200 g

3.
Add 1000ml water

4.
Add 200ml, 12N-HCl
and stir for 10 min.

5.
Add 1ml phenol-
phthalein alcohol
solution

6.
Add 95% of required
amount of 3N-NaOH

7.
Add remaining 3N-NaOH until
the color turns slight crimson

Fig. 5 Flow diagram of the chemical method by titration /2/

69

MOUNT PLEASANT LIBRARY
TEL. 051 207 3581 Ext. 3701 .

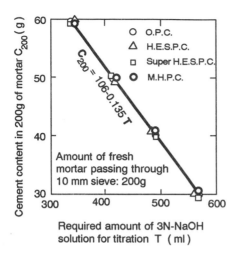

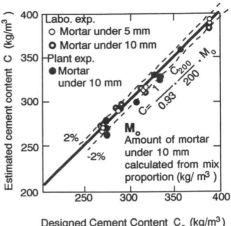

Fig.6 Amount of 3N-NaOH solution required for titration T vs. cement content in 200g of mortar C_{200} /2/

Fig.7 Designed cement content C_0 vs. estimated cement content C /2/

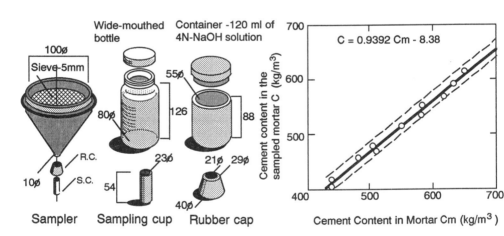

Fig.8 Tools employed for the prediction of cement content by the simple chemical method /3/

Fig.9 Cement content in sampled mortar C vs. designed cement content in mortar Cm /3/

B. The chemical method (2), The simple chemical method (3)--- Methods based on dissolving cement in hydrochloric acid and titrating with sodium hydroxide. The former is suitable for laboratory use as shown in Fig.5, while the latter is applicable to the site practice (refer to Fig.6). Relationship between the amount of sodium hydroxide needed for titration and the unit amount of cement is shown in Fig.7. As for the Portland cements, the relations are linear. Fig.8 shows the relationship between designed cement content and the estimated cement content by the method (2).

Since the hydrochloric acid employed in (1), (2) and (3) can dissolve lime components which is occasionally present in aggregates as well , an estimated cement content tends to show lager value. However in the case of using aggregates with less lime components, fairly well accuracy can be expected.

C. The sodium gluconate dissolution method (4) --- A method based on dissolving cement into sodium gluconate. This method was developed in order to avoid error in the application of hydrochloric acid, which tends to dissolve calcium carbonates in aggregates. It was first proposed by Y. Kasai et al./4/ for the test of hardened concrete, and then applied to fresh concrete by K. Kato et al./5/. Further improvement for on-site application is necessary. Fig.10

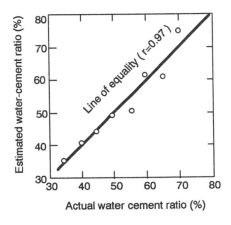

Fig.10 Actual water-cement ratio
vs. estimated water-cement ratio /5/

shows the relationship between an actual and estimated water-cement ratio from the result of K. Kato.

D. The specific gravity hydrometer method (5-1) and The volumetric weight method [5-2]--- Methods based on the specific gravity of the cement suspension. The former has been of in practice by K. Shiina /18/ and A. Shimizu /19/ and the latter has been applied to the Rapid Analysis Machine (RAM). These methods are likely to be affected by clay components in aggregates.

E. The color difference method (6)---A method based on the color differences of cement suspension colored by calcein. This method has not been fully studied.

F. The heating and breaking method (7) --- A method based on the rapid heating by microwave or other rapid drying method to break specimens into small pieces and the quantity passing through the 88μm sieve which are taken as cement/8/. More actual result is needed for this method.

G. The RAM method (8-1) /9/, The mix analysis method (8-2)/10/--A method based on mechanical washing to precipitate cement component in the applied water. A better accuracy and actual results are present in (8-1) and (8-2), but an expensive equipment is necessary.

Among above methods, (5), (6) and (8)) can be applicable directly to concretes, while (1), (2),(3),(4)and (7) deal with mortars separated from concretes. The on-site application is relatively easy in methods (1), (3) and (7).

2)Testing the unit amount of water

A. The drying method (9), The microwave heating method (10) --- A method based on heating and drying to determine the amount of water. The former includes the fryingpan method/11/,/12/ , infrared heating method /13/ and graphite plate heating method, while the latter is of induced heating/8/,/14/.

Fig.11 shows the relationship between the designed and measured water-cement ratio by microwave radiation/15/.

B. The weight difference method (11)--- A method based on the difference in weight of degassed mortar weighed in air and in water corrected by the weight of cement and aggregate under water calculated from mix proportion. This method has been successfully applied to the determination of water-cement ratio by M. Kanda/16/.

C. The volume difference method (12)---A method based on the difference in volume of degassed mortar or concrete by subtraction of calculated volume of cement and aggregate. This method is likely to introduce an error in measuring aggregate volume and is of less accurate.

D. The alcohol-hydrometric method (13)---A method based on changes in the specific gravity of water in fresh concrete by adding alcohol/17/.

E. The centrifugal dehydrator method for mortar(14-1), The centrifugal dehydrator method for concrete(14-2)---A method determining the unit amount of water by means of dehydrating by centrifugal machine/21/,/22/.

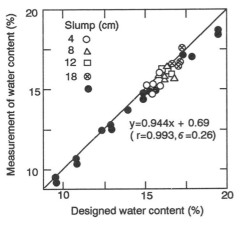

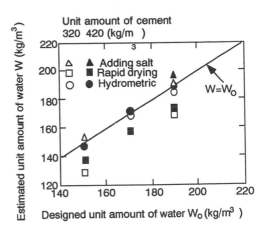

Fig. 11 Measurement of water content   Fig.12 Estimated unit amount of water
of wet-screened mortar by the microwave W vs. designed unit amout of water W_o /20/
heating method /8/

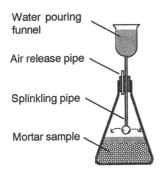

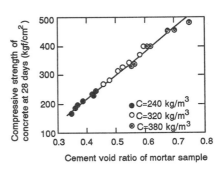

Fig. 14 Cement void ratio of a mortar
Fig. 13 Water pouring apparatus /16/   sample vs. concrete strength at 28 dys /16/

F. The salt concentration method (15)---A method based on changes in salt concentration by
adding salt. This method has long history and becomes a simple and useful method owing to
recent development of salt concentration sensors. A successful result using sodium ion meter
has been reported by A. Shimizu. The comparison of relationships between the designed unit
amount of water and the estimated water content for three testing methods is shown in
Fig.12/19/.

The methods (9) and (10) can determine the drying weight loss directly, while (11) , (12), (13)
and (15) are indirect method. The methods (12), (13) , (14-2) and (15) are applicable to
concretes, while (9), (10), (11) and (14-1) are for mortars without any results in concretes.

3)Testing the water-cement ratio

A. Hydrochloric acid dissolution method (**16**)---A method calculating water-cement ratio by a combination of cement content based on the hydrochloric acid dissolution method (1) and water content based on the weight difference method (11).

This method has actual results and is of good accuracy, but when fine aggregates in a mortar specimen contains lime component, it is hardly applicable since it utilizes hydrochloric acid.

Fig.13 shows a water pouring apparatus and Fig.14 shows the relationship between cement void ratio of mortar sample and the concrete strength at 28 days/16/.

B. The hydrometric and volume difference method (**17**)--- A method calculating water-cement ratio by a combination of cement content based on the specific gravity hydrometric method (5-1) and water content based on the volume difference method (12).

C. The hydrometric and centrifugal method (**18**)--- A method in combination with (5-1) and (14)/18/.

D. The centrifugal and washing method (for mortar)(**19-1**)/21/, The centrifugal and washing method (for concrete)(**19-2**)/22/---Methods combining the centrifugal dehydrator method(14) and the washing test method.

E. The mix analysis equipment method (**20**)---A method calculating water-cement ratio by a combination of cement content based on the washing test method (8) and water content based on the subtraction of cement and aggregate content from the weight of a fresh concrete/9/.

A special washing machine is necessary for method (20)/9/,/10/, and the measured water content is in general not so accurate.

F. The hydrometric and salt concentration method (**21**)---A method calculating water-cement ratio by combining the cement content based on the hydrometric method(5-1) and water content based on the salt concentration method (15).

A single concrete sample can be used throughout the test procedure. A simplified method applicable on site has been developed by A. Shimizu/19/, whose test procedure is shown in Fig.15. The salt concentration detector is shown in Fig.16, which measures the change in light deflection in accordance with the salt concentration from a drop of the filtered bleeding water of concrete. The relationship between the water-cement ratio measured on site and the compressive strength is shown in Fig.17, and the mix analysis is possible by measuring the under-water weight of aggregates.

G. The microwave heating and breaking method (**22**)---A method calculating water-cement ratio by a combination of cement content based on the heating -breaking method (8) and water content based on the microwave heating method (10).

The method (21) ,as well as the method (22), is based on an excellent combination of test principles since the same concrete sample can be supplied to the each test, and (21) is probably applicable as an in-situ test, while (22) is probably applicable if the specimen is extended to concrete other than mortar.

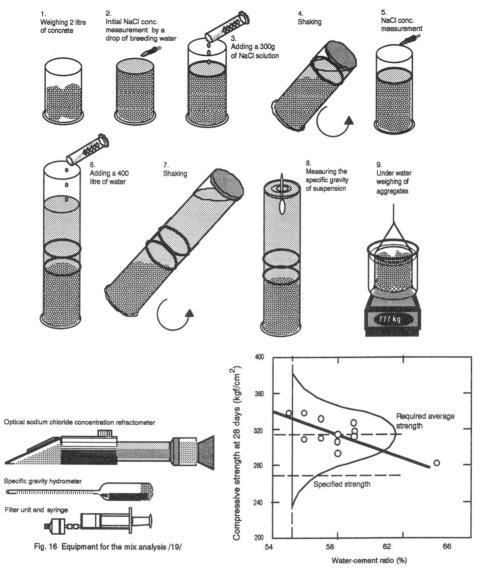

1. Weighing 2 litre of concrete

2. Initial NaCl conc. measurement by a drop of breeding water

3. Adding a 300g of NaCl solution

4. Shaking

5. NaCl conc. measurement

6. Adding a 400 litre of water

7. Shaking

8. Measuring the specific gravity of suspension

9. Under water weighing of aggregates

??? kg

Optical sodium chloride concentration refractometer

Specific gravity hydrometer

Filter unit and syringe

Fig. 16 Equipment for the mix analysis /19/

FIG. 17 Measured water- cement ratio vs. 28 days compressive strength from Shimizu, A.

Compressive strength at 28 days (kgf/cm^2)

Water-cement ratio (%)

Required average strength

Specified strength

H. The elastic attenuation method (**23**)---A method estimating water-cement ratio by measuring amplitude variation as an elastic attenuation of sine wave generated in a concrete. This excellent idea is first reported by K.Kanda et al. in 1979/23/ and now commercially available. The actual results in construction site, especially applications with superplasticizer, must be reported.

As a whole, the method (23) is quite smart thought the method (21) is also promising. The apparatus is shown in Fig.18, and the relationship between the measured amplitude value and the designed water-cement ratio are shown in Fig.19.

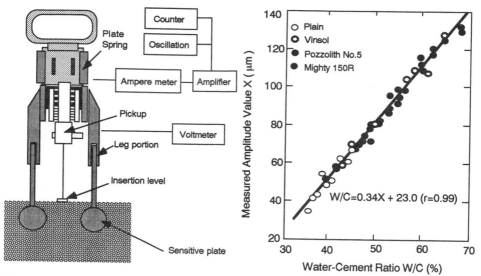

Fig.18 Vibration driver for the elastic attenuation method /23/

Fig.19 Water-cement ratio vs. measured amplitude value /23/

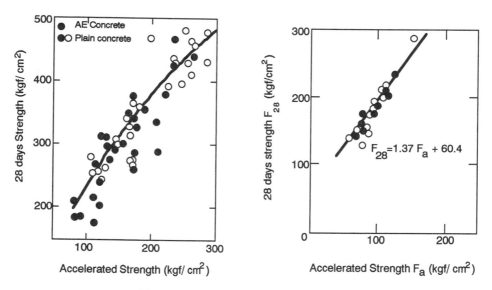

Fig. 20 Accelerated strength of concrete by the hot water method (55°C)vs. 28 days strength /26/

4)Accelerated strength test method using concrete and mortar

A. The heat of hydration method for 24 hours (**24**) ,The heat of hydration method for 48 hours (**25**)--- A method based on the curing in an adiabatic container making use of the heat of hydration.

No particular pre-curing is necessary, and the standard mold can be used for (24) and (25). Since the curing temperature is not high enough, the available strength after 24 hours is relatively smaller for the method (24), and the error in the prediction of 28 days strength is appreciable. A longer curing period as in (25) may be preferable to be more accurate.

B. The slightly hot water method (35°C)(**26**), The hot water method (55°C)(**27**),/26/,/27/   The hot water method (70°C) (**28**)/28/--- A method based on the accelerated curing in hot water .

The hot water curing method forms a main stream to obtain the accelerated curing strength and has many variations. The accelerated strength by the method (26), though one of the method of ASTM C-684/25/ is small as the curing temperature is 35°C for 24 hours. The relationship between the accelerated strength and the 28 days strength is shown in Fig.20, and this method (27) is specified both by JCI and AIJ and has considerable actual results. The compressive strength test can be carried out after 24 hours using simple restriction mold capping. The curing temperature of 55°C is almost safe enough from a scald. The methods with curing temperature of 70°C and 80°C, in which strength can be tested within 48 hours including pre-curing, has been reported. The method (28) is more accurate than the method (27), but longer time is needed and more dangerous with respect to a scald.

C. The boiling method (one day)(**29**), The boiling method (28.5 hours)(**30**)--- A method based on the accelerated curing in the boiling water.

Attentions must be paid to the boiling of specimens in the methods (29) and (30). A noticeable error in the test of accelerated strength within 3 to 4 hours is reported due to the shortage of curing period. The degree of restriction of the mold should be large enough. The method (30) specified in the ASTM C-684 needs only 28.5 hours for the test.

D. The electric oven method (**31**)/30/, The hot oil method (**32**)/31/--- A method based on the high temperature curing within the restricted and sealed mold.

The curing temperature of (31) and (32) exceed 100°C, and the mold should be particularly stiff, and monolithic with tapered specimens. The demolding is carried out using a piston. An electric oven is employed for the method (31), when a special care must be taken for the position of specimens in the oven, since the temperature distribution and the control capability varies according to the position.

The method (32) uses high temperature oil bath which can provide an uniform temperature distribution. The both of these two methods can be performed within one day. Attentions should be directed to a scald.

E. The rapid hardening method (33)/32/,/35/ ---A method based on the set accelerating agent applied to the wet-screened mortars. Fig.21 shows a diagram of the rapid hardening method by accelerated mortar strength. A comparison was made with respect to the age of test. It is shown in Fig.22 that a better accuracy can be obtained when tested in the age of 2 hours rather than 1.5 hours.

The effect of chemical admixtrure on the estimated compressive strength of concrete is shown in Fig.23, and the relationships between the accelerated mortar strength and the 7 days or 28 dyas compressive strength are shown in Fig.24.

The method (33) using set accelerating agents requires some skills, and it can evaluate the designed strength within two hours.

As a whole, hot water curing method of 55°C, (27) is reasonable if the specimen is to be tested next day, while a safe and high temperature curing with sealed, restricted mold is necessary for the concrete test within several hours.

Fig.21 Flow diagram of the rapid hardening method by accelerated mortar strength /32/

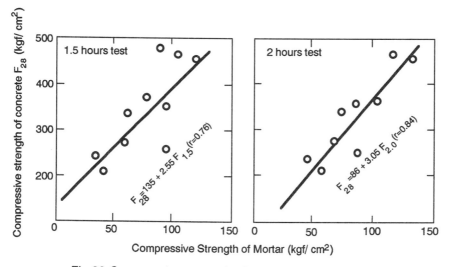

Fig.22 Compressive strength of mortar by accelerated testing vs. compressive strength of concrete at 28 days F_{28}/34/

78

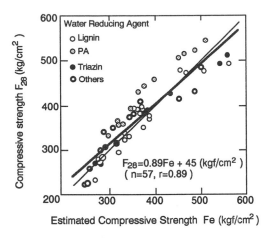

Fig.23 Estimated compressive strength of concrete by accelerated mortar strength Fe vs. compressive strength of concrete at 28 days F_{28} . /33/

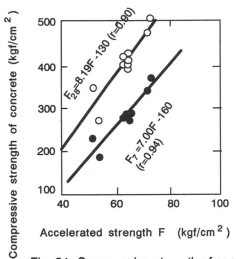

Fig. 24 Compressive strength of concrete vs. accelerated compressive strength F /36/

5 CONCLUDING REMARKS

Test methods of unit amount of water, unit amount of cement, water-cement ratio to check or predict the strength of concrete at 28 days are reviewed. The rapid and early age test methods have been generously studied for those ten years, and some reasonable and realistic methods are developed. It appears that the next stage to standardize them must be considered.

REFERENCES

A)Tentative standard test methods for the early age prediction of concrete quality, JCI, (1985)
B)A Collection of the Rapid, Early Age Test Method Predicting the Concrete Quality, AIJ, (1985)
C)Proc.Rapid test method for early age prediction of concrete quality, JCI, (1979)

The following references are mainly quoted from the above three major publications indicated as A,B or C.

/1/Kanda, M. A measuring method for water cement ratio in fresh concrete, Transaction of JSCE, vol.2, Part 2, 1971 (in English)
/2/ Kasai, Y. et al., A testing method for cement content of fresh concrete - By dissolving cement in hydrochloric acid and titrating with sodium hydroxide-, Transactions of the JCI, 1979 (in English)
/3/ Method of test for determination of designed unit cement content of fresh concrete with mortar sample by hydrochloric acid and sodium hydroxide, A, B
/4/ Kasai, Y. et al.,Examination and test method for cement content in hardened concrete by sodium gluconate, Review of the 40th general meeting, CAJ, 1986
/5/Kato. K. et al., Rapid evaluation of concrete strength by fresh concrete, Extended Abstracts, The 43rd Annual Meeting of CAJ, 1989
/6/ Method of test for water-cement ratio of fresh concrete by use of the specific gravity hydrometer and volumetric flask, A, B
/7/Nakagawa, M.,Miura, I., Kemi, T.,A study for the estimation of cement content in a fresh concrete, C
/8/ Tsunoda, S., Akashi, T.,A method measuring the water-cement ratio by microwave heating, C
/9/Method of test for cement content of fresh concrete by use of the automatic washing machine, A, B
/10/ Ohgishi, S.,Tanahashi, I., Matsuo, Tamura,Test apparatus for measuring constituent materials of fresh concrete, Trans. JCI.,5, (1983), pp53-60, (in English)
/11/ Method of test for water content of fresh concrete-The drying method, B
/12/ Kasai, Y., Matusi, I.,Method of test for water content of fresh concrete- The drying method, C
/13/ Nishizawa, A study for the applicability of a method predicting the cement content and water-cement ratio by means of washing machine and infrared dryer, C
/14/ BRI of Japan,Technical report on the test method for unit amount of water in a fresh concrete, Annual Report of the Technical development for improving durability, Technical Center for the Land Development, (March 1988)
/15/ Horikawa, K. et al., Comparative test for the quality evaluation of various fresh concretes, Proceedings of the JCI, vol.11, No.1, pp.159-164, 1989
/16/ Kanda, M. et al., A study on early estimation of concrete strength, Transaction of JSCE, vol. 5, 1973 (in English)
/17/ Nakajima, M. et al., A rapid analytical method of water content in fresh concrete (mortar and cement paste), C
/18/ Shiina, K., Toyoda, M., Measurement of water-cement ratio of fresh concrete by means of combining the hydrometric method and centrifugal machine method, Cement Gijutu Nenpo, 39, (1985), pp100-103
/19/ Shimizu, A. et al., Studies for simple test method predicting the water content in a fresh concrete, Part 1, Part 2, (1986), Part 3, (1987), Part 4, Part 5 ,(1988), Summaries of technical papers of annual meeting of AIJ
/20/ Marushima, M., Study on three methods for the determination of water content in fresh concrete, Summaries of Technical Papers of Annual Meeting of AIJ, 1989
/21/ Method of test for water-cement ratio of fresh concrete with mortar sample by use of the centrifugal machine and washing screen, A, B
/22/ Method of test for cement content and water-cement ratio of fresh concrete by use of the centrifugal machine and washing screen, A, B
/23/ Kanda,M. et al., A method for measuring water cement ratio in fresh concrete by using vibration, Concrete library of JSCE, No.3, June, 1984 (in English)
/24/ Soshiroda, T., Fujisawa, K, Accelerated strength tests of concrete, Cement Concrete, No.302, (1972)
/25/ASTM Standard Test Method for Making Accelerated Curing, and Testing Concrete Compression Test Specimens C-684
/26/ Soshiroda, T., Masada, Y., Accelerated strength test of concrete, C
/27/Method of accelerated strength test of concrete by use of specimens cured in hot water(55°C), A, B
/28/ Method of accelerated strength test by use of concrete specimens cured in hot water (70°C), A,B-supplement

/29/ Soshiroda, T., Fujisawa, K., Accelerated strength tests for an early age prediction of concrete quality, Cement Concrete, No.280, (1970)

/30/ Ozawa, M, Nihei, T. Early age prediction of the concrete strength by closed form, C

/31/Yoshikane, R., On the early age prediction of the concrete strength using sealed, restricted mold in high temperature curing, C

/32/ Ikeda, S., A rapid estimation method for concrete strength by accelerated hardening, Transactions of JSC, No.255, 1976, 11

/33/ Fukushima, N. et al., Test method for the rapid estimation of concrete strength by accelerated hardening,C

/34/Abe, R. et al., On early age evaluation of concrete on site by accelerated hardening method, C

/35/ Method of test for estimating the strength of concrete by rapid-hardening process using mortar from fresh concrete sample, A, B

/36/ Central Research Institute of the All Japan Ready-Mixed Concrete Association, A study on the comparisons of test methods of accelerated strength test for concrete, 1990

8 IN-SITU DETERMINATION OF ENTRAINED AIR IN FRESHLY MIXED CONCRETE USING FIBER OPTICS

F. ANSARI
Civil and Environmental Engineering, New Jersey Institute of
Technology, Newark, NJ, USA

Abstract
Quality control and condition analysis through non-destructive testing in concrete is
one of the important issues in concrete construction. In this paper, a methodology
for determining the distribution and amount of entrained air system in concrete is
presented. Based on the developed methodology, to facilitate acceptance or rejection
of the concrete at the construction site, a test method and apparatus is invented. The
distribution and percentage of entrained air in freshly mixed concrete will be measured using Fiber Optics. Reflected light intensity measurements will determine the
distribution of air bubbles in fresh concrete. The new Fiber Optic Airmeter's speed,
reliability and accuracy has been examined through repeated experiments against
conventional techniques.

1 Introduction and objectives

The way in which air entrainment increases the resistance of concrete to deterioration by frost is widely acknowledged and understood. Volumetric, and Pressure
methods are among the most widely used procedures for air content determinations
of freshly mixed concrete. For instance, in the Volumetric method [1,2] a known
volume of concrete is removed from the mix and mixed with water. Air is sepera-
ted from the slurry by agitation, and measured from the decrease in volume. In the
Pressure method (3) air which has been pumped to a predetermined pressure in a
compartment of known volume is released into a sealed container full of concrete. The
pressure volume relationship, Boyle's law, is then used to measure the amount of air.

In all the above-mentioned techniques experiments have to be performed in the
laboratory, and in-situ characteristics of the air entrained concrete may be quite
different from the laboratory measured amount. Moreover, by the time the laboratory
results are analyzed the concrete in question may have been cured and set. It is also
desirable to be able to measure the amount of entrained air at several locations in
a pavement slab while it is being placed in order to assure uniformity of the mix
throughout the pavement.

The main objectives of the research reported here are to develop a methodology for monitoring the amount and distribution of entrained air bubble system in concrete, and to construct a handheld portable device based on the developed methodology for use at the construction site.

2 Detection of air bubbles by fiber optics

An unconventional use of Optical Fibers which has been gaining in stature over the past few years is the development of Fiber-Optic sensors. Fiber-Optic sensors have been extensively employed in Aeronautics and Space applications where measurements of temperature, pressure, displacements, and flow rates at hostile environments are necessary [4-8]. The advantages of Fiber Optic sensors include high sensitivity and immunity to electromagnetic interference. Also, since no electricity is required at the location of the sensor, they can be safely used in any environment. Fiber-Optic sensors employ intensity or phase modulation of the light propagating through them in order to sense and measure the mentioned effects. In the particular application of Fiber Optic sensors to the entrained air content in freshly mixed concrete, a system based on the intensity of reflected light through different mediums is developed.

When a light wave travels from one medium into another, the indices of refraction in the two mediums determine the ratio in between the amount of refraction and reflection. If the light wave travels from a material with higher refractive index into a material with lower index, most of the light reflects back, whereas if the refractive indices are the same or almost the same, most of the light will enter the new medium (Fig. 1). The angle by which light ray strikes the interface between two mediums, the angle of incidence, also play an important role in controlling the amount of refraction and reflection. As shown in Fig. 1, this angle is measured with respect to the line perpendicular to the interface. Larger angles of incidence bring about more reflected rays, whereas sharper angles give rise to further refractions.

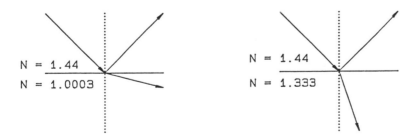

Fig. 1. Effect of refractive index on reflection and refraction

The simple concept of refraction and reflection when combined with fiber optics will result in a powerful measurement instrument for air entrained concrete. As shown in Fig. 2, if the light exiting end of a glass optical fiber (N = 1.44) meets an air bubble (N = 1.0003), most of the light will reflect back due to large difference in between the indices in these mediums.

DETECTION METHODOLOGY

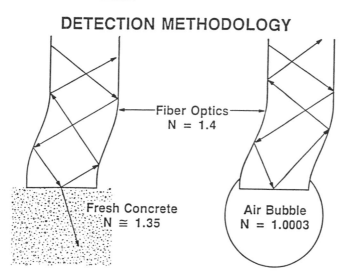

Fig. 2. Air bubble detection methodology in concrete

Therefore, measurements of the reflected light, during one of our experiments, showed high intensity (Fig.3). On the other hand, the same fiber tip when immersed in fresh concrete (Fig. 2), showed a large decrease in the intensity of the reflected light (Fig. 3). The glass air interface at which total internal reflection occurred was eliminated and the index of the glass was nearly matched by the fresh concrete. When this happens, the light is no longer totally internally reflected and most of the optical power is transmitted from the fiber end into the concrete. As shown in Fig. 3, a refractive index difference of 0.33 (air index = 1.0003, cement paste = 1.3) caused a large difference in the intensity of reflected light (about 6.8 Volts). Therefore, such a light reflecting fiber optic sensor offers great accuracy in high sensitivity measurements. Results shown in Fig. 3 also indicate sensor's quick reponse in recognizing the change in it's surrounding environment (less than 1/31 second).

2.1 System Configuration
Fig. 4 describes components of the hand held fiber optic sensor for the determination of entrained air in freshly mixed concrete. A battery powered diode laser emitting visible light at 670 nm wavelength is employed in transmitting optical signals. A silica glass multimode optical fiber delivers the light to the concrete. Depending on

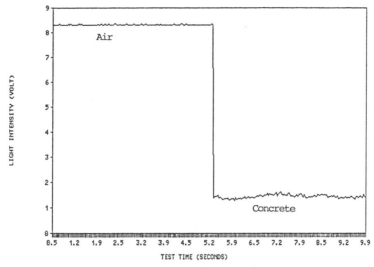

Fig. 3. Reflected light intensities in the air and concrete

the amount and number of air bubbles in the fresh concrete, light will reflect back into the fiber. The reflected signal is transmitted back to a photodetector. The reflected light intensity signal is changed to an electrical current in the photodetector and amplified through an amplifier. At this stage the voltage output of the amplifier

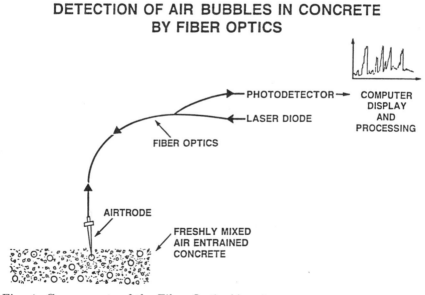

Fig. 4. Components of the Fiber Optic Airmeter

is converted to a digital signal via an analog to a digital converter. Real-time data is then transferred to a personal portable computer for data processing, and percent air determinations. As shown in Fig. 4, air bubbles are detected by reflected light signal pulses as the sensor is moved onward in fresh concrete.

2.2 System Operation
Steps involved in operation of the sensor are outlined as follows:

1. The sensor and the computer are turned on and the data acquisition program is triggered so as to collect data. A run duration of ten seconds at a sampling rate of thirty Hertzzare selected (these values were chosen through trial and error, other settings may also prove to produce optimum results).

2. Sensor's tip is immersed in concrete for ten seconds in order to initialize the system with respect to minimum intensity readings associated with concrete.

3. Sensor's tip is held in the open air in order to establish the maximum intensity readings due to air.

4. Once the system is initialized, sensor's tip is plunged into the fresh concrete and moved onward in order to sense the presence of air bubbles at different locations in concrete. The computer stops data acquisition after ten seconds. Real time plot of reflected light intensity versus time is displayed, and the intensity values are saved in a file for further processing.

3 Experimental Program

Experimental program consisted of measuring the air content of different batches of freshly mixed concrete by the Pressure, Volumetric, and present studies' method. A mix proportion of 1.1 : 1.5 : 1.5 : 0.5 (cement : sand : aggregate : water) by weight was employed for all batches, and only the amount of air entraining admixture was varied. However, in order to examine the response of the system to variations in water contents, similar mixes with water-to-cement ratios of 0.45 and 0.50 were also tested. Type III Portland cement conforming to ASTM C150 and ASTM No. 2 grade river sand passing through sieve No. 8 were employed. Maximum coarse aggregate size passing 9.5 mm and retained on No. 4 sieve was used. Air entraining agent used in the present study was manufactured by Master Builders (MB-VR) conforming to ASTM C260 and AASHTO M154 requirements. The dosage of MB-VR for different batches of fresh concrete ranged between 2 to 16 milli-liter per 45 kg cement used in the mix (0.41 to 3.6 fl.oz per 100 lbs. cement).

3.1 Air content from air bubbles

Conversion of air bubble data to a single value representing percentage of air, requires a calibration procedure at the beginning of measurements. Results shown in Fig. 5 consist of data collected by keeping the sensor stationary in the fresh concrete for a period of ten seconds, and data from measurements in the open air for another ten seconds. The area in between the two horizontal lines representing the response in the air and concrete (Fig. 5), correspond to 100 percent air content. At this point, the Airmeter is ready for air content measurements. Therefore, the sensor tip is plunged back into the concrete and moved onward for another ten seconds. Typical air bubble data collected in this way are shown in Figures 6 thru 10. Percentage of air is determined by calculation of the area under the curve representing air bubble data, and then dividing this value by the value calculated earlier representing 100 percent air.

The above-mentioned air content determination procedure, and the calibration procedures are programmed in a batch file so that all the calculations can be performed at a single computer command at the construction site. Air content measurements by the Fiber Optic Airmeter requires 110 seconds. This time duration includes calibration, data acquisition and computation time.

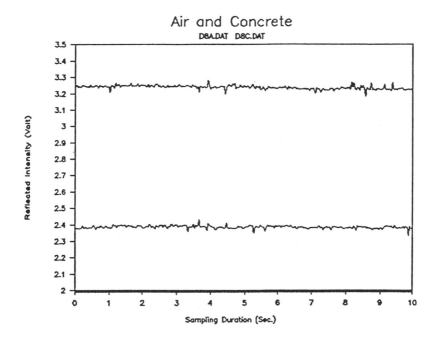

Fig. 5. Calibration curves prior to measurements

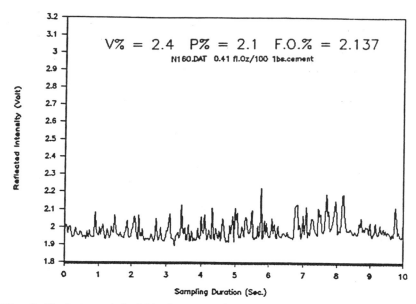

Fig. 6. Real-time air bubble data in concrete with 2 milli-liter admixture

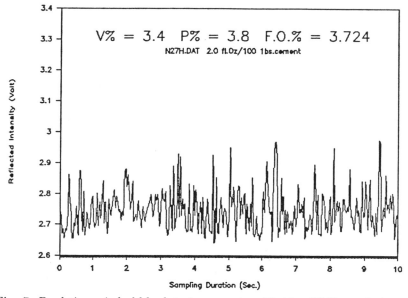

Fig. 7. Real-time air bubble data in concrete with 10 milli-liter admixture

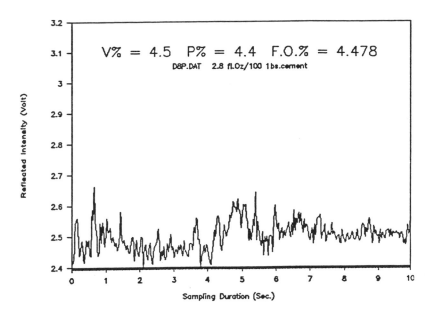

Fig. 8. Real-time air bubble data in concrete with 14 milli-liter admixture

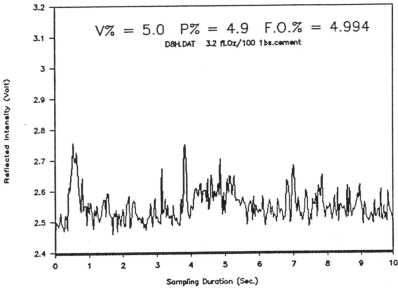

Fig. 9. Real-time air bubble data in concrete with 16 milli-liter admixture

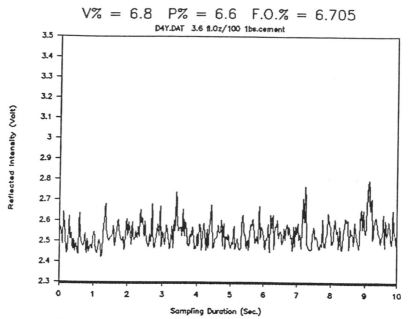

Fig. 10. Real-time air bubble data in concrete with 18 milli-liter admixture

Figures 6 thru 10 represent typical real-time computer plots of data acquired by the Fiber Optic Airmeter developed in this study. For each fresh batch of concrete, due to ease and speed of operations involved in using the Fiber Optic Airmeter, uniformity of air entrainment was examined by collecting data at three different locations in the concrete mixer. Values shown on top of the air bubble data in Figures 6 thru 10 represent air content percentages measured by the Volumetric, Pressure, and the Fiber Optic methods respectively. As these percentages reveal, results are comparable to a very good degree of accuracy. Concretes made in the laboratory were prepared in a small mixer, and therefore, repeated Fiber Optic measurements of the same mix revealed uniform dispersion of air bubbles in the mix. However, the system developed in the present study can be easily employed at a construction site so as the uniformity of large mixes of concrete during mixing or pouring can be determined.

4 Acknowledgements

This study was made possible by a grant from the National Research Council, Strategic Highway Research Program (SHRP), contract number SHRP-87-10011.

References

[1] Pearson, J.C., Volumetric Method For Determining The Air Content of Freshly Mixed Concrete, Proceedings ASTM, Vol. 44, 1944, 343-350.

[2] Menzel, C.A., Development And Study of Apparatus And Methods For The Determination of The Air Content of Fresh Concrete, Proceedings ACI, Vol. 43, 1947, pp. 1053-72.

[3] Klein, W.H., Walker,S., A Method of Direct Measurement of Entrained Air in Concrete, Proceedings ACI, Vol. 42, 1946, pp. 657-68.

[4] Davis,C.M., Fiber Optic Sensors: An Overview, Society for Photo-Optical Instrumentation Engineers (SPIE), Vol. 478, pp. 12-19.

[5] Baumbick, R.J., Alexander, J., Fiber Optic Sense Process Variables, Control Engineering, (March 1980), pp. 75-77.

[6] James, K.A., Quick, W.H., Fiber Optics: The Way to True Digital Sensors, Control Engineering, (February 1979), pp. 30-32.

[7] Ramprasad, B.S., Radhabai,T.S., Speckle - Based Fiber Optic Current Sensor, Optics and Laser Technology, (June 1984), pp. 156-158.

[8] Nakayama, T., Optical Sensing Technologies By Multimode Fibers, Society for Photo-Optical Instrumentation Engineers (SPIE), Vol. 478, 1984, pp. 19-27.

9 DETERMINATION OF THE AIR-VOID PARAMETERS IN FRESH CONCRETE

H. SIEBEL
Research Institute of the German Cement Industry,
Düsseldorf, Germany

Abstract
The air-void parameters of concrete are determined by the classical method on hardened concrete and with the new Danish DBT air-void measuring equipment on fresh concrete. The results are compared and evaluated.

1 Introduction

Concrete with a high resistance to freezing and thawing in the presence of de-icing salt must have an adequate content of small voids. With stiff road concrete it can generally be assumed that when using a permitted air-entraining agent and with correct preparation of the concrete, there will be a sufficient content of small voids if the total air content in the fresh concrete tested by the pressure method does not fall below the minimum value stipulated in the standards and specifications, e.g. 4 % for a maximum grain size of 32 mm. It has been found that this assumption is not valid when softer concretes or concretes with superplasticizers or plasticizers are used. It is then necessary to determine the air-void parameters, such as the micro air-void content L 300 (all pores up to 300 μm), the spacing factor and the air-void distribution, as well as the total air content. So far it has only been possible to determine these air-void parameters in hardened concrete. The newly-developed Danish equipment "DBT air-void measuring equipment" makes it possible to determine these parameters in fresh concrete.

2 Air-entrained concrete

The preparation and placing of air-entrained concrete does not generally present any difficulty when stiff concrete of the type used in road construction is involved. This has been confirmed by large numbers of investigations, see among others [1,2]. With these concretes it is sufficient to determine the total air content in the fresh concrete [2]. The increased use of soft air-entrained concrete with high resistances to freezing and thawing in the presence of de-icing salt, e.g. for bridge cappings or sewage treatment plants, and of air-entrained concrete to which superplasticizers or plasticizers have also been added, make it necessary to check the air void parameters. Several investigations have shown that the void system in soft concretes are displaced from small voids towards larger ones, especially when super-plasticizers and plasticizers are used. Fig. 1 shows this change in the air-void system. A rapid-hardening air-entrained concrete with a spread of about 32 cm was plasticized by the addition of 3 % superplasticizer (relative to the weight of cement) so that its spread became 55 cm. While there was a high proportion of small voids in the concrete without superplasticizer – the micro air-void content L 300 was about 2 Vol.-% – this fell to about 0.7 Vol.-% on addition of a superplasticizer as the smaller voids had combined to form larger ones. If these concretes had been tested only for total air content, then values of about 4 Vol.-% would have been obtained for both concretes and the change in the air-void system would not have been detected. In such cases it is therefore absolutely essential to check the air-void parameters. Testing the hardened concrete takes too long, is too complicated and comes too late for any correction during concreting.

3 "DBT air-void measuring equipment"

For quite a long time various scientists have been working on the determination of air-void parameters in fresh concrete [6]. At the moment it seems that a Danish equipment [7] (see Fig. 2) is one of the most advanced in the field. A wire basket

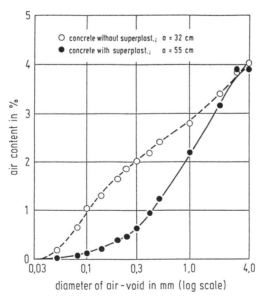

Fig. 1. Air content as a function of diameter of the air voids for concrete without and with superplasticiszers

(wire spacing less than 6 mm) is driven into the compacted fresh concrete with a percussion drill (see Fig.3). The mortar is collected in the wire basket up to a maximum aggregate size of 6 mm. This mortar is pressed into the body of a syringe of the type familiar from medicine. The mortar (about 20 ml) is then injected into the "DBT air-void measuring equipment". In the lower part of this equipment there is a special glycerine-type liquid above which there is water. The mortar is stirred with a rod so that the air bubbles pass from the mortar into the special liquid and then rise slowly. The air bubbles are caught in a bell (glass dish) suspended in the water.

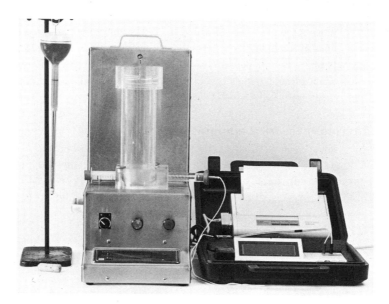

Fig. 2. DBT air-void measuring equipment

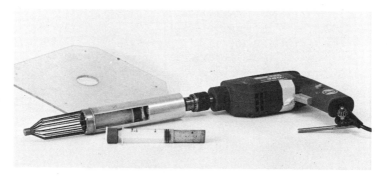

Fig. 3. Percussion drill with a wire basket

This bell is linked to a balance. In this way it is possible to determine the buoyancy of the bubbles. The large bubbles rise first, followed by the smaller ones, so the micro air-void content, spacing factor and air-void distribution can be determined using a computer program; the buoyancy of the bubbles is measured every minute over a 20 minute period.

4 Investigations

Investigations are now being carried out with this equipment at the German Cement Industry's Research Institute with financial support from the Federal Ministry of Transport. To this end concretes of varying consistencies are being prepared with both gravel and crushed aggregates. The quantity of air-entraining agent and the addition of super-plasticizer is also varied. The micro air-void content, spacing factor and air-void distribution are determined with the "DBT air-void measuring equipment" and compared with the parameters determined in the hardened concrete. The first test results for several soft concretes are shown in Fig. 4. The spacing factors determined with the equipment agree quite well with those determined in the hardened concrete. The investigations are being continued at the present.

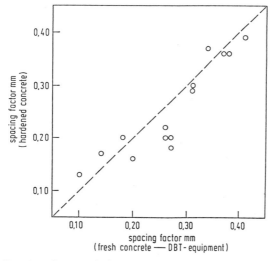

Fig. 4. Spacing factor of the hardened concrete as a function of the spacing factor of the fresh concrete

5 Preparation and placement of air-entrained concrete

Air-entrained concrete is a special concrete, and it must be prepared and placed very carefully. To ensure perfect quality it is advisable, even during the suitability testing, to determine not only the total air content but also the air-void parameters. If it were possible to carry out this test on fresh concrete using the new equipment, then any change in the air-void system could be detected quickly. For fairly large contract sections it could be advisable to use this equipment in the first concreting stages. Any change in the air-entraining agent or in the entire concrete composition could then be carried out successfully at an early stage. The DBT air-void measuring equipment must be tested further before it is possible to recommend such a test procedure at the construction site.

References

[1] Schäfer, A.: Frostwiderstand und Porengefüge des Betons. Beziehungen und Prüfverfahren. Deutscher Ausschuß für Stahlbeton, H. 167, Berlin 1965, S. 3/57

[2] Bonzel, J., und E. Siebel: Neuere Untersuchungen über den Frost-Tausalz-Widerstand von Beton. beton 27 (1977), H. 4, pp 153/157, H. 5, pp. 205/211 und H. 6, pp. 237/244; also Betontechnische Berichte 1967, Beton-Verlag, Düsseldorf 1978, S. 55/104.

[3] Superplasticizers and other chemical admixtures in concrete. Proceedings 3rd International Conference Ottawa, Canada 1989. Editor V.M. Malhotra. ACI publication SP 119. Detroit, Michigan 1989

[4] Siebel, E.: Air-void characteristics and freezing and thawing resistance of super-plasticized air-entrained concrete with high workability, in [3], pp 297/319.

[5] Whiting, D., und J. Schmitt: A model for de-icer scalling resistance of field concrete containing high range water reducers, in [3], pp. 343/359.

[6] Johansen, R.I.: Characterisation of the air-void system in concrete by a vibration test. Int. Symp. on Admixtures for Mortar and Concrete, Brussels 30. Aug. - 1. Sept. 1967.

[7] DBT Luftporenmesser. Dansk Beton Technik A/S, Hellerup, Bedienungsanleitung.

10 CHARACTERISTIC VALUES OF CEMENT DETERMINED IN THE CEMENT FACTORY AND AT A READY MIX CONCRETE PLANT: A COMPARISON

E. BIELAK
Wilhelm Dyckerhoff Institut, Wiesbaden-Biebrich, Germany

Abstract

According to german standards, a cement plant has to test twice a week the cement of any kind taken from the delivery silo. This last step of the quality control within the cement plant gives characteristic values for the water demand of the cement paste, the setting time and the strength of the standard cement mortar; it should describe the quality of cement.

These results are compared with the results of tests carried out with cement samples taken from a silo at a ready mix concrete plant. The samples from the ready mix concrete plant were taken once a week (average). The time period for the comparison of the characteristic values is nearly one year.

The scope of the paper is to show that the quality control procedures in a cement plant can describe the quality of the cement for a customer. The results will be discussed with respect to the reliability of the test procedures.

1 Introduction

Cement is in general used to produce concrete. About 50 percent of the cement manufactured in West Germany is customed by the ready-mixed concrete industry. Therefore there is a certain interest in the quality parameters of the ready-mixed concrete and the cement used. Besides a lot of other influences, the properties of concrete are more or less influenced by properties of the cement. One of these influences is outlined by the water-cement-ratio law (Walz, Powers) that combines the strength of a cement mortar and the water-cement-ratio of the concrete with the concrete strength. For the concrete strength and the workability of fresh concrete are assumed to be main parameters for the concrete quality (durability of the hardened concrete should be mentioned too), there is a certain interest in having knowledge about the cement parameters influencing these properties of concrete quality. Another criterion of quality estimation is the uniformity of the concrete or cement properties.

In Germany normally a producer of ready-mixed concrete does not test the cement received. He only can gain knowledge about the cement properties through the data from the quality control tests of the cement manufacturer. These data are received from tests carried out twice a week with samples from the delivery silo of the cement plant.

The aim of the reported work was to compare this data of the quality control procedure at the cement plant with the results of matching test with cement samples from a silo at the ready-mixed concrete plant. The comparison can show if the data from the quality control of the cement manufacturer can be representative for the description of the cement quality at a single customer.

2 Quality control procedures according to German standards

The German standard DIN 1164 regulates the test procedures to fulfill the quality control. According to Part 2 of this standard, a cement manufacturer has to test the composition and the properties of each cement type and strength class as follows:

daily:
>   setting time
>   soundness

twice a week:
>   loss of ignition
>   content of carbon dioxide
>   insoluble residue
>   content of sulphat (SO_3)
>   fineness of grinding
>   mortar strength

monthly:
>   determination of constituents
>   determination of the heat of hydration

The test procedures are described in other parts of DIN 1164 or in the European standard EN 196.

3 Scope of investigation

The investigation took place in 1988. Within another investigation programm, it was possible to take cement samples of one type, strength class and manufacturer from the silo at a ready-mixed concrete plant. It was a blast furnace slag cement HOZ 35 L according to German standard DIN 1164. The samples were taken over a period from January to October. The number of random samplings is about 60.

The samples were taken from the conveyor device from the silo to the batch mixer corresponding to a certain concrete mix.

These cement samples were tested at the central investigation and development laboratories of the Dyckerhoff AG in Wiesbaden. The investigation is concentrated on the characteristic values fineness of grinding, water demand of the cement, begin of setting, end of setting and the strength of the standard cement mortar at 2, 7 and 28 days.

The results of the tests carried out with the samples from the ready-mixed concrete plant are compared with the results from the quality control of the delivery cement plant. The number of those samples within the observed period is about 78. The values compared are produced from two different test laboratories.

Because it is not possible to assosiate the time of sampling at the cement plant with a certain sample at the ready-mixed concrete plant, the time of sampling at each place is taken into account.

In general the ready-mixed concrete plant got two to three times a week a delivery of 24000 kg of the cement under investigation. The average amount of shipped cement of this type from the cement plant is about 14000 to 15000 tons a month.

4 Results and evaluation

Table 1 gives a survey of the test results. The table shows the average of the characteristic values fineness of grinding, water demand of the cement, setting time and strengths of the standard mortar, their scatter (standard deviation) and the number of samples. The data is given for the test series from samples from the ready-mixed concrete plant (customer) and from the quality control procedure of the manufacturer (manufacturer), both within the observed period from January to October 1988.

These values (Table 1) show a satisfactory correspondence between the two data groups for the average as well as for the scatter.

It was tried to compare the averages and the scatters of the two data pools (customer vs. manufacturer) with statistical methods. The standard deviations were

Table 1. Test results

characteristic value dimension	fineness of grinding cm2/g	water demand cem. %	beginn of setting min	end of setting min	mortar strength at		
					2 days N/mm2	7 days N/mm2	28 days N/mm2
mean value — manufactorer	3746	28,3	197	269	13,9	31,7	49,5
mean value — customer	3685	28,4	164	228	12,6	31,6	50,8
standard deviation — manufactorer	181	0,4	16	24	1,1	1,6	2,0
standard deviation — customer	164	0,6	25	43	1,0	1,6	2,1
number of samples — manufactorer	78	78	78	78	78	78	78
number of samples — customer	63	57	57	57	62	60	62

compared with the F test and the mean values with the t-test. Table 2 shows the results of this statistical testing. The mean values of the characteristic values of the two data pools, quality control at the manufacturer and sampling at a ready-mixed concrete plant, are assumed not to be equal (with the exception of the water demand of the cement and the 7 day strength). It is remarkable that the mean values of the two data pools for the 7 day strengths should be statistical equal, the strengths for the other testing times not. Therefore the statistical test procedure should be discussed.

Table 2. Statistical tests

	characteristic value	fineness of grinding	water demand cem.	beginn of setting	end of setting	mortar strength at 2 days	7 days	28 days	
comparison of scatter F-test	F calculated	1,22	1,59	2,44	3,00	1,20	1,01	1,10	
	F table	1,65	1,48	1,48	1,48	1,52	1,50	1,48	
	judgement	equal	not equal	not equal	not equal	equal	equal	equal	
comparison of average t-test	t calculated	2,078	1,791	8,828	6,737	7,549	0,356	3,460	
	t table	1,977	1,983	1,987	1,989	1,977	1,978	1,977	
	judgement	not equal	equal	not equal	not equal	not equal	equal	not equal	

The used statistical procedures are valid for the case that the results are gained from a normal distributed (Gauss distribution) statistical data basis. This assumption is not fulfilled. During the period of observation, the quality of the cement is always regulated according to internal quality goals and with respect to the preproducts of the manufacturing process. This regulating process in combination with a significant scatter of the testing procedures for the different characteristic values at a relative low span of the quality can result in the not plausible results of Table 2.

To get a better idea whether the data from the quality control of the cement manufacturer are representative for a customers cement quality or not, we now look at the temporal course of the characteristic values of both samplings (manufacturer and customer).

This data, for the characteristic values grinding fineness, water demand of the cement, setting times, and strengths at 2, 7 and 28 days, are shown in Figures 1 to 6. The characteristic value concerned is plotted over a time axis for the sampling in the cement plant (manufacturer) as well as the sampling at the ready-mixed concrete plant (customer). The curves in these figures show the gliding average for the data. This gliding average is calculated from the last 5 test series.

These gliding average curves show a satisfactory correspondence. Especially in the comparison of the two data bases (manufacturer and customer sampling) trends are well reproduced. Not at all, the curves are not identical. The remaining difference can be attributed to the fact that the time of sampling at the cement plant and the ready-mixed concrete plant is not directly corresponding and that the data are gained from two different testing laboratories. Beyond this, each value has a testing scatter. This testing scatter should be taken into consideration. Therefore a span is calculated from the gliding average of each characteristic value (Figures 1 to 6) by adding/subtracting one time the standard deviation of testing scatter of the testing procedure concerned.

This standard deviation of the testing scatter (Table 3) was worked out within one other test series and statistically proved.

Table 3. Standard deviation of testing scatter

characteristic value dimension	fineness of grinding cm2/g	water demand cem. %	beginn of setting min	end of setting min	mortar strength at		
					2 days N/mm2	7 days N/mm2	28 days N/mm2
standard deviation s	21,6	0,12	6,1	11,0	0,64	0,97	1,00

Figures 7 to 12 show this span plottet against a time axis. These figures show, that within a practicable range the results of the quality control of the cement manufacturer can describe the quality at a customer (ready-mixed concrete plant) as long as statistical aspects are taken into consideration.

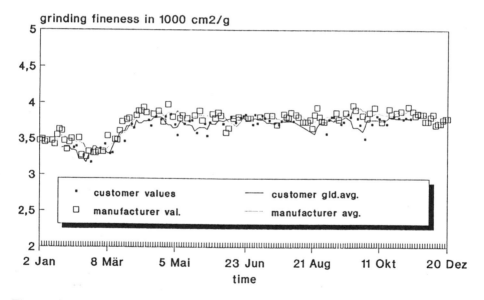

Fig. 1. Grinding fineness; comparison customer vs. manufacturer

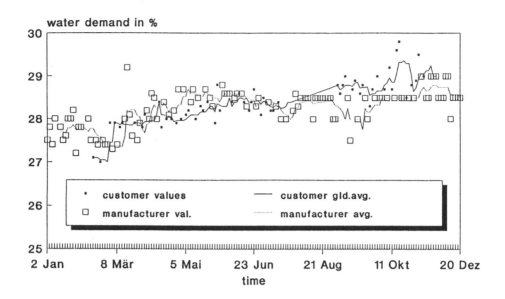

Fig. 2. Water demand of cement; comparison customer vs. manufacturer

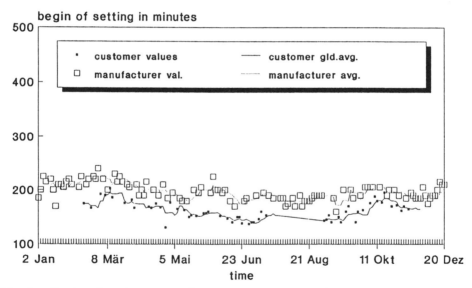

Fig. 3a. Begin of setting; comparison customer vs. manufacturer

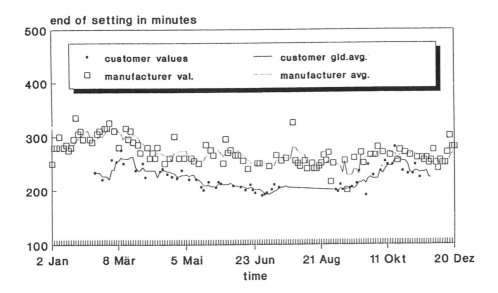

Fig. 3b. End of setting; comparison customer vs. manufacturer

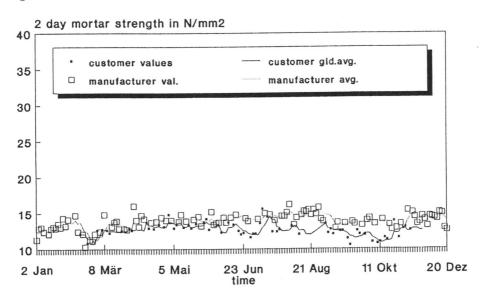

Fig. 4. 2 day mortar strength; comparison customer vs. manufacturer

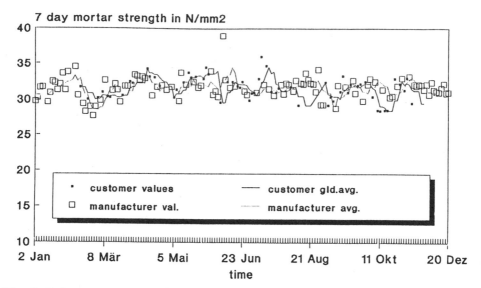

Fig. 5. 7 day mortar strengthy; comparison customer vs. manufacturer

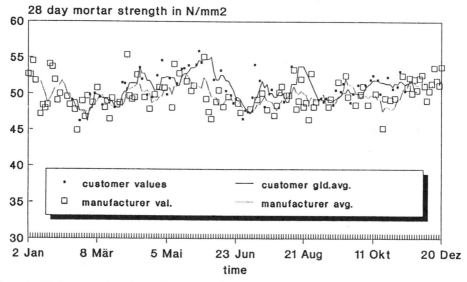

Fig. 6. 28 day mortar strength; comparison customer vs. manufacturer

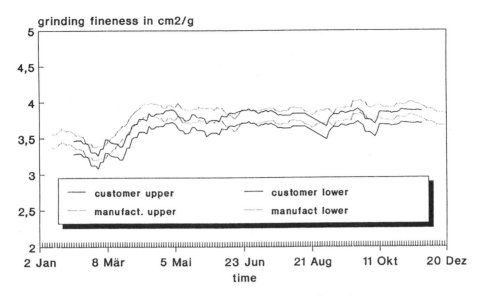

Fig. 7. Grinding fineness, avg ± s; span by average and testing scatter

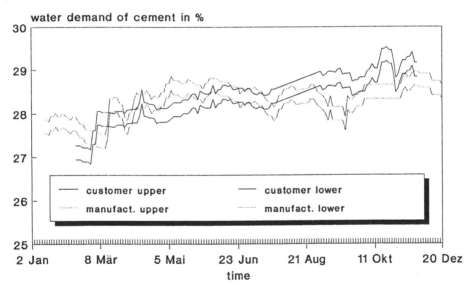

Fig. 8. Water demand (cement), avg ± s; span by average and testing scatter

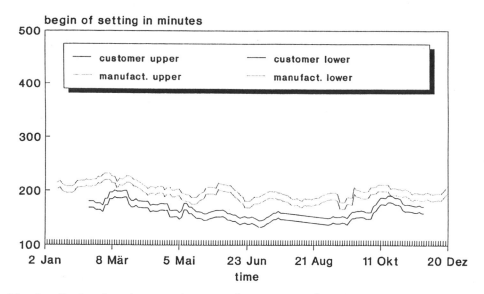

Fig. 9a. Begin of setting, avg ± s; span by average and testing scatter

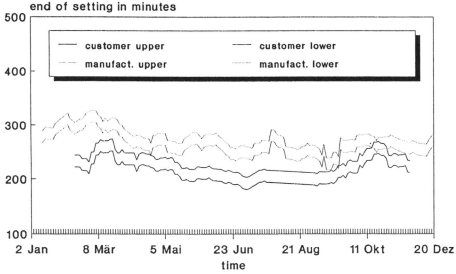

Fig. 9b. End of setting, avg ± s; span by average and testing scatter

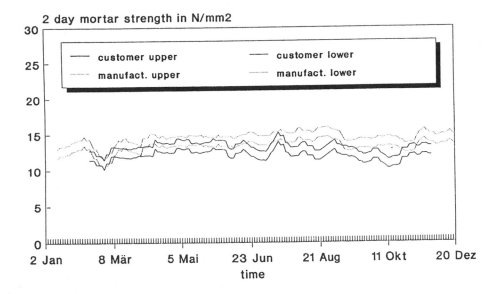

Fig. 10. 2 day mortar strength, avg ± s; span by average and testing scatter

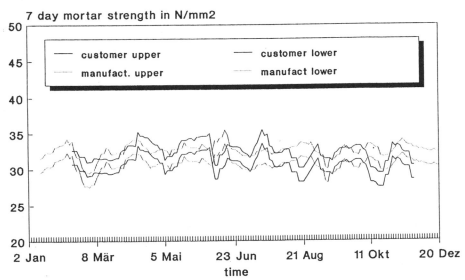

Fig. 11. 7 day mortar strength, avg ± s; span by average and testing scatter

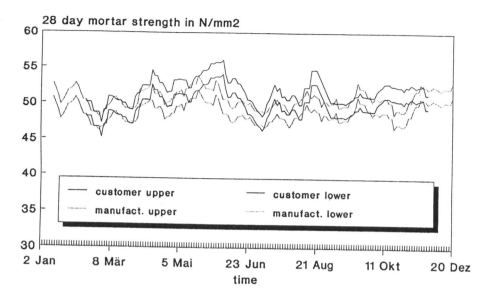

Fig. 12. 28 day mortar strength, avg ± s; span by average and testing scatter

5 Summary

Some characteristic values describing cement behavior gained from a sampling at a ready-mixed concrete plant are compared with the data from the quality control procedures at the cement plant. The observing period was nearly one year.

It was shown, that statistical methods cannot give a satisfactory answer, because the cement manufacturing is a controlled and regulated process. The given data basis from test of temporarily different samplings from this process cannot be assumed to be normal distributed and statistically random.

Calculated gliding averages that take the testing scatter of the testing procedures into consideration, are able to illustrate that the quality control data of the cement plant are representative for the cement quality at a certain customer.

The results also show that it makes no sense to withdraw conclusions about cement quality from single tests. The next step of investigation should find relations between cement quality parameters and the concrete quality within the small range of one cement type and strength class and one concrete, described by the strength class, workability and composition.

PART THREE
SETTING AND COMPACTION

11 SETTING AND EARLY STRENGTH GAIN OF CEMENT CONTAINING FLY ASH, SLAG OR LIMESTONE FILLERS

C.D. POMEROY
British Cement Association, Wexham Springs, Slough, UK

Abstract
This paper reports measurements of the setting times of Portland cements to which additions of fly ash, slag or limestone fillers were made. Data for Portland pfa cements and coarse ground cements are also included. The effects of cement type on the bleeding of concrete are also discussed. The relevance of low early age curing temperatures is considered.

1 Introduction

During recent years, many variations have been made in the formulations of hydraulic cements and some of these variations will affect the early age behaviour of concrete. In addition, fly ash and ground granulated blast-furnace slags and silica fume have been combined with Portland cements at the concrete mixer and this practice, too, affects the early age performance of concrete. The importance of these changes is not always obvious. One such effect relates to the setting and early strength gain of concrete. Such delays can be unwelcome on site if tasks can no longer be completed during a working shift.

Another effect is on the cohesivity of the concretes made from the different cements, some of which have a greater propensity to bleed or to promote greater segregation within the concrete.

The development of the provisional European Standard (pr ENV 197) "Cement: composition, specification and conformity criteria" introduced prospective changes to the formulations of the cements that have traditionally been used in the UK. The principal differences are the permissible inclusion of inert fillers of unspecified origin into cements, within prescribed limits (up to 5 per cent) and the specification of composite cements that contain fly ash, slag or inert fillers, sometimes in combination. Some of these factors have been studied as part of a very large programme of work carried out jointly between the UK Building Research Establishment, the British Cement Association and the cement makers. An interim report [1] on this programme was presented at a seminar at BRE on 28 November 1989 and some of the information in this paper comes from this joint programme.

2 Materials

When changes are made to the chemical composition of Portland cements, it is not unusual to find that other factors change simultaneously. This is particularly so when two components are blended or interground. Because the different materials will be of different hardness, there will be preferential degradation of the weaker component when a mix is interground. The resulting particle grading and the relative fineness of the two components is thus likely to differ from that obtained if each component is ground separately and then blended with the other. The cement maker thus is able to adopt a grinding and blending system to make a composite cement that will have preferential water retention and strength development properties.

In the work reported, cements based on different Portland cements combined with limestone, fly ash, slag or other additions have been used to make mortars and concretes of different mix proportions and certain aspects of early age performance have been measured.

Details of the individual experiments are given at the relevant parts of the text.

3 Measurements

The fineness of the cements was measured using the method described in BS 4550: Part 3: Section 3.3 (1978). The setting times were determined for cement pastes of standard consistence as described in BS 4550: Part 3: Sections 3.5 and 3.6 (1978). For compliance, the initial set shall be not less than 45 minutes and the final set not more than 10 hours.

Bleeding from fresh concrete was measured using the RILEM test procedure CPC 16, "Bleeding of concrete" [2]. Compressive strengths were measured on 100 mm cubes.

4 Setting times

Table 1 records the setting times for three OPC's to which small additions of fly ash (two sources) and a slag were added. The nominal percentage additions were 2.5 and 5. Although all of the cements complied with the setting requirements of the British Standards, a 5 per cent addition generally delayed initial and final setting by 20 (pfa) to 40 (slag) minutes. The lower percentage additions tended to show a smaller and inconclusive delay. The work was extended to include the addition of limestone fillers to the cement at levels of about 5 and 25 per cent. With the exception of the deliberate use of a poor quality lias that did not satisfy the requirements of the provisional European Standard for cements, the limestone additions advanced the initial setting times by up to 30 minutes and the final set by up to an hour.

Table 1.

Cement		Surface area fineness (m^2/kg)	Initial set (mins)	Final set (mins)
A1	OPC 1	392	100	140
AE2	OPC+2.5% pfa 1	389	125	180
AE5	OPC+5% pfa 1	389	130	180
A2	OPC 1	391	120	170
AD2	OPC+2.5% pfa 2	394	145	200
AD5	OPC+5% pfa 2	405	135	150
B1	OPC 2	354	150	240
BE2	OPC+2.5% pfa 1	357	130	220
BE5	OPC+5% pfa 1	357	160	215
B2	OPC 2	351	145	215
BD2	OPC+2.5% pfa 2	339	130	225
BD5	OPC+5% pfa 2	344	175	250
C	OPC 3	334	120	190
CS2	OPC+2.5% slag	340	130	220
CS5	OPC+5% slag	355	165	230

The effect was most marked for the higher percentage additions. The finenesses of all the cements with a limestone component were higher than the base OPC's used, the average increase being about 90 m^2/kg. Set accelerations of a similar magnitude were observed when OPC's were more finely ground in the manufacture of a rapid hardening Portland cement.

When 25 - 30 per cent of high quality fly ash (BS 3892, Pt 1) was combined with an OPC, the setting was delayed by about one hour (both initial and final set). In the two cases studied, the finenesses of the Portland pfa cement were similar to those of the parent OPC.

The question thus arises as to whether the changes in setting time are related to the presence of the filler, be it slag, fly ash or limestone, or whether they are simply a manifestation of the changes in fineness.

A possible explanation is provided by the work of Gutteridge and Dalziel, who have used XRD methods to measure the progressive hydration of the constituents of Portland cements in the presence of different additions, including the chemically inert rutile. They have convincingly shown that the presence of additions can stimulate hydration and this can be seen even after a few hours. Figures 1 and 2 show some typical results obtained by Gutteridge and Dalziel [3,4] for the hydration of the alite phases in an OPC either alone or in combination with pfa or the inert rutile (titanium dioxide).

It can be clearly seen that the presence of a secondary component has affected the hydration of the alite. Reactions of the other phases are also affected. Thus it is pos-

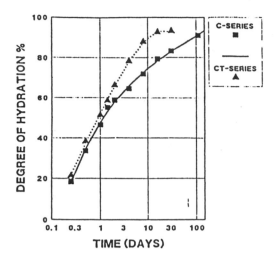

Fig. 1. The hydration of alite in a Portland cement (CT) and in the presence of rutile (C)

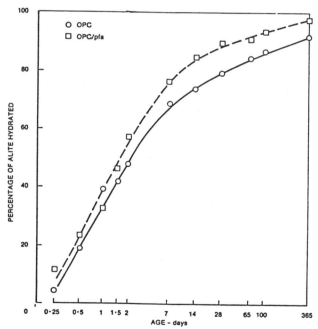

Fig. 2. The hydration of alite in an OPC and an OPC/pfa blend

sible that the early strength development and setting of cements is related not only to the cement chemistry and fineness but also to the presence of fine particles that can act as sites for the nucleation of hydration. However, there are further factors to consider, especially the water uptake or demand of the cements and their propensity to bleed. Ideally, a concrete mix must be workable and easily placed at as low a water:cement ratio as possible and these requirements are not always compatible.

5 Water demand and bleeding

Jackson [5] has reported the effects of limestone additions on the water required to make cement pastes of standard consistence. Again, the inclusion of a fine limestone has a large effect on the specific surface area of the cement, so that the explanation for any observed change is not straight-forward. Table 2 summarises the results which show that the inclusion of a limestone filler either has no effect or it reduces the water demand by a few per cent.

Other factors of possible relevance are the differences in cement density and in particle gradings, since the amount of water that is needed to impart a particular workability depends upon the free space between the packed cement grains. This can be seen from the tests on Portland pulverised fuel ash cements that contain about 30 per cent of fly ash. When a fly ash is blended with an OPC, the density is likely to be lower than that when the components are interground.

Presumably in the latter case, any hollow pfa particles are broken, thereby increasing the bulk density. For the blended fly ash cement, there was a reduction in water demand comparable with that for the limestone addition. With intergrinding, the water demand actually rose.

Table 2. The effect of additions of limestone on the water percentage required for standard consistence

Cement	OPC Surface area (m^2/kg)	% water	OPC + 5% limestone Surface area (m^2/kg)	% water	OPC + 16 - 28% limestone Surface area (m^2/kg)	% water
U	275	26.5	320	26.5	470	23.0
V	315	27.0	355	26.5	465	27.0
W	310	27.0	365	25.0	610	25.0
Y	265	28.0	325	28.0	410	28.0
G	395	26.5	510	26.5	530	26.5
F	445	27.0	465	26.5	580	26.0
H	345	26.5	505	26.5	515	26.0
Mean		27.0		26.5		26.0

It is impossible to give a satisfactory explanation as several factors change simultaneously - namely the cement composition, the fineness and grading, the particle shape and the physical nature of the exposed surface. However, the collective effects are of importance to the cement maker, as they provide one means of tailoring a composite cement to the needs of the customer.

6 The effect of fillers on water:cement ratio

One of the problems of applying the results from standard consistence testing is that the differences between cements can become masked by other factors, not least being the mix design. Brookbanks [6], using the data from the joint BRE/BCA filler programme, has shown that the w/c required for a nominal 60 mm slump concrete is as shown in Table 3.

Table 3. The effect of limestone fillers on the w/c for nominal 60 mm slump concretes

Cement content kg/m^3	% limestone in cement		
	0	5	16 - 28
225	W/C 0.92	W/C .89	W/C .88
300	.60	.60	.59
350	.52	.52	.52

It is seen that the reduction in water demand is most strongly reflected in the values for the leaner concrete mixes.

7 Bleeding

The other early-age factor studied was the relative susceptibility of the different cements to bleed. This is obviously of great practical significance. Rates of loss of bleed water were measured using the RILEM test method for a range of concretes that were based on a single 300 kg/m^3 cement content mix, using a range of cements of different fineness and limestone or fly ash content.

As seen in Figure 3, bleeding is closely related to the fineness or surface area of the cements. Probably particle gradings or packing densities are also important, though the results shown suggest that this would be a secondary effect. It will be seen that the cements with a high proportion of limestone do not bleed significantly, but as these are of very large surface area, the result is not unexpected.

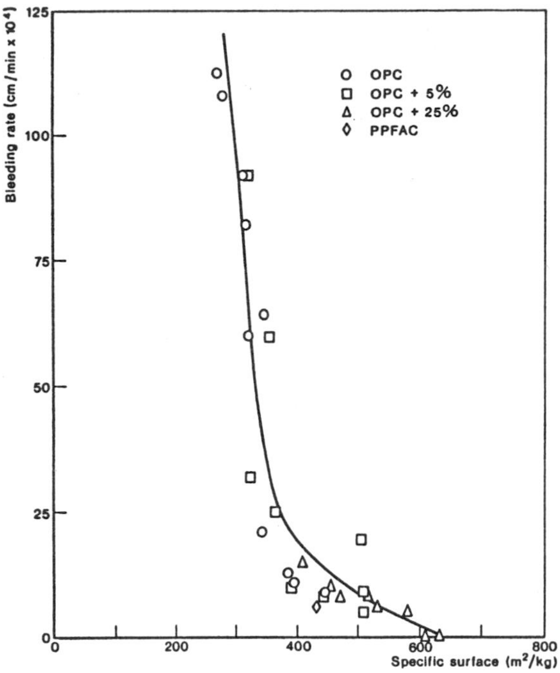

Fig. 3. Relation between bleeding rate of concrete and specific surface of cement used

8 Curing temperature

One of the most misleading test parameters is the laboratory temperature, especially in North European winter climates. Concrete performances are compared on the basis of tests carried out on samples made, stored and tested at temperatures close to 20°C. The average temperature on site in the UK, for all seasons and taking daily cycles into account, is much closer to 10°C. This, therefore, raises the question of the influence of temperature on early age performance. Livesey [7] reported some compressive strength measurements for 100 mm concrete cubes of a standard mix (300 kg cement, with the w/c adjusted to give a slump of approximately 60 mm) cured in water at either 20 or 10°C. Measurements were made at intervals from 1 to 28 days. A range of OPC's was used either as supplied or in combination with a limestone filler at about 5 and 25 per cent levels. Results for a coarse ground cement and a Portland pfa cement are also given. The results for all of the cements were comparable and are summarised in Table 4. It will be seen that by 28 days, the cements containing a limestone filler had attained comparable strengths irrespective of the curing temperature, whilst the coarse ground and Portland pfa cements were still given lower strengths at 10°C. However, at the more relevant 1 day curing, temperature has a very profound effect with strength at 10° being only about one third of that reported at 20°C.

117

Table 4. The effect of curing temperature on the strength gain of 100 mm water-cured concrete cubes

	1 day		28 days	
	20°C	10°C	20°C	10°C
Ordinary Portland cements	11.2	3.5	41.2	40.0
OPC + 5% limestone	10.2	3.5	37.7	38.3
OPC + 25% limestone	7.3	2.5	37.2	37.5
Coarse ground OPC	6.5	1.5	37.5	34.0
Portland pfa cement	7.5	2.5	36.0	33.0

The coarse ground cement, the Portland pfa cement and those with high (25-30) percentages of limestone additions were slowest in gaining strength.

It is interesting to note that at 28 days the cements that contained limestone, both at the 5 and the 25 per cent levels, reached higher strengths at 10°C than at 20°C, even though their early strength gains were greatly retarded.

A further programme of work has been undertaken at the BCA by Hobbs [8] who made and cured concretes using OPC alone or in combination with a high quality (BS 3892: Pt 1) pulverised fuel ash or a slag (BS 6699) at 5°C and 20°C. The mixes used are given in Table 5 and a summary of the results in Table 6.

Table 6 and Figure 4 show the sensitivity of composite cements containing pfa or slag to the curing temperature. This is particularly noticeable for the slag/OPC blend which at 5°C only reached 56 per cent of the 28 day 20°C cured value, whereas with OPC alone, the strength attained at 28 days at the lower temperature exceeds that at 20°C.

Table 5. Concrete mix designs for study of effect of temperatures on strength gain

Series I	OPC kg/m^3 (c)	Addition kg/m^3 (a)	w/c+a	Aggregate kg/m^3
OPC	367	–	0.48	1869
OPC + pfa	237	129	.44	1865
OPC + 2 pfa	284	152	.36	1808
OPC + slag	183	183	.47	1865
Series II				
OPC	278	–	0.65	1925
OPC + pfa	183	98	.57	1950
OPC + 2 pfa	220	118	.47	1892
OPC + slag	140	140	.61	1937

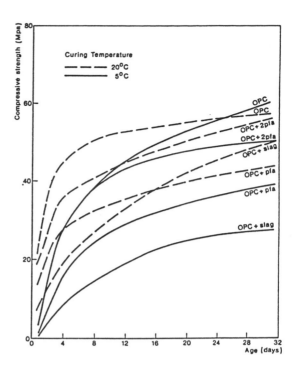

Fig. 4. The effect of curing temperatures on strength of OPC and OPC/PFA blends in concrete

Table 6. The effect of curing temperature on the compressive strength (MPa) of concrete

	OPC		OPC+ pfa		OPC+ 2 pfa		OPC+ slag	
Age (days)	5°C	20°C	5°C	20°C	5°C	20°C	5°C	20°C
SERIES 1								
1	4.1	26.6	1.6	14.9	(14.7)*	20.4	1.0	8.4
3	23.4	42.5	11.6	25.2	23.2	33.9	6.9	16.3
7	36.9	49.4	23.3	31.6	36.1	39.6	13.6	24.7
14	47.1	53.1	30.5	36.7	44.4	46.6	20.8	36.1
28	58.3	57.0	37.8	42.6	49.7	54.3	27.0	48.1
					* 2 days			
SERIES 2								
1	–	12.9	–	7.6	–	10.7	–	3.3
2	8.3	18.9	4.7	14.0	6.6	17.2	2.6	7.4
3	11.9	23.6	9.3	17.6	12.5	21.0	3.3	10.7
7	20.2	29.5	14.4	21.5	20.0	27.4	7.6	16.2
14	26.4	31.4	19.1	27.1	26.7	32.9	12.5	21.5
28	32.9	35.2	23.2	31.2	30.9	36.3	16.7	29.8

Another way to consider these early age effects of curing temperature is to compare the time it takes with the different cements to reach a given strength. For example, to reach 20 MPa at 20°C, it takes 1 day for the OPC and OPC + 2pfa and 4 days for the Portland blastfurnace slag blend. At 10°C the OPC and OPC + 2 pfa take 3 days while the OPC + pfa takes 6 and the Portland slag blend takes almost two **weeks**.

9 Discussion

The water demand and retention of cements and the mortars or concretes made with these are crucial factors that affect not only the rheology of the mixes and the facility with which they can be placed, they also affect the sensitivities of the concretes to bleed and to lose water during the curing period and these effects are reflected in the gains in strength during the early life of the concretes. The prevailing environment is an interactive parameter, with low temperatures and low humidities (not studied here), accentuating the different performances between the cements. Basically the message is that those cements most susceptible to bleeding and with the higher water demands are usually the cements most susceptible to inadequate curing. It follows that some cements should not be recommended for certain uses (e.g. the manufacture of thin sections of poorly cured concrete) even though they may be perfectly acceptable for large elements or where premature drying is impossible (e.g. foundations).

10 Conclusions

Some of the sensitive characteristics of cement have been considered both in relation to the early age properties and setting, and to the consequent effects of these properties on strength development. It is clear that the different performance characteristics cannot be considered in isolation and that improved early age properties regarding the cohesivity of cements and the retention of water are desirable attributes that will be reflected in later age performance.

References

[1] Performance of limestone-filled cements: Report of joint BRE/BCA/ Cement Industry Working Party (BRE, to be published).

[2] RILEM draft recommendation: Bleeding of Concrete: Materials and Structures (1983), Vol. 16, No. 91, pp 51-52.

[3] Dalziel, J. A., Gutteridge, W. A., C&CA Technical Report No. 560 (1986)

[4] Gutteridge, W. A., Dalziel, J. A., Filler cement: 'The effect of the secondary component on the hydration of cement. Advances in cement research.' (in the press).

[5] Jackson, P. J., 'Manufacturing aspects of limestone-filled cements' as reference [1].

[6] Brookbanks, P., 'Properties of fresh concrete' - as reference [1].

[7] Livesey, P., 'Strength development characteristics of limestone filled cements' - as reference [1].

[8] Hobbs, D. W., Private communication.

12 ULTRASONIC MEASUREMENT FOR SETTING CONTROL OF CONCRETE

N.G.B. van der WINDEN
INTRON Institute, Houten, The Netherlands

1 Introduction

1.1 Preamble

Ultrasonic pulse velocity measurements are used extensively as a non-destructive test method to establish the in-situ strength of hardened concrete. This method used in combination with drilled cores and rebound tests gives reliable information regarding the quality of the concrete used. Ultrasonic pulse velocity measurements are also used to locate cracks and non homogeneous areas in concrete.

Studies carried out by several researchers showed that the ultrasonic pulse velocity measurements can also be used to establish the setting characteristics of fresh concrete. Researchers who have been active in this field are Whitehurst, Elvery, Casson and Domone, Byfors and others.

The writer of this paper developed a method to measure the development of ultrasonic pulse velocity in fresh concrete. The method was developed in order to be able to control concrete mixes and to produce a constant flow of concrete with equal early properties. Properties such as setting time and available working time were of the utmost importance for the project at hand, which was a large offshore project where more than 2000 linear meter of wall had to be slipformed in one go.

The initial study of the method showed that the ultrasonic pulse velocity measurement gives much more information about the behaviour of concrete in the first 24 hours after mixing than the traditional methods such as the Vicat measurements of initial and final setting time. Initial and final setting time have no direct relation to the behaviour of concrete and gives only information about the behaviour of cement mortar mixes.

In the following chapters the results of the study mentioned are given as well as a comparison between the results of this study and other research work. This comparison showed that measurements carried out in different parts of the world become almost equal once the results are corrected by using correction factors developed by Byfors.

1.2 Setting and hardening of concrete in the first 24 hours after mixing

Once cement is mixed with water a chemical reaction starts. As a result of this chemical reaction cement lime becomes cement stone. This cement stone bonds aggregate particals together to concrete. Additives can influence this chemical reaction but, in general, do not prevent it.

Due to the chemical reaction, the mortar, which behaves more or less like tixotropic fluid, transforms slowly into a non-workable claylike substance and later into stone. During the claylike phase the concrete is very vulnerable to cracking. The tensile strain capacity is minimal. In case the concrete is demoulded during the claylike phase, the surface may be damaged although the concrete is basically stable. In the phase following the 'claylike' phase, properties like compressive and tensile strength develop slowly. The modulus of elasticy, however, develops rapidly.

By measuring the ultrasonic pulse velocity, one measures more or less the development of the modulus of elasticity and the development of the Poisson ratio.

For the ultrasonic pulse velocity, the following equation is valid:

$$v^2 = \frac{E_{dyn}}{m} \frac{1 - \mu}{(1 + \mu)(1 - 2\mu)}$$

in which:

v   = ultrasonic pulse velocity in m/sec.
E_{dyn} = dynamic modulus of elasticity in N/m^2
m   = mass of the concrete in kg/m^3
μ = Poisson ratio

Due to the fact that the Poisson ratio of hardened concrete is constant and that the relation between dynamic and static modulus of elasticity hardly varies, the ultrasonic pulse velocity measurement can be used to establish the strength of hardened concrete. Several researchers have shown that the increase of the ultrasonic pulse velocity in the first 24 hours after mixing gives a good picture of the changing setting and hardening characteristics of concrete.

2 Ultrasonic testing of fresh mixes

Study carried out by N.v.d. Winden for the Dunlin A project

2.1 Preamble

During the preparation for the slipform work of the caisson of the Dunlin A offshore platform it became apparent that the standard methods available for the testing of fresh concrete mixes would not give sufficient information about the behaviour of the concrete in the first 24 hours after placing.

Nevertheless, this information was of the utmost importance because the slipform

123

speed would only be 1m per day and the concrete would therefore remain inside the slipform for almost 24 hours.

The writer of this paper developed in conjunction with the research department of the Netherlands Cement Industry a simple system to measure the ultrasonic pulse velocity in fresh concrete. The set up and the results of the investigation are described in the following paragraphs.

2.2 The apparatus

The arrangement of the equipment is shown in Fig. 1. The transmitting head sends signals at regular intervals through the concrete. These signals are registered at the receiver. Because a large part of the already limited input energy is destroyed by scattering in the concrete and does not register at the receiver, it is necessary to fix the sensors as securely as possible to the concrete.

For measurement of cured concrete the normal method is to use a contact adhesive. For measurement of uncured mixes a continuing reliable bond is somewhat more difficult. In the test set-up developed, one of the sensors rested on the concrete, under its own weight, the contact being maintained even if there was bleeding or drying shrinkage. The other sensor, by means of a hole in the base of the test container carried part of the content. The ideal spacing of the sensors was found to be 10 centimetres.

In order to measure the transit time of the pulses, the ultrasonic 'Pundit' apparatus was connected to an analog digital converter and the analog output of this apparatus used to drive a plotter. During investigations it became apparent that continuous measurement was not desirable, because the constant stream of ultrasonic

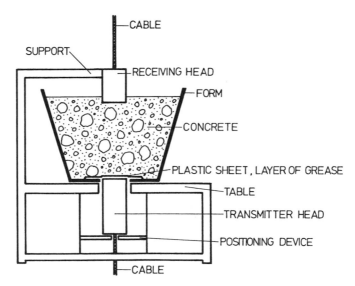

Fig. 1. Arrangement of equipment for vertical measurements

vibrations had an influence on the behaviour of the mix. Therefore the ultrasonic measuring apparatus, the converter and the plotter were activated, using a time clock, for 15 seconds every twenty minutes. In this way a stepped graph was produced, which is easily readable and gives a lot of information for a small amount of paper (Fig. 2). As could be expected, the measurements were sensitive to external vibrations. The disturbing influence of these was alleviated by making the set-up as vibration resistant as possible.

The method described above does not measure directly the setting behaviour of the concrete mix. It is also impossible to find by theory the relationship between the results from the ultrasonic equipment and the properties of practical importance for the concrete mix and concrete. It is therefore necessary to make an emperical interpretation of the results.

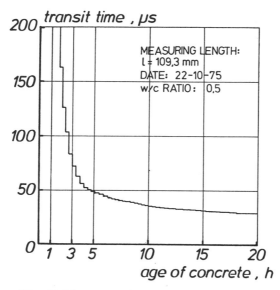

Fig. 2. The stepped graph from the plotter

2.3 Investigation of the practical meaning of the results

The usage of the measurements was of primary importance for ANDOC in defining the end of workability, the stability of the fresh concrete during slipforming and the striking time. Comparison with the conventionally measured compressive strength of the cubes was not possible. Therefore more practical comparisons were sought.

The workability was the most simple to control, in spite of the fact that the end of workability, the initial set, is a very subjective decision. For our investigation, a number of practical criteria to define the limits of workability were set out.

These criteria were:

a) The concrete mix must quickly return to its original shape after deformation due to a force on a vertical face. This test was carried out simply by means of a flexible plastic bucket. When the mix no longer followed the flexing of the bucket when picked up and put down, the workability had ended. Objections to this were (a) For each measurement a new bucket had to be used. Therefore the same sample was not used continuously; (b) The conclusion was dependent on the person who carried out the test; and (c) Because the test was not carried out continuously, there was chance the important changes in the rheological characteristics of the mix could be missed.

b) Control of the stiffness using a vibrating needle. As the concrete mix set, the area of influence of the vibrating needle was smaller. At a given moment, the needle left a hole in the concrete. The concrete mix had become unworkable. This test had the same objections as (a).

c) The forming of a construction-joint. For this test the cube moulds were filled to half height and the concrete mix prevented from drying out. At fixed periods the moulds were filled. Cubes were then split at 7 days and an indication of the reduction in the splitting strength found. Vibration of the concrete using a needle was not possible in this test because of disturbance of the bonding layer in the very small mould.

With the help of these checks it was shown that the end of the workability was defined by the area where the propagation speed increased from 1 km/sec to 1.5 km/sec.

All tests were initially carried out on the standard mix used by ANDOC, which consisted of 375 kg Dutch Portland blast furnace cement, class B, per cubic metre, 38 % Rhine sand, 62 % sea dredged aggregate, 0.4 % superplasticiser manufactured by Tillman, water cement ratio of 0.45 % and slump of 14-18 cm.

Study of the behaviour of mixes in practice has shown that, provided that the cicumstances are close to those of the laboratory, the values found in the laboratory are also applicable in practice. Two test slipforms on full scale were carried out on the site and this provided an excellent opportunity to check the phenomena found in the laboratory. During the first slipform test, the effect of the quantity of retarder on a slipformed wall was checked, and it was found that the results from the laboratory tests were directly applicable. With a sound propagation speed less than 1.5 km/sec then the stability of the wall was insufficient to stand unsupported. This arose mainly because of the fact that vibrations in the concrete were caused by the compacting of higher layers. During the second test concrete mixes with differing types of plasticisers were investigated. For these tests also the laboratory results were directly applicable. With the help of these and other tests it was found that, as long as the ultrasonic pulse velocity is greater than 1.5 km/sec, the concrete has a sufficient stability such that the lateral support from a shutter is superfluous, unless it is required for other reasons, for instance, due to wind-loading.

If the sound velocity is as high as ± 3 km/sec, then the concrete is in a stage where normal shutters can also be removed. When the sound velocity is above 3.5 km/sec, then one is in the area where normal tests can be used.

2.4 Results from tests carried out by ANDOC

The ultrasonic measurement method was used by ANDOC for two routine investigations:

a) Investigation of the properties of cement in combination with additives.

From each boat load of cement a sample of 25 kilos was taken with which a test mix of concrete was made. This mixture had the same composition as that produced at the batching plant, including additives. From this mixture the slump and flow-table rates were determined and cubes were made. Besides this a sample was placed in the ultrasonic apparatus to investigate the workability and setting speed during the first 24 hours.

b) Investigation of the effect of different additives on the stiffening and setting of the concrete mix.

The plasticising additive which was used for the lower part of the platform was found to have a too great retarding effect on the concrete mix to be used for sliding of the four towers. Therefore it was necessary to investigate other plasticisers which had little or no delaying effect. The ultrasonic measurement method was found especially useful for this, primarily because it was found that each additive produced differing setting characteristics. From these and other investigations a number of interesting points came to light. These can be illustrated using the following examples.

Example 1 Influence of the water/cement-ratio on the setting characteristic of concrete mix

From Fig. 3 it can be clearly seen that a reduction in the water cement ratio gives a definite increase in the stiffening and setting.

Example 2 Setting characteristics of different cement types.

From Fig. 4 the differences between cement types can be clearly seen. It should be noted that these are only examples and not definitive rules for the different types of cements. The variations found by us in cements of the same specification over an eighteen month period show that the boundaries of the areas in which the setting characteristics of the different cements fall, overlap each other; the trend shown is correct.

127

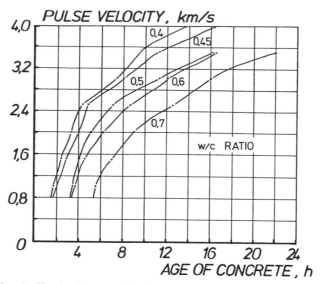

Fig. 3. Hardening speed related to water/cement-ratio

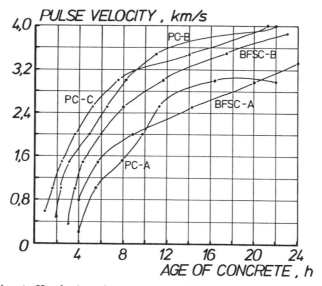

Fig. 4. Hardening characteristics of different types of cement
(w/c-ratio = 0.45; 375 kg cement per cubic metre

Example 3 The influence of environmental temperature and the initial temperature on the setting.

The examples quoted here are for two different concrete mixes with different additives. It should also be mentioned that our laboratory was a site laboratory where maintenance of a constant temperature was more difficult than in a research laboratory. In the graphs of Fig. 5 it can be seen that the ambient temperature has more influence on the setting than on the workability. The reduction of the initial mix temperature give a lengthening of the period of workability; this fact has been known for a long time.

Example 4 Influence of retarders (Fig. 6) and accelerators.

By means of retarders and accelerators the setting chracteristic of concrete can be visibly altered. Because of the required workability, ANDOC made more use of retarders than accelerators. It was found that practically all retarders are partly setting retarders and partly hardening retarders.

If for example, one wishes to keep concrete workable 24 hours to prevent construction joints, this effect should be borne in mind. Another effect that should be noted in this connection is the bleeding of water out of the surface. This effect was not registered by the ultrasonic equipment.

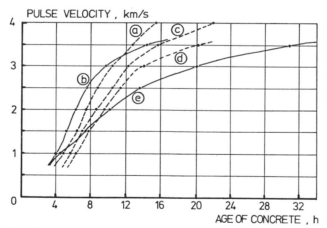

Fig. 5. Hardening of concrete mixtures at different temperatures (w/c-ratio of all mixtures 0.45, 375 kg of blast furnace cement, type B, per cubic metre

temperature	a	b	c	d	e
initial, °C	20	12.5	11	12	13.5
ambient, °C	20	20	20	10	10

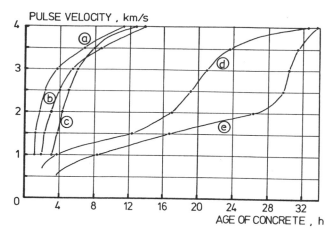

Fig. 6. Influence of accelerators and retarders on the hardening (mix with 375 kg blast furnace cement type B/per cubic metre and w/c-ratio 0.45)

a 0.5 % Cl-free accelerator
b plain mix
c normal quantity of retarder

d large amount of retarder
e 1 % super retarder

Example 5 The influence of different sorts of plasticisers on the setting characteristics. (Fig. 7) has a visible effect on additives. There are additives which at 20°C have practically no retarding effect and that with lower temperatures give an increasing retardation. It is therefore recommended to always investigate the influence of additives at temperatures above and below 20°C.

With all the foregoing it is hoped that the possibilities offered by the investigation of fresh concrete mixes using ultrasonics have been illustrated. Of necessary we had to work with a laboratory not expressly designed for such an investigation.

2.5 Reproduction

For all investigations it is necessary to know if the results can be reproduced. Reproduction of results is never a simple matter where concrete is concerned, especially where test conditions can scarcely be maintained constant. The number of variables concerned when testing concrete is legion and this all makes it more difficult to determine whether results are reproducible. Nevertheless, we are convinced that the reproduction of results is good and that the system can be used in less favourable circumstances. This can be shown using an example (Fig. 8) where a number of test results are shown that came from tests that were carried out to check the properties of cement delivered to ANDOC.

Five test results are shown of which four lie in the area within which 70 % of

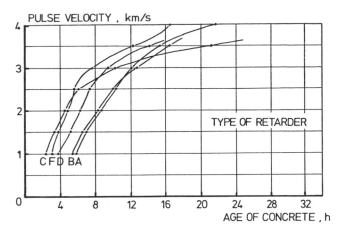

Fig. 7. Influence of plasticiser on the hardening (mix with 375 kg blast furnace cement, type B, per cubic metre and w/c-ratio 0.45)

the results lie and one which falls obviously outside. The result from this last test is not due to deviation of the measuring system, the reduced workability time was also clearly noticeable in practice.

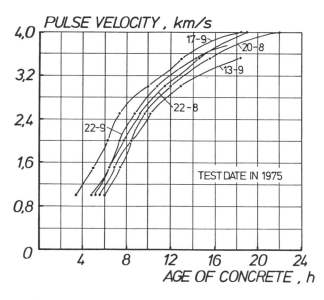

Fig. 8. Results of some tests on cement (mixture with 375 kg blast furnace cement, type B, per cubic metre, w/c-ratio 0.45 and 0.6 % retarding plasticiser

3 Ultrasonic testing of fresh mixes

Summary of studies carried out by others.

3.1 Preamle
In 1984 a working group of the Dutch 'Study society for concrete technology' (Stutech) carried out a study of the available information regarding ultrasonic testing of fresh concrete. The writer of this paper was a member of this study group. During the study it was found that although several researchers in different parts of the world had been working on this subject only a limited number of papers were available.

The following parameters were investigated:

a) ultrasonic pulse velocity compressive-strength

b) aggregate-cement ratio

c) cement type

d) ambient temperature.

In relation to the investigation reported in chapter 2 the investigations of Byfors, Elvery and Ibrahim and Casson and Domone are important because their results are comparable to the mentioned investigation (see Figures 9 to 12).

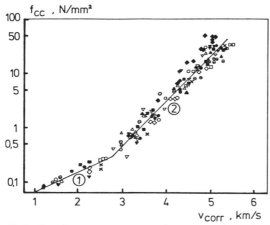

Fig. 9. Relation between compressive strength and pulse-velocity corrected for the air contents and maximum aggregate size according to expression (6)

1) $f_{cc} = 266 \cdot 10^{-2} \cdot exp(0871 \cdot v_{corr})$

2) $f_{cc} = 150 \cdot 10^{-3} \cdot exp(1933 \cdot v_{corr})$

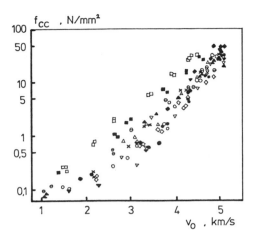

Fig. 10. Relation between compressive strength and uncorrected pulse velocity

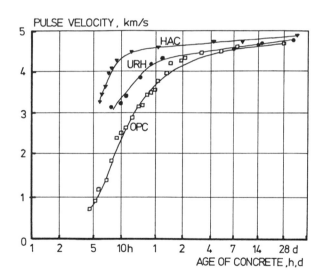

Fig. 11. Pulse-velocity development for concretes made with three different types of cement (series IV tests)

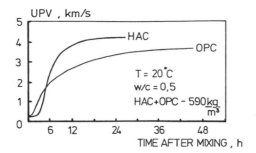

Fig. 12. Difference between OPC and HAC as shown by UPV measurements

When the test results of these investigations are compared, it seems that they all find different results for concrete mixes that are more or less comparable. Byfors investigated a number of parameters and concluded that the air content and the max. size aggregate can be pin-pointed as factors causing these differences.

The Byfors equations are:

Air content: $\Delta V = 0.05\ (A_o - 2)$
            A_o = air content   (A)

Max aggregate size: $\Delta V = 0.42\ \ln \frac{32}{d_{max}}$ (B)
d_{max} = max. aggregate size.

Finally

$$V_{corr} = V_o + 0.05\ (A_o - 2) + 0.42\ \ln \left(\frac{32}{d_{max}}\right) \qquad (C)$$

Byfors states that these corrections are valid for ultrasonic pulse velocities higher than 3 km/sec. On Appendix 1 the results of three investigations are plotted.

Line 1   Investigation by v.d. Winden
        Mix design:
        375 kg Portland blast furnace slag per m³.
        w/c-ratio 0.45
        max. size aggregate 31.5 mm
        mix temperature 15 - 20°C.

134

Line 2   Investigation of Elvery and Ibrahim
          Mix design:
          375 kg o.p.c./m^3
          w/c-ratio 0.45
          max. size aggregate 19 mm
          mix temperature 18°C.

Line 3   Investigation of Casson and Domone
          Mix design:
          340 kg o.p.c./m^3
          w/c-ratio 0.5
          max. size aggregate 4,76 mm
          mix temperature 20°C.

For all these investigations the same Pandit apparatus has been used.

The differences in ultrasonic pulse velocities as shown in Fig. 13 could come from differences in:

a) cement quality

b) mix design

c) temperature

d) used method of measurement

When, however, the results are corrected with the Byfors equation, the test results come very close together as shown in Fig. 14. For this case it is assumed that the air content of all mixes has been the same.

In view of the fact that different investigators find more or less the same results in tests carried out in different places in the world and in different time periods, the ultrasonic pulse velocity measurement gives a reliable test method for testing fresh concrete. Setting and initial hardening of fresh concrete can be investigated with this test method and gives much more information about the stiffening and setting phenomena than any other test methods available until now. Therefore, I am of the opinion that this test method should be checked more thoroughly and that international standards should be developed to describe the test method.

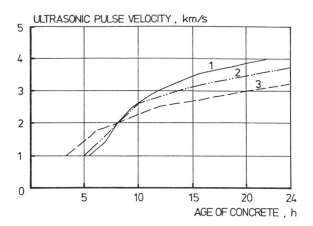

Fig. 13. Ultrasonic pulse velocity vs. age of concrete (1 v.d. Winden; 2 El-very/Ibrahim; 3 Casson/Domone

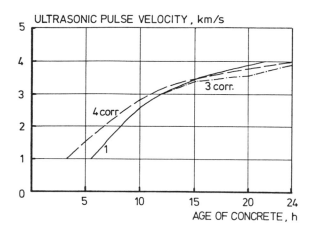

Fig. 14. Ultrasonic pulse velocity vs. age of concrete corrected acc. eq. (C)

References

[1] Voorschriften Beton, 1974, Dutch code of practice for concrete

[2] Report no. 5: Niet destructief onderzoek van beton, deel 1, non-destructive testing of concrete, part 1, published by CUR (Commissie voor uitvoering van Research)

[3] Report no. 18: Niet destructief onderzoek van beton, deel II, Non-destructive testing of concrete, part II, published by CUR

[4] Report no. 33: Niet destructief onderzoek van beton, deel III, Non-destructive testing of concrete, part III, published by CUR

[5] Report no. 69: Niet destructief onderzoek van beton, deel IV, Non-destructive testing of concrete, part IV, published by CUR

[6] R.H. Elvery and L.A.M. Ibrahim, Ultrasonic assessment of concrete strength at early ages Magazine of concrete research, vol. 28, nr. 97, December 1976

[7] CUR report no. 5, Non destructive testing of concrete 1953

[8] Whitehurst, E.A. Use of the soniscope for measuring setting time of concrete; ASTM proceedings, vol. 51, 1951, pp. 1166-1183

[9] N.v.d. Winden Investigations carried out

[10] Jones, R. A review of the non-destructive testing of concrete. Symposium of non-destructive testing of concrete and timber on 11 and 12 June 1969 London, organised by The Institution of Civil Engineers. Jones points out that there are two ways to measure the ultrasonic pulse velocity

[11] Facaoaru, I. Non-destructive testing of concrete in Romania; Lecture on the same symposium as indicated under [5]

[12] Pimenov, V.V., Grapp, V.B. Ultraschall zur Kontrolle der Herstellungstechnologie des Betons; **Bauplanung-Bautechnik 26 jg** Heft B. August 1972.

[13] Gensel, J. Ultraschall Untersuchungen und Erstarrenden Bindemitteln, **Baustoffindustrie**, 1969, nr. 11, page 380-382; 1970, nr. 1, page 19-2

[14] Casson, R.B.J., Domone, P.J.J., Ultrasonic monitoring of the early age properties of concrete, Symposium concrete at early ages, Rilem, Paris, 6-8 April 1982

[15] Byfors, J. Pulse velocity measurements for indication of the compressive strength at early ages, Symposium concrete at early ages, Rilem, Paris, 6-8 April 1982

[16] Gorur, K. Summary of a lecture on 5-10-1981, Research Fellow Technical University of Delft, about an investigation carried out to study cement hydration with microwaves

13 RADIOMETRY OF COMPACTION OF FRESH CONCRETE IN-SITU

A. HÖNIG
Brno Technical University, Brno, Czechoslovakia

Abstract
Nuclear gauges for density and moisture measurement are compared. Radiometric density gauges have been developed at Brno Technical University for measurement of the density of fresh concrete by the compaction process.

1 Comparison of portable nuclear gauges

Radiometric determination of density and moisture of soils and building materials has been carried out successfully and many types of instrument are available. The development of those methods and instruments has reached such a stage that it has become essential to review the results and experiences of their use and to inform practising engineers of the current possibilities and advantages of radiometric determination of density and moisture. On the other hand it is necessary to indicate to the manufacturers of such instruments and equipment the requirements of the users [1].

2 Radiometry of density and moisture

The radiometric measurement of density is based upon the transmission and attenuation of gamma radiation and the principle of photon scattering in the materials being tested. When employing the gamma radiation transmission and attenuation method, the radiation source and the detector are located on opposite sides of the test material. The result of the measurement is the average density of the materials between the radiation source and the detector. When employing the gamma radiation back scatter method, the gamma radiation source is located on the same side of the material under test or in the material under test. Between the source and the detector is a shield which prevents the direct passage of the radiation from the source into the detector. The detector counts only the photons that have been scattered in the materials, predominantly by the Compton effect. The sources of gamma radiation most frequently used are cesium Cs 137 with an energy of E = 0.66 MeV and cobalt Co

60 with an energy of E = 1.25 MeV. The radiation detectors are either Geiger-Müller tubes (GM-tubes) or scintillation detectors which are connected with a counter. The result of the measurement is the count rate which has to be transformed with the aid of a calibration relationship into density. In modern instruments the calibration relationship is programmed into the evaluation unit and the results of the measurement are specified in an analogue or digital manner in kilograms per cubic meter.

The radiometric determination of moisture is based on the principle of moderation of fast neutrons - predominantly on hydrogen atoms. As a source of fast neutrons an isotope is used which emits α particles and target elements as beryllium. The most frequently employed radiation sources are Ra 226, Am 241, Pu 239 and Po 210. Of these the best is Am-Be mixture which produces a sufficient number of neutrons and the secondary gamma radiation is low.

The volume of the material required for the measurement is a function of the moisture content. The more moisture the material contains, the smaller is the volume needed for slowing down the neutrons. The slow neutron detectors are connected onto a counter. The moisture value is read off either from the calibration curve on the basis of the count rate, or it is indicated directly, in an analogue or digital manner, in kilograms per cubic meter. In materials which have varying density, the determination of moisture is finalized by the simultaneous measurement of density. The output quantity will then also be the absolute or relative moisture content.

3 Basic types of detection gauge for density and moisture measurement

Radiometric gauges consist as a rule of two parts. They have a common counter and separately various detection gauges. The different types of detection gauge utilize the scattered and/or direct attenuation of gamma-rays or the slowing down of neutrons and have various geometries of the radiation source, detector and shielding. These are illustrated below.

3.1 Depth detection gauges (for depths from 0.6 to 20 mm):

a) One-way detection gauge for the measurement of density (Fig. 1a),

b) two-way detection gauge (reduces the influence of the chemical composition of the material - Fig. 1b),

c) detection gauge for the measurement of moisture (Fig. 1c).

3.2 Surface detection gauges (up to a depth of 0.15 m):

d) detection gauge for the measurement of density (Fig. 2d),

e) detection gauge for the measurement of moisture (Fig. 2e),

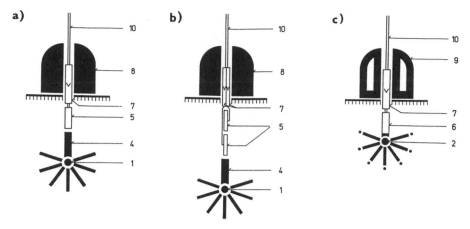

Fig. 1. Radiometric depth gauges

1 gamma radiation source
2 fast neutrons source
3 source of neutrons and gamma-rays
4 shielding between source and detector
5 gamma radiation detector
6 slow neutron detector

7 electronic circuit
8 gamma radiation shield
9 neutron radiation shield
10 cable
11 air gap frame
12 collimator

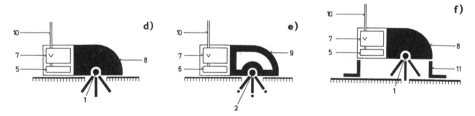

Fig. 2. Radiometric surface detection gauges

f) detection gauge for the measurement of density with an air gap (ratio of the counting rate immediately on the surface of the material and the counting rate with the air gap - Fig. 2f).

3.3 Detection gauges utilizing direct attenuation of gamma radiation (transmission method):

g) detection gauge for measurements in a narrow beam (the narrow beam is attained by shielding the radiation source, collimator and shielding the detector, the attenuation is exponential - Fig. 3g),

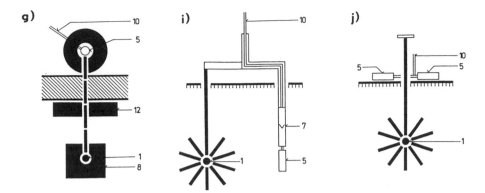

Fig. 3. Radiometric gauges for density measurements (transmission method)

h) detection gauge with a single-channel analyzer, in which is measured only the photon attenuation of a single energy - as a rule the energy of the peak of the spectrum. Such an arrangement is possible in any type of detection gauge that utilizes the effect of direct attenuation,

i) lysimetric detection gauge (fork gauge) for depths from 0.2 to 1.0 m (Fig. 3i),

j) stabbing detection gauge for depths from 0.20 to 0.80 m (Fig. 3j).

3.4 Combined radiometric detection gauges

k) depth detection gauge, in which the density and moisture measurements are combined (Fig. 4k),

l) surface detection gauge, in which the density and moisture measurements are combined (Fig. 4l)

m) detection gauge, in which one quantity (generally moisture) is measured by the e) method and the second quantity (generally density) by the j) method. In doing so either the radiation source or the detector is inserted into the material. This is a combination of a surface detection unit and a probing detection unit (Fig. 4m).

n) detection unit, in which the possibility of measuring density as well as moisture both for depth and surface measurements with a single detection unit are combined

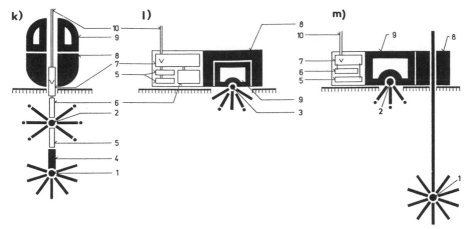

Fig. 4. Combined radiometric detection gauges

4 Compaction measurement

To measure the density of fresh concrete during and after compaction only gauges of type (j) can be used.

Radiation density meters developed at the Technical University in Brno measure the density of fresh concrete mixes, hardened concretes, bulk soils, and gravels. They are also useful for determining the density of the placed concrete used as biological shields in nuclear power sites, experimental nuclear reactors, particle accelerators, technical and medical irradiators, and also in the inspection of compacted subgrades in the construction of roads and motorways, earth dams, bridge concretes, etc. [2].

The operation of radiation density meters is simple [3]. Radiation density meters VUT in the form of a single gauge consisting of a detector and a counter comprise one mechanical unit.

4.1 Description of density meters
A significant design of the radiation density meter series is the VUT model IV with an analog output (Fig. 5.1) [3,4].

The parts are indicated by numbers as follows:

1 radionuclide Cs 137 in operating position

2 hollow steel needle

3 space for Geiger-Müller counter

4 lead shielding

5 connecting tube

6 safety control value

7 electronic equipment

8 meter

9-13 push-buttons for selecting density measurement ranges and standard

14 start push-button

15 signal for converter operation

16 signal for sliding out of radionuclide

17 holders

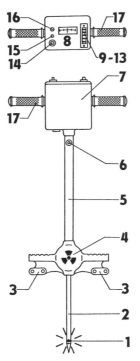

Fig. 5.1 Radiation needle-type density meter VUT model IV, with an analog output

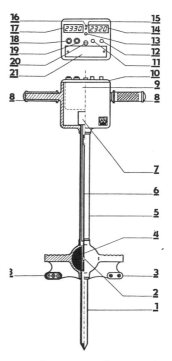

Fig. 5.2 Radiation needle-type density meter VUT model VI, with digital output

143

A continuation of the development of radiation density meters at the Technical University of Brno is the VUT model VI with a digital read-out of the measured quantity. It links up with the proven type with an analog output and is an entirely new instrument which employs the most up-to-date components [5].

The schematic arrangement and location of the functional components of the VUT radiation density meter, model VI with a digital output is shown in Fig. 5.2.

1 Hollow steel needle facilitates movement of radio-nuclide up to a distance of 200 mm into the surface of the cast concrete.

2 Radionuclide Cs 137 with a maximum activity of 320 MBq.

3 Detectors comprise 6 Geiger-Müller counters. They are located in groups of three in the tread-on part of the density meter.

4 The shielding of the radiator is accomplished by a lead sphere of diameter 92 mm.

5 The hexagonal tube connects the bottom density meter casting with the upper casting.

6 The radionuclide carrier and the radiator advancing mechanism facilitate the shifting of the radionuclide from the shielding cover into the hollow needle and back.

7 Body.

8 The handles provide easy handling of the density meter during measurements.

9 The electronic evaluation unit is fixed on TS boards and fitted with MOS digital integrated circuits.

10 Front panel, on which are located all the controls, signalling and display elements.

11 Push-button SOURCE. When depressed, the radiator is lowered to the tip of the hollow needle. After reaching the end position, the radionuclide shift mechanism motor is automatically switched off. When depressed again, the radionuclide returns into the shielding container.

12 Push-button START. When depressed, the measuring cycle is started

13 The glow lamp indicates the operation of the converter.

14 The display unit AVERAGE displays the average density value from two, four or eight measurements.

15 The pilot lamp signals a low-voltage of the battery pack. When it lights up, the battery pack must be replaced.

16 Pilot lamp of alarm signalling of radionuclide position. It begins to flicker as soon as the radiator leaves its shielded position.

17 The display unit MEASUREMENT displays the density of the measured material.

18 The selector switch RANGE makes possible density corrections in relation to the composition of the measured material.

19 The selector switch AVERAGE serves to switch off the supply voltage for the electronic evaluation unit and also as a switch for selecting the number of measurements for averaging.

20 Inspection input. The signal from an external pulse generator is connected here for checking the electronic evaluation unit.

21 The battery pack contains 10 NiCd storage batteries.

4.2 Principle of measurement

The scatter and absorption of gamma photons in the measured material is evaluated. The number of recorded photons decreases exponentially with an increase in density in accordance with the relation:

$$N = B \cdot N_o \cdot exp - (\mu_m \cdot \rho \cdot d) \tag{1}$$

where   N   rate after transmission of radiation through material $[s^{-1}]$
   N_o rate without irradiated material $[s^{-1}]$
   μ_m mass attenuation coefficient $[m^2/kg]$
   d   thickness of irradiated material $[m]$
   ρ density of the material $[kg/m^3]$
   B   build-up factor for the rate $[l]$.

With an increasing density the time required for filling the register set to a constant number of pulses is also increased.

During the measuring cycle which is limited by the filling of the register, is filled also the decadic counter. The frequency of the signal which is recorded by the counter, is variable and a function of time

$$f_m = F(t) \tag{2}$$

The relationship shows the exponential dependence of the recorded pulse rate on density. The frequency and its dependence on the period of measurement have been designed in such a way that the output decadic counter registers directly the value of the density of the material. This is facilitated without the switching of measuring ranges for densities in the range 2,100 - 2,900 kg/m^3 corresponding to the densities of concrete mixes. For measuring densities lower or higher than those given above, and also for determinations of special types of concrete, e.g. containing baryte aggregates, the range selector switch has to be employed.

MOUNT PLEASANT LIBRARY
TEL. 051 207 3581 Ext. 3701.

4.3 Technical parameters

Weight	15 kg
Dimensions	1050 x 430 x 150 mm
Range of measurement	1700-4000 kg/m^3
Accuracy of measurement	For 95 % of the results the accuracy is better than or equal to 2 %
Radioactive source	approx. 100 MBq Cs 137
Averaging	selectable out of two, four or eight measurements
Resolution	10 kg/m^3
Power supply	10 pcs NiCd storage batteries with a capacity of 4000 mA/h in battery pack
Accessories	storage battery charger 220 V/12 V shipping case spare pack with batteries
Weight of shipping case	18 kg
Dimensions of shipping case	1060 x 490 x 205 mm

4.4 Long-term and temperature stability

The stability of the density meter was determined by a standard method [1].

The daily and long-term stability was determined at the laboratory temperature of +20°C. The density meter with charged batteries was placed on the first day onto a calibration specimen of plain concrete, $\rho = 2330$ kg/m^3. The source was lowered to its operating position. Measurements were recorded in groups of five. The battery pack was always replaced when the pilot lamp signalled a low battery voltage.

Two hundred measurements were carried out on the first day. On each of the following days 100 measurements (20 groups) were made. The period of each measurement was given by the filling of the input register with 8192 pulses equivalent to 41 seconds. For each group were calculated the average and the group range. The average value as well as the group range were always related to the mean value calculated from the first hundred measurements which were carried out on the first day (2333 kg/m^3).

On the first day the following were determined: mean value from the first hundred measurements and its standard deviation. From this standard deviation were determined the probability limit of appearance for 95 % of the measurements and for 99.0 % of the measurements. For data from the second and subsequent days the mean values from all the measurements were plotted.

The method of testing at +40°C, +30°C, +10°C, ±0, −10°C was analogous to what was employed for long-term stability. To make possible an assessment of the change in the read-off values caused by temperature deviations from the laboratory temperature of +20°C, the statistical characteristics in percent were calculated, again related to the mean value of 2333 kg/m^3 which was obtained from the first 100 measurements during the daily stability test.

4.5 Measuring procedure

The hollow needle of the density meter is stuck into the fresh concrete in such a way that the bottom surfaces of the casting parts of the detectors rest on the concrete surface. Between casting and measurement on the concrete no air gap is allowed. The hollow needle must be pressed into the concrete perpendicularly to the horizontal joint of the concrete, so that both arms of the detectors are arranged symmetrically on the surface.

The RANGE selector switch is adjusted in accordance with the assumed density of the measured material. Position 1 corresponds to the density of plain concrete with a dry density of 2300 kg/m^3. The density meter is switched on by turning the selector switch to AVERAGE from the OFF position. The glow lamp lights up and on the display unit appears an undefined read-out. In the selector switch positions marked K the storage contents for averaging are determined. Before selecting the number of measurements for averaging, the selector switch must be set to position K, and then the number of measurements for averaging is selected.

By depressing the SOURCE push-button, the radiator is lowered into the tip of the hollow needle. The push-button remains locked in the depressed position and the signal lamp flickers. The motor switches off automatically after the end position is reached.

When the START microswitch is depressed, the density meter begins to record the gamma photons which impinge onto the detectors. This is indicated by a flickering of the last 0 position on the measurement display unit; if this is the second or further determination for averaging, the last 0 position on the AVERAGE display unit continues to flicker. After the input counter has been filled, the zero ceases to flicker and MEASUREMENT display unit shows the density. The START push-button can now be immediately depressed again. When this is done, on the first three positions of the MEASUREMENT display unit remain the data from the previous measurement and the last 0 position flickers in the course of the measurement. The AVERAGE display unit operates as follows: during the first measurement out of the present number remain displayed the density of the previous determinations. During the second and further measurements, respectively, the data in the first three positions are extinguished and during the filling time of the input counter, 0 begins to flicker. After the selected number of determinations the AVERAGE display unit shows the average density. If the average value of the selected number of measurements does not reach the required density, the concrete has not been compacted sufficiently.

After the measurement, the repeated depression of the SOURCE push-button returns the radionuclide into its shielding container (the push-button remains in its undepressed position). After the radiator has returned into its cover, the alarm pilot lamp ceases to flash. Finally the AVERAGE button is switched to the OFF position. Thus the instrument is switched off and the radiator is securely located in its shielding container. After a longer period of measurement the battery pack must be removed from the density meter and the batteries charged. After charging, the battery pack is stored in the shipping case.

4.6 Calibration of the instrument

The density meter was calibrated on range 1 on a calibration line made up from bilaterally ground stone and concrete plates of density 1950 - 3000 kg/m^3. Further ranges were calibrated in accordance with customer requirements. Each point of the calibration relationship is obtained from ten independent radiation measurements and the entire calibration relationship consists of a minimum of seven calibration points.

A check of the correct operation of the instrument is always carried out before the beginning and after the end of a shift, or if doubts concerning its correct operation arise. The density meter is checked either with the aid of a source which remains in its shielding container, or with the aid of a pulse generator whose signal is fed into the INSPECTION INPUT socket.

4.7 Biological shielding tests

Tests were carried out on the biological protective wall of two betatrons 41 MeV of Onkological Institutes, two linear accelerators, two experimental nuclear reactors and five nuclear reactors out of the total number of eight reactors of the nuclear power stations in Jaslovské Bohunice, Dukovany, Czechoslovakia, and Pakš in Hungary.

The advantages of radiometric tests are their speed of approximately 100 seconds for one test location compared to 1 - 2 hrs in classical tests, the immediate availability of test results and the possibility of increasing the density of insufficiently compacted concrete.

It was found that during the inspection on the walls up to a thickness of 100 cm one operator was sufficient, for larger wall thickness two operators were required. Repairs of insufficiently compacted locations could be carried out by a single worker with an internal vibrator. The determined values were processed statistically. For example, during density inspections of baryte concrete in each wall layer approximately 90 locations were tested in two measured ceiling layers were included 187 and 192 locations. In all 1740 locations were inspected, with two determinations at each point this represented 3480 measurements. The resultant density was even after revibration only at 34 locations i.e. 1.95 % slightly lower than 2800 kg/m^3. The homogenity factor K = 0.93 places the concrete of this shield into the laboratory quality class.

5 Conclusions

A radiometric inspection of concrete density ensured the required density of the placed shield concrete. From the economic point of view this could lead to a reduction of the required thickness of the shielding walls and ceilings.

If the proper thicknesses of concrete shields are maintained by applying operational radiometric inspection safety of nuclear reactors is enhanced.

References

[1] Hönig, A., Pospíšilová, L., Kablena, Pl, and Habarta, J.: "Commercial portable gauges for radiometric determination of the density and moisture content of building materials", A comparative study, International Atomic Energy Agency, Vienna, 1971

[2] Czechoslovak Standard ČSN 73 1375 "Radiometric testing of density and moisture", Publishing House of the Bureau for Standardization and Measurements, Prague, 1973

[3] Hönig, A.: Radiometric testing of density and moisture, Inženýrské stavby, 31, pp 11-17, 1983

[4] Hönig, A.: Inspection of fresh concrete in situ of biological shields of nuclear reactors, Stavby jadrovej energetiky 2, pp 1-6, 1983, Enclosure of the journal Inženýrské stavby, 31, 1983

[5] Hönig, A., Šupčík, J., New generation of a radiation density meter with a direct fresh concrete density read-out, Inženýrské stavby, 31, pp 389 392, 1983

14 MEASUREMENT OF WATER CONTENT IN FRESH CONCRETE BY USE OF NUCLEAR PROBES AND OTHER METHODS

P. van den BERG
Hollandsche Beton Groep (HBG) n.v., Rijswijk,
The Netherlands

Abstract

To find a rapid alternative for measuring the water/cement-ratio in fresh concrete, HBG and CEMIJ (cement manufacturer) made a test program with nuclear measuring equipment (in use for measuring the density and water content in soils) in comparison with two drying methods: drying with a frying pan on a gascooker and drying with a micro-wave oven. Series of mixes with two types of cement, with water/cement-ratios between 0.40 and 0.70 and with several admixtures have been tested. The results of the tests will be presented.

1 Introduction

One of the most important durability factors of concrete is the permeability of the hardened concrete. The permeability of concrete is the easiness of penetration of gases, water and the dissolved chemicals in water. Reduction of the permeability, being the same as an increase of the resistance, is possible by reducing the pore system of the cement matrix.

Reduction of the pore system can be achieved by decreasing the content of mixing water in relation to the cement content. Either less water together with the same cement content or more cement together with the same water content will increase the quantity of hydration products and thereby reduce the pore system. Also, a lower water/cement-ratio results in more resistance of the concrete or with other words, a concrete with a lower water/cement-ratio will be more durable.

In the Dutch Regulations for Concrete Technology (VBT 1986) more attention is given to the durability of concrete. Table 1 shows the exposure classes in relation to the environmental conditions, mentioned in these regulations.

For the several exposure classes of Table 1 requirements are given for mix design to realize a minimum resistance against the environmental actions. These requirements are shown in Table 2.

Important for the durability of the concrete will be the control of the water/cement-ratio during concrete production.

Table 1. Exposure classes related to environmental conditions

Exposure class	Environmental conditions	
1	dry	
2	humid	
3	humid with frost and de-icing agents	
4	sea-water	
5 a	aggressive	mildly
b		moderately
c		highly
d		very highly

Table 2. Requirements for mix design related to the exposure class

exposure classes	1	2	3	4	5a	5b	5c	5d
max. w/c-ratio for:								
- reinforced concrete	0.65	0.55	0.55	0.55	0.55	0.50	0.45	0.45
- prestressed concrete	0.60	0.55	0.55	0.55	0.55	0.50	0.45	0.45
min. cement content in kg								
- aggregates between A-B	280	280	280	280	280	300	300	300
- aggregates between A-C	280	300	300	300	300	300	300	300
min. air content in % *)								
- aggr. max. size 63 mm	-	-	3.0	3.0	-	-	-	-
- 31.5 mm	-	-	3.5	3.5	-	-	-	-
- 16 mm	-	-	4.0	4.0	-	-	-	-
- 8 mm	-	-	5.0	5.0	-	-	-	-
cement type	-	-	-	hc	-	-	src	src

Note *): no requirements with w/c-ratio < 0.45.

In the regulations for concrete technology several methods are mentioned for measuring the water/cement-ratio in fresh concrete:

- method 1   : by weight control of the materials

- method 2a : by Thaulow method

- method 2b : by drying samples of fresh concrete

- method 3   : by solution method

Two methods are favourite in the laboratories of the ready-mixed concrete production plants: method 1 and 2b. For method 1 the water/cement-ratio is calculated with the registrated figures of cement and water per batch, together with the moisture contents of the aggregates. Therefore the aggregates must be dried. After drying and calculating the moisture content of the aggregates, the total water content is known and the water/cement-ratio is to be calculated. For method 2b the water/cement-ratio is calculated by drying samples of fresh concrete for determining the water content of these samples, together with the registrated figures of the materials per batch.

For water/cement-ratio control a number of tests is required, dependent on the exposure class of the concrete (see Table 3).

Table 3. Number of water/cement-ratio tests

Exposure class	Number of w/c-ratio tests
1 and 2	1 per 40 m^3, min. 1 per day max. 3 per day
3, 4 and 5	1 per 40 m^3 min. 1 per day max. 6 per day

The problem with this number of tests is the time needed for each test. For both methods a minimum time of about one hour is required to give a water/cement-ratio result. By making the number of tests required, one man should be doing only water/cement-ratio tests during each production day.

Practice for most of the batching plants is making once a day a water/cement-ratio test by method 1 or 2b, and during the rest of the day control of the consistency of the fresh concrete by making slump tests. Together with control of the registrated figures of the weighed materials of all batches an acceptable control of the water/cement-ratio is reached, but only during production of one mix design per day. During production of several alternative mix designs together a more rapid method is desired. To find an acceptable alternative, HBG and CEMIJ started last year a test program with three methods of water/cement-ratio control:

- nuclear method

- microwave oven method

- frying pan method

2 Description of the methods

2.1 Nuclear method

The nuclear method is in use both for measuring the moisture content in different materials and for compaction control of soils. For density and moisture control of soils and for bitumen content control of rolled asphalt HBG is familiar with nuclear equipment, particularly HWZ as the road construction company of the Group. The principal of the water detection by a nuclear probe is based on the reduction of the velocity of high-speed neutrons.

High-speed neutrons, transmitted by an Am-241/Be (American-Beryllium) source, will be reduced in velocity by repeated collide with hydrogen atoms. The other, only heavy, atoms do not have any influence on the velocity of the neutrons. The low-speed (so-called "thermal") neutrons can be measured by a Li (Lithium) scintillation counter built into the probe. Incoming thermal neutrons produce pulses which are counted by a micro-processor. Because of the relation between the number of hydrogen atoms, only present as water in aggregates and concrete, and the number of low-speed neutrons in a specified time, the water content of concrete is to be calculated by the micro-processor. This is only possible after calibration with some samples with known water content.

For these tests the BIMO is used, which is especially developed for bitumen control, existing of a box of $0.40 \cdot 0.40$ m^2 with a height of 0.90 m, divided in two parts and a microprocessor for operation and measuring. The upper part of the box is to place a sample of fresh concrete inside a special mould and the bottom part of the box includes the nuclear source in its position of rest.

For a measuring cycle the source, situated in the middle of the box, must be lifted into the centre of the concrete sample. Directly after placing the source in the measuring position counting of the low-speed neutrons is started. Within a period of 16 minutes 10 measurements will be made, and each new result is the average value of all foregoing measurements.

During the tests a measuring time of 16 minutes is chosen, based on experience of HWZ by using this equipment by asphalt control. After finishing the tests, the results after 16 minutes have been compared with the results after 8 minutes (by using the registrated figures of 5 measurements). The several results are given as BIMO-8 and BIMO-16.

2.2 Microwave oven

The microwave oven is introduced as a possible rapid alternative for the drying test. In these tests the oven is used for drying fresh concrete samples like method 2b. The mircowave oven is a professional Philips AKB 115, with a capacity of 1400 W. Drying of samples of about 2.5 kg, placed in dishes of glass, did not be any problem.

2.3 Frying pan

Drying fresh concrete samples with the frying pan method is to control this method with the theoretical figures of the mixes and to compare these results with the other methods.

3 Testprogram

For these tests several mixes have been constructed. A range of water/cement-ratios between 0.40 and 0.70 is chosen because the limits of the limits of the Regulations of Concrete Technology for ready-mixed concrete are 0.45 and 0.65.

Mixes of 30 kg of fresh concrete have been constructed in the CEMIJ laboratory. For each mix a constant cement content of 330 kg per m^3 is chosen. Two types of cements are tested: a blast furnace cement class A with 25 % Portland cement clinker and a Portland flyash cement with 70 % of Portland cement clinker. The most important data of the mix designs are given in Table 4.

Table 4. Mix designs

water/cement-ratio		0.40	0.46	0.52	0.58	0.64	0.70
cement content	kg/m^3	330	330	330	330	330	330
water	kg/m^3	132	152	172	191	211	231
aggregates	kg/m^3	1977	1924	1871	1821	1768	1715
density	kg/m^3	2439	2406	2373	2342	2309	2276
water content in %		5.41	6.32	7.25	8.16	9.14	10.15

To test the influence of admixtures on the nuclear results in relation to the other methods a number of mix designs have been constructed with some different admixtures, as shown in Table 5.

Table 5. Mix designs with admixtures

water/cement-ratio		0.40	0.52	0.52	0.40	cement type	
cement content	kg/m^3	330	330	330	330	bfc	pfc
water	kg/m^3	132	172	172	132		
plasticizer	kg/m^3	4.0	-	-	-	*	-
retarder	kg/m^3	-	4.6	-	-	*	-
air ent.agent	kg/m^3	-	-	0.4	-	*	-
silica slurry	kg/m^3	-	-	-	40	-	*
total w/c ratio		0.406	0.530	0.526	0.460		

Note: bfc = blast furnace cement, pfc = Portland-flyash-cement

All the mixes have been tested twice and each mix of 30 kg is divided in smaller parts for the several tests as shown in the next figure.

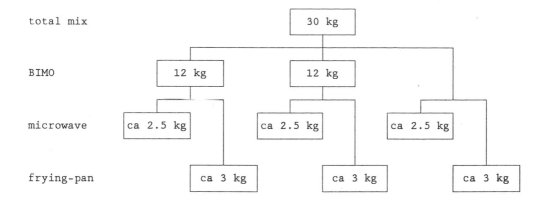

This means per mix: 2 nuclear results, 3 microwave results and 3 frying pan results.

4 Test results

4.1 Calculation of water contents

The water content per sample is measured with the several methods. For the drying tests (by microwave oven and frying pan) the water content is calculated by dividing the measured loss of weight after drying by the total weight of the wet sample. For the nuclear test method a calibration diagram is used. The relation between the water content and the registrated counts is appointed before measuring the samples. The number of the counts depends both on the water content of the samples and the water content of the surrounding air. With a minimum of 3 samples with specified water content a calibration diagram can be made, however, the influence of the environment must be taken into account by making zero-tests without samples in the box.

With special software, developed by the manufacturer of the BIMO and being a part of the microprocessor, all measured counts of the calibration samples and also the counts of the zero-test and the reference test (a sample with an unchangeable water content) are used for calculation of the calibration diagram. Registrated counts of other samples can be translated in water contents by using this calibration diagram. To keep the calibration diagram up to date every day will be started by making a zero test and a reference test.

4.2 Calculation of the water/cement-ratios

The water/cement-ratio of each sample is calculated by using the water content result and the specified cement content per mix. The relation between the specified water content per mix design and the water/cement-ratio is used as a calibration diagram (see the data of Table 4 and Fig. 1). Deflecting results have been interpolated between the given data.

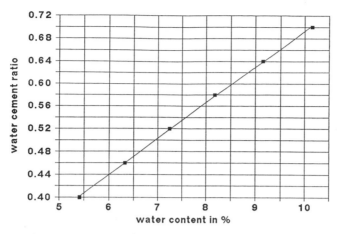

Fig. 1. Relation between w/c-ratio and water content

4.3 Regression analysis

The results per method for the several mix designs have been compared with the theoretical water/cement-ratios. For example Fig. 2 shows the BIMO results after 8 minutes. This comparison is made for all tests without admixtures together, for the tests only with blast furnace cement, for the tests only with Portland flyash cement and for all tests with the admixtures. In all these cases the diagrams are similar to Fig. 2. The average value of the results is calculated with a linear regression analysis and the calculated regression line is also given in Fig. 2.

To compare the test results of the several test methods, Fig. 3 shows the corresponding regression lines. The regression lines of BIMO-8 and BIMO-16 are very close together as well as the regression lines of magnetron and frying pan. The regression data of the results in respect of the regression line are given in Table 6.

Table 6. Regression data

testmethod	microwave	frying pan	BIMO 8 min.
correlation coefficient	0.931	0.935	0.941
gradient	0.876	0.860	0.982
constant factor	0.043	0.055	0.010
standard deviation	0.021	0.020	0.022

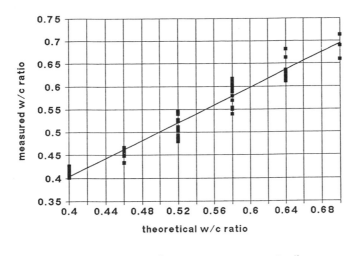

Fig. 2. Test results of BIMO after 8 minutes

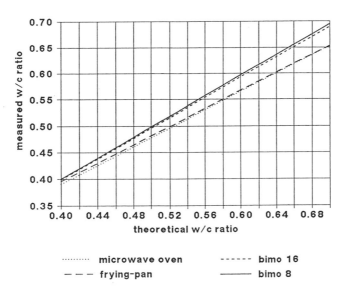

Fig. 3. Regression lines of the test methods

The figures of Table 6 for the several methods are nearly the same. The only difference is the gradient of the regression line of the nuclear method with respect to the other methods. The regression line of the BIMO is practically equal with the theoretical line. The other methods will give lower regression lines.

There is no difference observed between the two types of cement, corresponding to the expectations, because both cement types are missing hydrogen atoms. The test results of the mixes with admixtures give the same relation, but only after calculating the water part of the admixtures as added water (also a higher water/cement-ratio shown in Table 5 as "total w/c").

Comparing the results of the BIMO to the drying results of the microwave oven show nearly the same figure as before (see Fig. 4), with a standard deviation of the results to the regression line of 0.024.

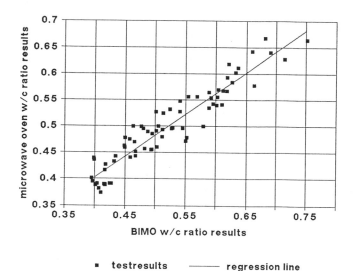

Fig. 4. Relation of w/c ratio results between BIMO and microwave oven.

4.4 Testing time

The time required for doing a test is very different for the three methods. For the frying pan a drying time of about 45 minutes is necessary. Using the microwave oven the samples could be dried within twenty minutes. The drying time for both methods depends on the size of the sample and the water/cement-ratio of the fresh concrete.

Fig. 5 and 6 show the influence of the size of the sample and the water/cement-ratio on the drying time by microwave oven.

The measuring time for the BIMO can be limited to 8 minutes. This is a remarkable reduction in time with respect to the traditional methods.

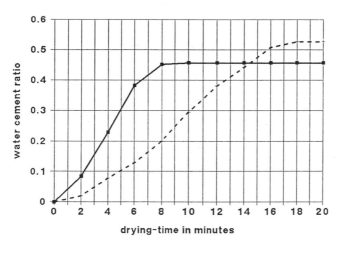

Fig. 5. Influence of sample on drying time by microwave oven

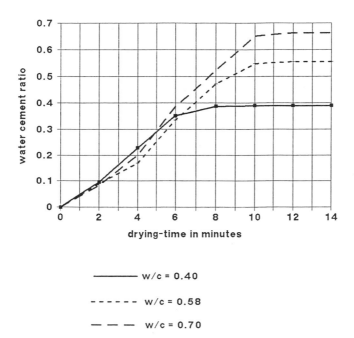

Fig. 6. Influence of w/c ratio on drying time by microwave oven

5 Safety aspects

With reference to the test results the nuclear method could be seen as a good and very rapid alternative for measuring the water/cement-ratio in fresh concrete. However, the fright for radiation is an important handicap for the introduction of this method in practice. Therefore it is important to demonstrate the safety of the use of a nuclear gauge.

At first, working with nuclear equipment is only allowed by special trained operators under supervision of the Röntgen Technical Service (RTD). A licence, given by the government, is required to use a nuclear gauge. At second, the radiation level of this kind of probes is very low. To demonstrate the low radiation level, a calculation is made for the following situation:

- test equipment is placed on a distance of 2 m of the operators desk without any complimentary safety measure,

- caused by a misunderstanding of the operator, the nuclear source is placed in the measuring position instead of the protected restposition during the whole day.

The radiation values for the BIMO close to the surface have been measured by the RTD during the tests:

- in restposition   - 0.4 mRad/hour
- in operating position   - 2.4 mRad/hour

The radiation equivalent dose for the operator is calculated during a working period of 220 days per year and should be 5.1 mS. On this moment the maximum allowed equivalent dose for operators is appointed on 50 mS, which means that the calculated value in this very exceptional circumstance should only be 10 % of the maximum value.

In accordance with the licence for the use of the nuclear gauges, everyone working with or close to the gauge must carry a personal control batch. These batches are under monthly control of the RTD and everyone will be informed of his current radiation equivalent dose.

6 Summary and conclusions

- In a test program with several mix designs one familiar method and two alternative methods have been studied to find a reliable and rapid testing procedure for measuring the water/cement-ratio in fresh concrete according to method 2b of the VBT 1986. Drying concrete samples by a microwave oven and by a frying pan and measuring the water content of concrete samples by a nuclear probe have been tested. The results of the tests have been compared with the theoretical figures of the in laboratory constructed mix designs.

- The results of the several methods indicate the same accuracy. A standard deviation of about 0.02 for all methods is calculated.

- The gradient of the regression lines of the several methods is different. The BIMO results are very close to the theoretical values, but both the microwave oven results and the frying pan results give a lower regression line.

- A reduction of testing time by half can be achieved by the microwave oven.

- A remarkable reduction is possible by using a nuclear probe. A testing time of 8 minutes should be sufficient by using the BIMO.

References

[1] Mitchel, T.M. (1975) Nuclear gage for measuring the cement content of plastic concrete, Public Roads Vol. 38, no. 4

[2] Hofstra, A. (1985) Dichtheid en vochtgehalte van vers beton nucleair bepaald, publication Wegen

[3] Nucleaire meetmethoden in de wegenbouw, Mededeling 30, Stichting Studiecentrum Wegenbouw

[4] Baron, J.P. Détermination de la teneur en eau des granulats et du béton frais par méthode neutronique, **Rapport de recherche LPC**, no. 72, Ministère de l'équipement et de l'aménagement du territoire

[5] Scheurer, A. (1989) Moisture Measurement Systems and their Application in Concrete, **Betonwerk + Fertigteil-Technik**, Heft 3

15 SETTING TEST ON CONCRETE FOR SLIPFORM METHOD

J. NICOLAY
Wilhelm Dyckerhoff Institute, Wiesbaden-Biebrich, Germany

Abstract
Dependent on the height of the sliding formwork, young concrete leaves the form-work after 8 - 10 hours. Then the young concrete has to support its own weight and the surface of the concrete still must be finishable. The mix design used PC 35 F, fly ash and a plasticizer with retarding effect. In preliminary tests the temperature curve of a concrete specimen was automatically recorded and setting was defined as a temperature increase of 1 K due to hydration.

Hardening laboratory tests on ready mixed concrete taken from site exhibited si-milar temperature curves. Mainly the hardening characteristic of the design concrete was appropriate for the planned progress of construction.

1 Problem

1.1 Silo construction

The Dyckerhoff company intended the construction of a supplementary cement silo in its plant Wiesbaden-Amöneburg. For the construction of the outer and inner cylinder shells and the 3 partition walls the slipform method was used.

The outer diameter is constant = 18.60 m
the total height is 52.00 m
the wall thickness is 0.80 m in its lower part
the wall thickness is 0.30 m in its upper part.
The silo capacity is 4,800 tons.

1.2 Slipform method

The slipform method is the most efficient construction technique for high-rise shells with partially constant cross sections. The height of the slipform and the slipping rate both determine the age of young concrete when leaving the formwork.

The dead weight of the formwork, the platform and its load is transferred into climbing rods by jacking and clamping devices. The climbing rods stand on the har-dened foundation concrete and do not stress the new construction.

1.3 Concrete requirements

Structural design required a B 35 strength class. Cross sections and positioning of the reinforcing bars indicated a maximum aggregate size of 32 mm. For efficient and foolproof workability, the "standard" class of consistence was chosen (flow diameter 450 ± 30 mm). The silo walls will be exposed concrete. In order to get a durable concrete, a minimum cement content ≥ 300 kg/m^3 and a maximum water/cement-ratio ≤ 0.60 were specified by the owner. A plasticizer with retarding secondary effect had to be used. Plasticizers cause the gain of about 5-10 l/m^3 water content without the loss in strength and durability of the water itself. In addition to these "conventional" requirements, the slipform method demands 2 supplementary properties of the young concrete:

1.3.1 Loading by own dead weight

The young concrete, when leaving the formwork, has an age of about $125/14 \approx 9$ hours

with   1.25 m   = the height of the formwork
          0.14 m/h   = average slipping rate.

The stress of the young concrete due to its own dead weight is:

$$\sigma = 2400 \cdot 1.25 \cdot 10^{-5}$$
$$= 0.03 \text{ N/mm}^2$$

with   2400 kg/m^3 = the density of the concrete

The stress level is low. But lateral strain also must remain negligible.

1.3.2 Suface finishing

The young concrete, when leaving the formwork, must not be too much hardened. The concrete surface is not smooth and closed. It exhibits marks and traces of slipping. In this age the young concrete still must be apt for finishing. The surface will be smoothed with a float by manpower (Fig. 1 to 3). An additional filling coat is prohibited. Even wetting by additional water is adverse to durability.

2 Marginal conditions

2.1 Ready mixed concrete

The producing ready mixed concrete plant was situated nearby the job site. Therefore delivery time is negligible. A concrete acceptance rate of 6 m^3/h was intended. Therefore a truck mixer was poured out in 1 hour and the average placing time was an 1/2 hour.

2.2 Temperature of fresh concrete

The slipping construction periods were scheduled in September and November 1989.

Fig. 1. First slipping rates

Fig. 2. Concrete surface leaving the formwork

Fig. 3. Smoothing with a float by manpower

In the meantime the floor was built by precast elements.

For the first period the fresh concrete temperature was assumed to be 20°C. Indeed the weather was sunny and the fresh concrete temperature was 20 to 25°C.

For the second period the fresh concrete temperature was assumed to be 10°C. This was the minimum temperature the concrete producer guaranteed otherwise he has to warm up the concrete by using hot mixing water. Indeed the temperature of the fresh concrete was 10 to 15°C, but the ambient temperature decreased below 0°C.

3 Initial tests

3.1 Conventional tests

The conventional tests had to check the requirements of German Standard DIN 1045, §7.4.2.1, e.g. workability over the delivery period of 1 hour, and the final strength according the strength class B 35.

3.2 Testing method: Removal of formwork
Setting of concrete must be reached to remove the formwork. Setting time of a cement paste is determined according to German Standard DIN 1164/Part 5 or European Standard EN 196-3 using the Vicat-apparatus and penetrating with a needle.

The "Preliminary Guideline for retarded concrete" determines the end of workability by using an internal vibrator. When the vibrator is pulled out and its immersion point ceases to flow together, then the end of workability is reached.

A combination of both methods is imaginable: Setting of concrete is reached when an internal vibrator no longer immerses in a young concrete.

This method would be unfavourable because of

- the required quantity of undisturbed concrete

- the impossibility of automation

Therefore we defined the setting time of a concrete mix when a rise in temperature due to hydration of 1 K has taken place. A temperature rise due to mixing energy and to solution heat is not taken into consideration and the measurements first start an 1/2 hour after mixing.

A temperature rise of 1 K corresponds to the following heat of hydration:

2,580 kJ /cement content of 1 m^3
(615 kcal/cement content of 1 m^3)
Calculated with a cement content of 300 kg/m^3
8.6 kJ /kg cement
(2.0 kcal /kg cement)

This heat quantity is about 2 ... 3 % of the heat after a 28 days-hydration period. It is assumed that this small heat quantity is already a significant criterion of the hardening process of the mix.

3.3 Testing procedure
Period of measurements: Starting point was 30 minutes after mixing to simulate an average delivery time of ready mixed concrete.

Temperature variations: It was intended to vary the temperature of fresh concretes and of storage conditions respectively between 20 and 10°C. The constituent materials were stored at 20 and 10°C about 24 hours before mixing.

Testing procedure: 30 minutes after mixing, a 15 cm cube with a 1 cm polystyrene form was filled with fresh concrete and compacted. A thermocouple was put in the middle of the concrete and the cube was closed by a cap of polystyrene too. Each specimen was stored in a conditional space at 20 and 10°C respectively.

The temperature curves were recorded automatically.

3.4 Results of the initial tests
Two mixes were designed and tested, with variations in cement-type and fly ash addition:

- 300 kg/m³ PZ 35 F + 50 kg/m³ FA + 0.5 % o.c. plasticizer
- 340 kg/m³ HOZ 45 L   0.5 % o.c. plasticizer

Constituent materials:

PZ 35 F-, HOZ 45 L-DIN 1164, Dyckerhoff plant Amöneburg
Fly ash Safa Weiher II PA-VII 21/118
BV 90 Tricosal plasticizer with retarding secondary effect
quartz-sand and -gravel in grading according to German Standard DIN 1045 A/B 32.

Table 1. Test results: Time up to temperature rise = 1 K

Initial test mix containing	water content (1/m³)	T concr.	T stor.	time (1K) h	compressive strength after	
		°C			12 h 24 h N/mm^2	
300 kg/m³ PZ 35 F 50 kg/m³ FA	164	20	20	9	0,5	6,0
0,5 % o.c. plast.	162	14	10	15	-	1,1
340 kg/m³ HOZ 45 L	169	20	20	9	0,6	4,2
0,5 % o.c. plast.	165	14	10	15	-	1,2

3.5 Evaluation of test results

The both conceptions PZ 35 F + FA and HOZ 45 L have the same setting times at 20 and 10°C respectively and very similar early strength results after 12 and 24 hours. Because of more experience with PZ 35 F + FA, the conctractor took his choice for this conception.

With regard to the risk of a severe drop in temperature in the month of November this conception allows a change PZ 35 F to PZ 45 F. A change HOZ 45 L to PZ 45 F cannot be taken into account because of the significant colour variation of both cements.

3.6 Additional initial tests without plasticizer

It was intended to use the secondary retarding effect of the plasticizer during construction. Therefore 2 additional initial tests without any retarding plasticizer were carried out.

Table 2. Test results: Time upt ot temperature rise = 1 K

Initial test without plasticiser	water content (1/m³)	T concr. °C	T stor.	time (1K) h	compressive strength 24 h N/mm²
300 kg/m³ PZ 35 F	169	18,5	20	4	5,9
50 kg/m³ FA	167	11,6	10	15	0,7

4 Building construction

4.1 Work progress
The work progress was normal and the concrete in construction exhibited the expected way of setting.

The slipping rates were the following:
Average: 14.6 cm/h, time in formwork 125/14.6 = 8.6 h
Minimum: 12.5 cm/h, time in formwork 125/12.5 = 10.0 h
Maximum: 18.8 cm/h, time in formwork 125/18.8 = 6.6 h

A comparison between the test results and the practical requirements of the job site shows good applicability. We believe the test method is correct and useful.
The building supervision uses a most simple test excactly to the point: Previous to the next slipping rate, a thin steel rod is sticked into the fresh concrete. Considering the height of the formwork = 1.25 m rod may penetrate only less than 0.85 m and the lower 0.40 m must have a sufficient resistance against penetration.

4.2 Temperature curves of ready mixed concrete
During construction of the silo we had opportunity to compare:

- the work progress itself with

- setting time results of the initial tests

- temperature curves and setting times of samples taken from ready mixed concrete.

Because of some variation in the slipping rates and in the fresh concrete temperatures, the dosage of plasticizer was varied in the range 0.25 - 0.50 - 0.70 % of cement (Fig. 4 and 5).

Fig. 4. Concrete placement

Table 3. Test results: Time up to temperature rise = 1 K

Ready mixed concrete 300 kg/m³ PZ 35 F + 50 kg/m³ FA	Water [x] content (1/m³)	T concr.	T stor.	time (1K)	compressive strength 24 h
		° C		h	N/mm²
+ 0,25 % o.c. plast.	166	21	20	5,5	6,9
+ 0,50 % o.c. plast.	167	23,8	20	7	7,3
+ 0,50+0,20 % plast. [xx]		22	20	9	-
+ 0,70 % o.c. plast.	168	23,9	20	8,5	4,1

x) by drying   xx) subsequent dosage of 0.2 % plast. in the laboratory

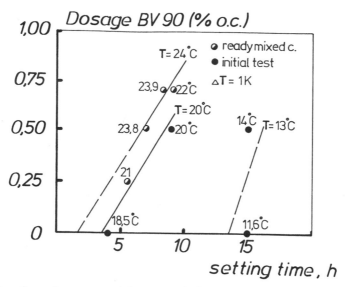

Fig. 5. Setting time of concrete with: 300 kg/m³ PZ 35 F + 50 kg/m³ FA + retarding plasticizer, dependent on - plasticizer dosage, - temperature of fresh concrete.

The results: Time up to temperature rise = 1 K, measured in laboratory on samples taken from the job site, exhibited

- a comparable range of setting time with the initial test results

- a significant influence of plasticizer dosage on setting time.

This influence on setting time was used by variable dosages of plasticizer, in particular in combination with fresh concrete temperatures about 25°C, during the first period in September.

In the second period in November, temperature was very low, even below 0°C. For better using the heat of hydration, the upper part of the silo was sheltered by canvas (Fig. 6 and 7). Moreover, the cement content was raised up to 350 kg/m³.

In this way, work progress was not hindered. The laboratory setting tests were no more valid because of the change in cement content and ambient temperature. But by means of the practicable test to prove the penetration resistance of the concrete within the formwork, slipping rates were controlled in a most reliable way.

Fig. 6. The top is sheltered by canvas

Fig. 7. Concrete placement

5 Conclusions

- The criterion temperature rise $\Delta T = 1$ K of a concrete mix is correct to determine its setting time.

- Temperature measurements are well applicable, they may be recorded automatically.

- Setting reactions of concrete are well predictible if
  - temperature of the fresh concrete
  - ambient temperature conditions
  are simulated realistically.

- The criterion $\Delta T = 1$ K is approved to compare
  - laboratory mixes to ready mixed concrete
  - test results to practical requirements: earliest strength and good finishing.

- The test method is only useful in case of initial test. During the construction period it is easier and nearer at hand to pierce a thin rod into the fresh concrete: The lower 40 centimeters of concrete within the slipform must not be piercable.

- The mix design containing ≥ 300 kg/m^3 PZ 35 F and a water/cement-ratio $\leq$ 0.60 was sturdy in terms of workability, setting, earliest and final strength.

  Also in terms of durability, the owner expects a maintenance-free concrete construction.

PART FOUR

EARLY AGE STRENGTH DEVELOPMENT

16 EARLY AGE STRENGTH DEVELOPMENT

R.N. SWAMY
University of Sheffield, Sheffield, UK

Abstract
This paper presents a brief and critical review of the available non-destructive and semi-destructive in-place test methods for estimating the strength of concrete, particularly at early ages. The implications of these test methods are discussed, with special reference to their variablity in practice and sampling. Available test results indicate that these methods can be used with confidence to give a reasonable degree of accuracy, although there is scope for more detailed studies of some of the parameters influencing these results. Satisfactory early age strength development can help to evaluate and enhance quality of construction but should not, however, be viewed as a guarantee of continued strength development or long term durability.

INTRODUCTION

In-situ testing during construction by in-place non-destructive or semi-destructive methods is now recognised to be an essential part of design and construction, particularly when limit state design methods are adopted. These techniques have several practical applications: in new construction, they can be used to determine safe stripping time, prestress release or post-tensioning time, to monitor strength development, particularly when concreting in cold weather, to check serviceability conditions or compliance and acceptability criteria, to ensure construction safety and generally to estimate the quality of construction and potential durability. In existing structures, these test methods can be used to evaluate the structure, detect imperfections, weaknesses and deterioration, and in post—mortem examinations of structural distress or structural failures. However, to be widely used in practice, such test methods need to give not only consistently reliable and reproducible results, but they should also be simple, quick and easy to operate, and robust and convenient enough to withstand site operations.

It is well known that the strength of the concrete in an actual structure is less than that obtained from standard cured control specimens (1). Concrete in a structure differs from that in a control specimen in terms of method of placing and its effects, compaction effort, temperature differentials, curing, rate of drying, moisture/relative humidity gradients, etc. - almost all factors that effect significantly the overall and long term strength, stability and safety of the structure. From an engineering point of view, strength is very much an essential parameter in design, and is also a reasonably reliable indirect indicator of durability, such as resistance to carbonation. The most important aim of in-place testing is thus often to establish a reliable estimate of the strength of the concrete in the structure, particularly at early ages. Prediction of the early age strength of concrete during construction has also special relevance and significance in the current climate of growing concern and awareness of possible causes of concrete deterioration in service. In-place testing can therefore help to assess and enhance the quality of construction.

Early age strength development, and its determination should, however, be seen in a proper perspective. Changes in the nature of cement and its ability for continued long term strength development, as well as resistance to sources of deterioration related to external environment or internal agencies, clearly warn us to exercise caution and engineering judgement in correlating early age strength, or indeed any early age property, to total guaranteee of continued strength development or long term durability.

Further, since strength cannot be measured directly by non-destructive methods, all these tests can only measure some other property which can then be related to strength. The accuracy of the strength prediction thus obviously depends directly on the degree of correlation between the strength of the concrete and the property measured by the test. It is, therefore, important to recognise and understand what the test measures, how it is related to strength and if there are factors other .than strength which can influence the test results.

METHODS OF STRENGTH ASSESSMENT

The methods available to estimate concrete in-situ strength by non-destructive or partially damaging tests can be broadly classified as follows:

- Surface hardness tests
- Penetration resistance tests
- Ultrasonic pulse velocity tests
- Impact-echo method
- Maturity tests
- Pull-out tests

- Break-off tests
- Combined methods

Of all these tests, the pull-out tests and break-off tests are probably more related to strength than others, but even then all the methods require some calibration charts to relate the measured value to strength.

Surface hardness tests

Surface hardness tests consist essentially of impacting the concrete surface in a standard manner, using a given mass activated by a given energy, and measuring the rebound or size of indentation. The Schmidt Hammer is probably the most well-known of all surface hardness tests. Since the underlying principle of this test involves both impact loading and stress-wave propagation, both strength and stiffness of the concrete which is impacted, adn the losses of energy due to mechanical friction in the instrument all have a significant effect on the measured value. The test result is therefore influenced by a number of factors such as

aggregate type
local site conditions at point of impact
moisture condition
degree of carbonation
curing conditions, age and size of specimen
surface texture, and
orientation of instrument during test.

The other points to be remembered are that the rebound hammer test probes only the top surface layer of the concrete (and therefore is unrepresentative of the interior of the concrete); and only the concrete in the vicinity of the plunger influences the rebound value.

Repeatability of results

The repeatability of the test and the variability of the concrete in the structure, i.e. the "within-test" variation, can be defined by the standard deviation or the coefficient of variation (CV) of the test results. The precision statement of ASTM C805 specifies the within-test standard deviation as 2.5 rebound numbers. Two sets of research data reported from Canada (2) and the U.K. (3) give standard deviations of 2.4 and 3.4 rebound numbers respectively. Since the rebound number is a measure of the concrete strength and stiffness, there appears to be a trend of increasing standard deviation with increasing rebound number, but the CV does not appear to have any clear trend with increasing rebound number. In general, the repeatability of the rebound hammer is about 10% CV.

Penetration resistance tests

Penetration tests determine the pentration resistance of concrete, and therefore, like the rebound hammer tests, they also involve the initial kinetic energy of a device or probe driven into the concrete, the energy absorbed by the concrete, and frictional losses. The Windsor Probe is probably the most well known commercial system available for this type of test. Unlike the rebound or other surface hardness tests, however, the Windsor probe system measures the quality of the concrete below the surface, and local surface conditions, such as moisture content and surface texture, do not however affect the test results. For concretes from 5 to 70 MPa cube compressive strength, the penetration depth varies from about 15–50 mm. The depth of penetration decreases with increasing compressive strength but not in proportion.

The probe penetration system uses different types of probes for normal weight and lightweight aggregate concretes, and operates at two different power levels depending on the strength of the concrete. The system measures the exposed length of the probes, although the fundamental relationship is between concrete strength and depth of penetration. The variability of the system thus depends directly on the conditions encountered by each probe. As the probe energy is constant, the resistance to penetration through the concrete depends on the type of aggregate, whether the probe meets a large or small piece of aggregate, whether the aggregate is fractured or not, or the aggregate distribution in the vicinity and whether the probe penetrates the matrix (Fig. 1). The Windsor probe allows either a single or a group of three probes to be fired into the concrete.

Repeatability of results

The variability of the probe penetration test results depends on the number of probes used to evaluate the average results (2, 3). Two sets of three probes in a triangular pattern seem to give consistent results with low variability (4). With an average of six results, the coefficient of variation at all ages up to 360 days has been found to be well within 4% for both normal weight and lightweight concrete as well as for different types and sizes of aggregates up to 20 mm (4). The hardness of the aggregate does not appear to have any significant effect on the variability of results at early ages. The harder aggregate tends to show marginally higher standard deviation and CV.

The most important thing to remember is that with commercial systems, the calibration charts relating exposed length of probes to standard cured cylinder strengths are based on the hardness of the aggregate, as determined from a standard mineral scratch test on Moh's scale of hardness. Extensive

tests (4) show that these charts invariably over-estimate
the actual cube strength, both wet and air dry cured, and
this is true for both normal weight and lightweight concrete
up to one year old, as well as for different sizes and types
of coarse aggregates. The strength-probe length relationship
is dependent on the type of concrete and curing regime (but
independent of the concrete mix), and the relationship valid
for new concrete (aged up to 28 days) is not valid for older
concretes. Factors other than the hardness of the aggregate
alone are involved in the penetration resistance of the probes,
factors not considered in the usual strength charts provided
with the equipment.

The real efficiency of a non-destructive test can be evaluated
only when the test is applied to assess in-situ concrete of
low strength and of concrete affected by environmental
conditions. Some typical test data obtained from reinforced
concrete slabs 1800 x 900 x 130 mm containing OPC concrete
or HAC concrete are shown in Tables 1 and 2 (5). Slabs S1
and S2 were compacted on a vibrating table, and tests were
carried out on the finished surface (S1) or mould face (S2).
Slab S3 was compacted by tamping with a steel rod several
times. Slabs with HAC were either cast with hot water 50°C
(S4), cold water 15°C (S5) or cured in hot air at 38°C (S6).
Each result in the Tables is the average of three probes.
The data show that it is possible to obtain low variability
of probe results. It is also interesting to see that the
increase in exposed probe length due to mould face or bad
compaction was less than 4% with OPC concrete. The probe
results over-estimated the 3 day OPC strength by as much as
65%; with HAC concrete, the results show reasonably good
agreement with strength range 35-40 MPa, but the probe is
clearly unable to predict high concrete strengths.

Ultrasonic pulse velocity method

The ultrasonic pulse velocity test primarily determines the
propagation velocity of a pulse of vibrational energy through
concrete by measuring the time of travel of the ultrasonic
wave through the concrete. Since the pulse velocity is directly
related to the material elastic modulus and density, and since
compressive strengh is also related to these two material
properties, the pulse velocity is often related to compressive
strength. However, many factors influence pulse velocity –
factors such as age of concrete, moisture content, mix
proportions, presence of cracks and voids and presence of
reinforcement such that no unique correlation exists between
strength and pulse velocity. Nevertheless the method has some
positive advantages in the sense that it does penetrate the
full thickness of the member, and it is very versatile since
measurements can be taken quickly, repeatedly, and with care
and skill, accurately. It has also the advantage of being
used with other test methods.

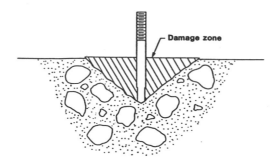

Fig 1. Probe penetration test : fracture zone

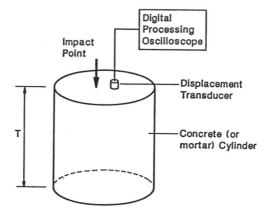

Fig 2. Impact-echo test : schematic arrangement

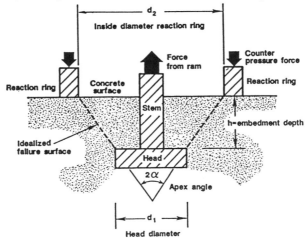

Fig 3. Pullout test : schematic arrangement

Table 1. Windsor probe test results for OPC concrete

Slab No.	Age (days)	Exposed probe length (cm)	Manual strength (N mm⁻²) Cube	Cube strength (N mm⁻²) Wet	Dry	Probe length Standard deviation (cm)	Coefficient of variation (%)
S1	3	5.232	23.0	14.0	12.8	0.229	4.37
	28	4.648	28.6	27.6	20.9	0.020	0.44
	122	4.877	35.5	29.5	22.2	0.051	1.05
	184	4.826	33.9	29.5	20.7	0.070	1.47
S2	3	5.461	26.4	18.6	15.2	0.086	1.58
	28	4.775	32.4	31.4	23.4	0.046	0.96
	122	4.928	36.9	35.5	22.9	0.116	2.36
	184	5.004	39.3	35.8	22.9	0.067	1.37
S3	3	5.385	25.3	16.3	15.8	0.028	0.52
	28	4.750	31.5	31.1	24.9	0.141	2.97
	122	4.851	34.8	35.3	24.7	0.168	3.46
	184	4.801	33.1	34.1	22.5	0.069	1.43

Table 2. Windsor probe test results for HAC concrete

Slab No.	Age (days)	Exposed probe length (cm)	Manual strength (N mm⁻²) Cube	Cube strength (N mm⁻²) Wet	Dry	Probe length Standard deviation (cm)	Coefficient of variation (%)
S4	3	4.953	37.6	39.3	44.9	0.117	2.36
	28	4.953	37.6	40.3	44.7	0.051	1.03
	122	4.801	33.1	33.1	35.4	0.025	0.53
S5	3	5.410	51.5	77.6	77.3	0.104	1.93
	28	5.588	56.9	93.3	86.7	0.082	1.45
	122	5.563	56.1	88.5	65.9	0.028	0.50
S6	28	5.525	54.9	91.2	89.1	0.041	0.74
	122	5.474	53.4	82.4	62.1	0.013	0.23

Repeatability of results

Pulse velocity results are relatively insensitive to the normal heterogeneity of concrete, so that it is possible to obtain extremely low within-batch coefficient of variation. Several laboratory studies show average CVs less than 2%. Strength-pulse velocity relationship is essentially exponential (6), so that even when the pulse velocity correlates well with strength at early ages, it is insensitive to even major increases in strength at later ages. A relationship established at early ages is therefore not applicable as the concrete matures. Relationships established using laboratory-cured specimens cannot be confidently used to evaluate strength development in a structure. Similarly, relationships established for sound concrete cannot be used to predict the strength of deteriorating concrete.

Impact-echo method

The impact-echo method, like the pulse velocity method, is also based on transient stress wave propagation to measure early age strength development, but has the advantage that only one side of the concrete surface needs to be accessible (Fig. 2). In this method, the concrete is subjected to a point impact, and the surface displacement, at a point adjacent to the impact, is monitored. From the measured displacement waveform and the thickness of the concrete, the compression wave (P-wave) velocity is calculated. Changes in the velocity with age can then be related to the development of mechanical properties as the concrete matures. As for the pulse velocity method, both density and elastic properties of the concrete influence the P-wave velocity, and factors which influence these two properties can therefore be expected to affect the strength - P-wave velocity relationship.

Pessiki and Carino have recently reported an initial feasibility study to examine the use of the impact-echo method to develop strength-velocity relationship (7). It was found that at early ages up to about 3 days at normal temperature, the relationship was independent of curing temperature and water-cement ratio. However, a change in the aggregate volume content had a significant effect on the relationship, whereas hot-cured specimens showed lower strength than room-cured specimens.

In these tests the earliest measurements were made at about 12 hours after mixing. Measurements taken up to 7 days showed that for a given concrete mixture, cured at room temperature, the sensitive region of the strength-velocity relationship lasted for about 3 days, i.e. for about 60% of the 28 day strength. Since this covers the strength and age required typically for safe removal of formwork or the transfer of prestressing, the impact-echo method may have very useful applications in estimating strength development at early ages.

Maturity method

Since the cement-water chemical reaction is exothermic, two basic factors influence the strength development of concrete, particularly at early ages, namely, the rate of hydration reaction and time. The extent of hydration is a function of the thermal history of the concrete so that a method of strength assessment based on the combined effects of temperature and time would appear to be ideally suited to concrete. The maturity method does precisely this by quantifying these combined effects by a "maturity" value from a knowledge of the thermal history of the concrete and a so-called "maturity function". The development of a strength-maturity relationship then enables the strength of a particular concrete mixture to be expressed as a function of its maturity. The effect of different curing conditions on strength can then be readily evaluated through such relationships and the thermal history of the samples.

Several mathematical expressions of the maturity function relating the influence of temperature on the rate of strength development have been proposed and reviewed (8, 9). Two expressions - the linear Nurse-Saul function and the Arrhenius exponential function, the former being suitable when concrete is cured between 10 and 30°C, and the latter, over a larger range - have found widespread usage and will be discussed elsewhere in this workshop. It will be obvious that to utilise this method in practice and estimate the in-place strength requires two activities - establishment of the strength-maturity relationship for the concrete in the structure and monitoring continuously the temperature of the in-place concrete to define its maturity (i.e. temperature-time factor).

A very recent study points out that it is unlikely that the maturity concept will lead to satisfactory prediction of strength because the rate of hydration of cement increases more rapidly than in proportion to temperature (10). Nevertheless, the error involved can be acceptable for prediction of early age strengths. Further, because cement hydration and concrete hardening are essentially retarding processes, maturity meters that simulate the hydration of cement, such as evaporation of a liquid from a capillary tube, may be preferable than electronic meters for early age strength prediction.

Repeatability of results

The repeatability of the maturity method depends on the instrumentation used to record the temperature history of the concrete. Electronic maturity meters are available, but it is important to ensure that its maturity function is applicable to the concrete in the structure. Maturity meters which reflect the retarding nature of cement hydration such as those incorporating the evaporation of a liquid from a capillary tube, may have a more rational basis for the maturity concept and be more advantageous in predicting early age strength of concrete (11). Maturity computed from actual temperature readings of the concrete may give a better and lower repeatability of the maturity values than commercial instruments.

Pullout tests

The pullout test measures the failure load required to pull out from concrete a specially shaped metal insert with an enlarged head which has been cast into concrete. The test involves static loading of the concrete, and therefore, unlike the rebound hammer and probe peneteration tests, some sort of stress analysis of the fracture zone, whether elastic or non-linear, can be carried out. There is general agreement that the fracture zone in the test is subjected to a highly non-uniform three-dimensional state of stress; but whether the ultimate failure is governed by crushing of the concrete or by a combination of shear/tension inclined fracture is still subject to considerable argument and debate (Fig. 3).

The result of the pullout test is very much dependent on the nature of the concrete in the immediate vicinity of the conic frustrum defined by the insert head and reaction ring; thus only a small volume of the concrete is involved in the test. On the other hand, the test produces a well-defined fracture zone and measures a static strength property; it is, therefore, possible to develop an empirical correlation curve between pullout strength and the compressive strength of concrete. However, the test needs preplanning the location of the inserts on the formwork, and unlike other in-situ tests, it cannot be performed at random after the concrete has hardened. There are modified forms of the pullout test, the internal fracture test, which can be carried out on existing construction by drilling a hole into which either normal pullout inserts or split-sleeve assemblies/wedge anchors are inserted which are then pulled out (12-14).

Repeatability of results

Published data on the within-test variability of the pullout test indicate that the standard deviation tends to increase with increasing pullout load, whereas the CV shows no such trend (15). The latter could thus be used as a measure of the repeatability of the pullout test. Analysis of the pullout test data reported from different laboratories show the within-test average CV ranging from 4.0% to 10% with one high value of 14.8% for apex angle 54° to 70°, embedment depth 25-50 mm, maximum aggregate size 10-25 mm and a wide range of natural aggregates including one lightweight aggregate (16). The Lok test with a 25 mm embedment and 62° apex angle shows a CV of 4.1 to 15.2% with an average of 8% (17): NBS results show the following CV (18):

Variable		Range of CV%
Apex angle	: 30-86	5.6 – 18.7
Embedment	:	6.5 – 14.0
Aggregate size	: 6-18mm	3.3 – 10.6
Aggregate type	:	5.6 – 16.8

A wide variety of test data seem to indicate that the variability of the pullout test is influenced by the ratio of the mortar strength to coarse aggregate strength and by the maximum aggregate size. Repeatability is improved when the maximum aggregate size is small, and aggregate and mortar strengths become similar (19).

The variability of a modified form of the pullout test based on the internal fracture test carried out on RC slabs 1800 x 900 x 130 mm with low strength or deteriorating concrete, is shown in Table 3 (5). The tests showed that the surface damage caused by the test was independent of the type of concrete, concrete mix and age of concrete. For OPC concrete, the average CV was about 12%, whereas with HAC concrete this value was about 14% for low strength concrete (30-40 MPa) and about half this value for high strength concrete (75-95 MPa). The results on the mould face (slab S2) were about 6-11% higher due to differences in the concrete density, the effects of compaction, the aggregate-matrix bond and the matrix strength. With this type of modified test where an expanding bolt is inserted into a drilled hole, the variability is higher because of the number of factors involved such as the localized nature of the test, the characteristics of the concrete surface, the imprecise load transfer mechanism, variations in the depth of embedment of the bolt and the characteristics of the hole.

Break-off tests

This is a partially destructive test and consists in deter-mining the flexural strength of concrete on a core with the rupture zone located at about 70 mm from the surface. The method consists in forming a circular slit in the concrete, and applying a transverse force to fracture the resulting core. In fresh concrete, plastic forms are inserted into the concrete as it is cast, and removed after curing. In hardened concrete, a diamond drill is used to cut the required slit. The force required to rupture the core is assumed to be a measure of the flexural strength of the concrete, and can

Table 3. Variability of pull-out test results

Slab No.	Age (days)	Pull-out force* (kN)	Wet cube strength+ (N mm−2)	Pull-out standard deviation (kN)	Pull-out coefficient of variation (%)	Average coefficient of variation (%)
S1 OPC	3	2.72	14.0	0.45	15.3	
	28	3.78	27.6	0.27	7.3	11.5
	122	3.72	29.5	0.51	13.7	
	184	3.70	29.5	0.36	9.7	
S2 OPC	3	2.88	18.6	0.45	15.4	
	28	4.21	31.4	0.53	12.6	11.9
	122	4.10	35.5	0.54	13.3	
	184	4.05	35.8	0.25	6.3	
S3 OPC	3	3.01	16.3	0.47	15.5	
	28	4.36	31.1	0.56	12.7	12.9
	122	4.30	35.3	0.47	10.9	
	184	4.26	34.1	0.53	12.4	
S4 HAC	3	2.79	39.3	0.36	12.8	
	28	2.82	40.3	0.52	18.4	13.8
	122	2.68	33.1	0.28	10.6	
	184	2.60	28.2	0.35	13.5	
S5 HAC	3	6.16	77.6	0.33	5.4	
	28	7.13	93.3	0.56	7.9	7.6
	122	6.53	88.3	0.53	8.1	
	184	6.18	77.0	0.55	8.9	
S6 HAC	3	6.29	84.1	0.51	8.1	
	28	7.17	91.2	0.41	5.7	6.7
	122	6.56	82.4	0.42	6.5	
	184	6.50	82.5	0.41	6.3	

*Average of eight tests   +Average of three cubes

be correlated to compressive strength. Tests reported in the literature show CV of 7 to 14% (20).

A modified form of this test is the pull-off test in which a circular steel probe is bonded to the surface of the concrete by means of an epoxy resin adhesive (21). The test can be carried out on uncored or partially cored specimens. The force required to pull-off the probe is a measure of the tensile strength of the concrete which can then be correlated to compressive strength. Results of early age tests using a portable apparatus have shown CV of 16.8%, 11.3% and 13.6% at 1 day, 3 days and 7 days (21).

Variability of early-age results

There is only limited laboratory and field data reported in the literature which are specifically desinged to examine the variability of in-place testing at early ages. Carette and Malhotra (22) have reported on the within-test variability at ages of 1-3 days of various in-situ tests, and their ability to predict early age strength development. They used four methods - penetration resistance, pulse velocity, rebound hammer and two types of pullout tests. The tests were performed at 1, 2 and 3 days on plain concrete slabs, 1220 x 1220 x 300 mm in size. Four concrete mixes with nominal cement contents of 250-350 kg/m^3 were used. Their results are shown in Table 4. The results show that all in-situ tests, with the possible exception of the rebound hammer which had the highest variability with a CV of about 12%, can predict the early age strength development of concrete with a reasonable degree of accuracy.

Pullout testing using "finger-placed" inserts to determine the strength development of concrete during the construction of a box culvert requiring about 11,500 m^3 concrete has also been reported (23). The data obtained from this study appeared to be normally distributed at early ages, although the within-test variation was rather high (Table 5). It was thought that this variability could be reduced by improvements in the equipment, technique of placing and extracting the inserts.

Comparative data on within-test variability of various in-place test methods have also been reported by Yun et al (24). Their results are presented in Table 6.

Sampling in practice

To gain maximum useful information from in-situ non-destructive testing, careful planning of both the number of tests and their location in the structure are required. The sample size to be used for a particular situation will depend on the variability of the test method, the maximum allowable error

Table 4   Early age variability of in-situ tests

Test	No of tests	CV% Av.4 mixes		
		1D	2D	3D
CANMET pullout	4	8.3	7.0	6.1
Commercial pullout	10	7.9	9.9	7.9
Penetration resistance	6	4.9	5.8	5.3
Ultrasonic pulse velocity	4	0.2	0.5	0.4
Rebound hammer	20	13.0	10.6	11.8

Table 5   Early age variability of finger-placed pullout tests

Variability	2D	4D	7D
S.D: Overall, kn	4.2	3.3	4.4
S.D.: Within test, kn	2.1	2.1	3.4
C.V: Overall, %	22.4	15.3	17.5
C.V: Within test, %	11.2	10.0	13.4

Table 6   Early age variability of tests, C.V.%

Test	25mm agg. Concrete			40mm agg. Concrete		
	1D	3D	7D	1D	3D	7D
Pullout	15.8	9.1	14.2	16.1	21.7	15.5
CAPO	-	-	-	17.6	27.0	19.0
Pulse velocity	0.5	0.4	0.5	0.8	0.6	0.8
Rebound hammer	13.2	11.9	8.8	11.0	11.4	11.6
Probe	11.7	13.8	13.8	13.1	16.3	11.9

and the CV. ACI Committee 228 suggests the following typical values of CV (%) as a guide in selecting the number of tests (16):

Rebound hammer	–	10
Probe penetration	–	5
Ultrasonic pulse velocity	–	2
Pullout	–	8

The following numbers of tests, it is suggested, would ensure the same degree of confidence for the average in-place test result as the average cylinder strength (16)

Rebound hammer	–	12
Probe penetration	–	3
Ultrasonic pulse velocity	–	5
Pullout	–	8

CONCLUSIONS

Results published in literature give considerable hope and confidence in the use of non-destructive and semi-destructive in-place tests to predict the early age strength development of concrete with reasonable degree of accuracy. However, avaiable data are limited, and there is need to examine in detail the effects of many parameters both in laboratory and field testing. New promising techniques such as the impact-echo method, and the wider use of combined methods involving the rebound hammer and the ultrasonic pulse velocity should extend the scope of available methods for practical applications. Two aspects of the in-situ tests, namely, the effect of large variations in temperature of the in-place concrete at early ages and the role of mineral and chemical admixtures deserve further careful study. Using pre-established relationships should be tolerated only with a clear knowledge of the implications and engineering judgement.

REFERENCES

(1)   Maynard, D. P., Davis, S. G. The strength of in-situ concrete, The Struct. Eng., 52, (1974), no. 10, pp. 369-374.

(2)   Carrette, G. G., Malhotra, V. M. In-situ tests: variability and strength prediction at early ages, ACI Publ. SP-82, (1984), pp. 111-141.

(3)   Keiller, A. P., Preliminary investigation of test methods for the assessment of strength of in situ concrete, Tech. Rep. No. 42.551, Cement and Concrete Association,

(1982), 37 pp.

(4) Swamy, R. N., Al-Hamed, A. H. M. S., Evaluation of the Windsor probe test to assess in situ concrete strength, Proc. Instn. Civ. Engrs., Part 2, 77, (1984), pp. 167-194.

(5) Swamy, R. N., Ali, A. M. A. H., Assessment of in situ concrete strength by various non-destructive tests, NDT International, 17, (1984), no. 3, pp. 139-146.

(6) Swamy, R. N., Al-Hamed, A. H. M. S., The use of pulse velocity measurements to estimate strength of air-dried cubes and hence in situ strength of concrete, ACI Publ. SP-82, (1984), pp. 247-276.

(7) Pessiki, S. P., Carino, N. J., Setting time and strength of concrete using the impact-echo method, ACI Materials Journal, 85, (1988), no. 4, pp. 389-399.

(8) Malhotra, V. M., Maturity concept and the estimation of concrete strength - a review, CANMET Inf. Circular No. IC 277, (1971), 43 pp.

(9) RILEM Committee 42-CEA, Properties of concrete at early ages - state-of-the-art report, RILEM Materials and Structures, 14, (1981), no. 84, pp. 399-450.

(10) Chengju, G., Maturity of concrete: method of predicting early-age strength, ACI Materials Journal, 86, (1989), no. 4, pp. 341-353.

(11) Hansen, A. J., COMA - meter, the mini maturity meter, Nordisk Betong, 25, (1981), 3 pp.

(12) Bungey, J. H., Concrete strength determination by pull-out tests on wedge anchor bolts, Proc. Inst. Civ. Engrs., Part 2, 71, (1981), pp. 379-394.

(13) Chabowski, A. J., Bryden-Smith, D. W., Assessing the strength of in situ Portland cement concrete by internal fracture tests, Mag. of Conc. Res., 32, (1980), no. 112, pp. 164-172.

(14) Ash, J. E., Assessment of in-situ concrete and its application to HAC beams, Concrete, 11, (1977), no. 3, pp. 24-25.

(15) Stone, W. C., Carino, N. J., Reeve, C. P., Statistical methods of in-place strength prediction by the pullout test, ACI Journal, 83, (1986), no. 5, pp. 745-755.

(16) ACI Committee 228, In place methods for determination of strength of concrete, ACI Materials Journal, 85, (1988), no. 5, pp. 446-471.

(17) Krenchel, H., Peterson, C. G., In situ pullout testing with Lok test – ten years experience, Nordisk Betong.

(18) Stone, W. C., Giza, B. J., Effect of geometry and aggregate on the reliability of the pullout test, Concrete International : Design and Construction, 7, (1985), no. 2, pp. 27-36.

(19) Bocca, P., Application of pullout test to high strength concrete strength estimation, RILEM Materials and Structures, 17, (1984), no. 99, pp. 211-216.

(20) Dahl-Jorgensen, E., Johansen, R., General and specialized use of the break-off concrete strength testing method, ACI Publ. SP-82, (1984), pp. 293-308.

(21) Long, A. E., McC. Murray, A., The pull-off partially destructive test for concrete, ACI Publ. SP-82, (1984), pp. 327-350.

(22) Carette, G. G., Malhotra, V. M., In-situ tests: variability and strength prediction of concrete at early ages, ACI Publ. SP-82, (1984), pp. 111-141.

(23) Vogt, W. L., Beizai, V., Dilly, R. L., In situ pullout strength of concrete with inserts embedded by finger placing, ACI Publ. SP-82, (1984), pp. 161-175.

(24) Yun, C. H., Choi, K. R., Kim, S. Y., Song, Y. C., Comparative evaluation of nondestructive test methods for in-place strength determination, ACI Publ. SP-112, (1988), pp. 111-136.

17 MATURITY FUNCTIONS FOR CONCRETE MADE WITH VARIOUS CEMENTS AND ADMIXTURES

N.J. CARINO
National Institute of Standards and Technology,
Gaithersburg, MD, USA
R.C. TANK
Narmada Hydro Project, Gujarat, India

Abstract
A model is proposed for estimating relative strength gain of concrete
under isothermal curing conditions. The key feature of the model is the
relationship between curing temperature and the rate constant for relative strength development. The strength gain of seven concrete and
mortar mixtures under three curing temperatures was studied. It was
found that a simple exponential function may be used to describe the
observed variations of the rate constant with curing temperature. It is
shown that the relative strength development of concrete can be estimated
from its temperature history using parameters determined experimentally
from tests of isothermally-cured mortar specimens.

1 Introduction

At early ages, it is difficult to estimate accurately the in-place strength of concrete based solely on the age and the strength-gain properties
measured under standard conditions. The reason is because early-age
strength is strongly influenced by the in-place temperature history.
During the late 1940's and early 1950's the idea was developed in Europe
that the strength development of concrete could be related to a quantity
called "maturity" [1-3]. Maturity attempts to account for the combined
effects of time and temperature on strength gain and is evaluated from
the in-place temperature history of the concrete. The "maturity concept"
proposed by Saul [2] states that samples of the same concrete mixture
will have equal strength if they have the same maturity value, irrespective of their temperature histories.

Three steps are involved in applying the maturity method to estimate
in-place strength: (1) establish the strength-versus-maturity relationship for the concrete mixture; (2) compute the in-place maturity value
based on the measured temperature history; and (3) estimate the in-place
strength using the strength-maturity relationship and the in-place maturity value. Use of the method requires an adequate supply of moisture
to sustain hydration during the period that strength is to be estimated.
A standard practice for using the maturity method has been adopted by
ASTM [4].

The maturity value is computed from the temperature history using a "maturity function." A variety of such functions have been proposed [5,6], but only two are widely used. The first is the so-called Nurse-Saul function:

$$M = \Sigma \ (T-T_o) \ \Delta t \tag{1}$$

where
    M = temperature-time factor (often called "maturity"),
    T = average temperature of concrete during the time interval Δt, and
    T_o = datum temperature (temperature below which there is no strength gain).

The Nurse-Saul maturity function is based entirely on empirical observations, and it has been found to be adequate only under certain conditions [5]. When different samples of a given concrete experience dissimilar early-age temperatures, Eq. (1) does not represent accurately the effect of curing temperature on strength development. Nevertheless, the Nurse-Saul function is widely used because of its simplicity.

In 1977, a maturity function was proposed [7] which is based on the well-known Arrhenius equation. The function is used to compute the "equivalent age" at a reference temperature according to the following equation:

$$t_e = \Sigma \ \left(e^{-Q(\frac{1}{T} - \frac{1}{T_r})} \right) \ \Delta t \tag{2}$$

where
    t_e = equivalent age at the reference temperature,
    T = temperature of the concrete during the time interval Δt, degrees Kelvin,
    Q = a constant (to be discussed later), degrees Kelvin, and
    T_r = reference temperature, degrees Kelvin.

The maturity function based on the Arrhenius equation has been found to be superior to the Nurse-Saul function, and it has been adopted in the practices of some Nordic countries. However, Eq. (2) is also empirically based and there are questions about the appropriate values of the constant Q [5].

Since the late 1970's, the U.S. National Bureau of Standards[a] (NBS) has studied the maturity concept. In the early work, the Nurse-Saul function was used [8,9]. In agreement with others, it was found that Eq. (1) approximated the combined effects of time and temperature on the development of various mechanical properties of concrete. However, a later study showed that under certain conditions, the Nurse-Saul function lead to consistent errors [10]. This finding was in agreement with previous critics of the maturity concept [11]. As a result, NBS embarked on a new study [12] of the maturity concept to find explanations for these discrepancies and to examine alternative approaches for computing maturity.

---

[a]Name changed in 1988 to the National Institute of Standards and Technology (NIST).

The results of the NBS study identified the inherent assumption behind Eq. (1) and explained why Eq. (2) is a more accurate maturity function [5]. It was also concluded that, contrary to the maturity concept, a given concrete does not possess a unique strength-maturity relationship, because the early-age temperature history affects the limiting strength. However, it was proposed that there is a unique relationship between maturity and relative strength (ratio of strength to limiting strength). A procedure was developed to determine experimentally the maturity function that would result in accurate estimates of relative strength gain for a given concrete mixture.

The key feature of the NBS procedure is to evaluate the temperature dependence of the rate constant governing strength development of concrete. The rate constant is the initial rate of relative strength development during the acceleratory period following setting. The value of the rate constant was determined by statistical analyses of strength gain data under constant temperature curing. In the data analyses, it was assumed that strength gain could be described by a hyperbolic strength-age curve. It was also suggested that the temperature dependence of the rate constant could be evaluated with sufficient accuracy by testing mortar cubes rather than concrete cylinders.

The NBS research produced a rational procedure to analyze the effects of curing temperature on strength development. However, the research program was limited to one cementitious system (ASTM Type I cement with no admixtures). Therefore, the present study was carried out to expand upon the initial work by considering a range of concrete mixtures used typically in construction. This expanded study is described in the Ph.D. thesis of the second author [13], and this paper summarizes the results.

1.1 Objective and scope

The goal was to develop a reliable model to quantify the effects of curing temperature on the strength development of concrete. To achieve this goal, the study had the following objectives:

- to verify the hyperbolic strength-age model for the development of strength under constant curing temperature;
- to examine the relationship between the rate constant and curing temperature for various concrete and mortar mixtures;
- to determine whether mortar specimens can be used to obtain the parameters to estimate relative strength gain of concrete specimens.

To achieve these objectives, the study examined the effects of the following variables on the strength development of concrete and mortar mixtures:

- Type of cement (ASTM Type I, II, and III),
- Curing temperature (10, 23, and 40 C),
- Water-cement ratio (W/C) (0.45 and 0.60),
- Admixtures (accelerator and retarder), and
- Pozzolanic additions (fly ash and ground blast furnace slag).

194

2 Experimental procedure

2.1 Mixture proportions and test specimens

The concrete mixtures were proportioned using the ACI trial proportioning procedure [14]. The coarse aggregate was crushed limestone with 19-mm nominal maximum size, and the fine aggregate was natural river sand. The cement-to-sand proportion of the mortar representing the concrete was selected to ensure that the rate constants for mortar mixtures would be similar to those of the corresponding concrete. Based on experiments, the optimum cement-to-sand proportion of mortar was found to equal the proportion of cement-to-coarse aggregates in the concrete it represented [13].

A drum mixer was used to mix concrete and a bench-top paddle mixer was used to mix mortar. Mixing and specimen preparation were carried out at room temperature, and separate batches were mixed for each curing temperature. Cylindrical concrete specimens were molded using 102 by 203-mm plastic molds. Cubic mortar specimens were molded using 51-mm cube steel molds.

2.2 Curing

The three constant curing temperatures were approximately 10, 23, and 40 C. Within one hour from the start of mixing, the freshly molded specimens were transferred carefully to water baths maintained at the intended curing temperatures. Efforts were made to ensure that the initial temperatures of the mixtures were close to the intended curing temperature, because the initial temperature significantly affects the initial rate of strength development. Therefore, initial mixing temperatures were controlled by using hot or cold mixing water. The initial temperature of each fresh concrete and mortar batch was within 5 C of the desired curing temperature.

Specimen temperatures were monitored using thermocouples placed at the center of a cube or a cylinder from each batch. A data logger was used to measure temperatures at one-minute intervals and print the daily average temperatures. The concrete and mortar specimens reached the water bath temperatures within 2 hours after the start of mixing.

A maximum deviation of 3 C above the intended curing temperature was observed for the specimens cured in the high temperature bath due to the early-age heat of hydration. However, the deviation was less than 3 C for specimens cured at 10 C and room temperature because the heat of hydration evolved more slowly and was absorbed by the water baths. In subsequent data analyses, the curing temperature of each batch was taken as the weighted average specimen temperature up to the time of the third strength test.

2.3 Strength testing

All mortar cubes were removed from their molds at the time of the first strength test, but the concrete cylinders were kept in their molds until the day of testing. Three cubes and three cylinders were tested in compression at seven different ages. Sulfur capping compound was used with the cylinders. The testing schedule was selected so that specimens cured at different temperatures were tested at approximately the same maturities. The age at time of test was considered from the time when the the mixing water was added.

3 Data analysis

3.1 Strength development model

It was assumed that, under isothermal curing, the strength gain of the concrete and mortar specimens could be represented by the following hyperbolic equation:

$$S = \frac{S_u \, k_T \, (t - t_o)}{1 + k_T \, (t - t_o)} \tag{3}$$

where

S = compressive strength at time t,
S_u = limiting compressive strength at infinite age,
k_T = rate constant for strength development at the curing temperature T, day^{-1}
t = curing age at temperature T, days, and
t_o = age when strength development is assumed to begin, days.

The basis of Eq. (3) has been explained independently by Carino [5] and by Knudsen [15]. Least-squares regression analysis [16] was used to determine the values of S_u, k_T, and t_o for each mixture.

3.2 Rate-constant-versus-temperature relationship

The relationship between the rate constant and curing temperature is the key to developing the correct maturity function. The suitability of three functions in representing this relationship was examined.

Linear Function -- The linear relationship between the rate constant and curing temperature is

$$k(T) = \beta \, (T - T_o) \tag{4}$$

where

$k(T)$ = rate constant function, day^{-1},
T = curing temperature, C, and
β, T_o = regression constants.

The regression constant T_o represents the temperature at which the rate constant is zero, and is analogous to the datum temperature in the Nurse-Saul maturity function (Eq. (1)).

Arrhenius Function -- The rate constant function based on the Arrhenius equation is expressed as:

$$k(T) = A \, e^{\left(\frac{-Q}{T} \right)} \tag{5}$$

where

A = regression constant, day^{-1},
Q = regression constant, Kelvin,
T = absolute curing temperature, Kelvin.

Exponential function -- The following exponential function, suggested in Ref. [17] as an alternative to the Arrhenius function, was also examined:

$$k(T) = C\, e^{(B \cdot T)}$$ (6)

where

  C = regression constant, day^{-1},
  B = regression constant called the "temperature sensitivity factor," 1/C, and
  T = curing temperature, C.

The regression constants in Eqs. (5) and (6) were obtained by ordinary least squares analysis using natural-logarithm transformations of the equations.

3.3 Relative strength gain

The authors have proposed[b] that a unique relationship exists between relative strength gain and equivalent age for a given concrete mixture. Relative strength is the ratio of strength to the limiting strength when the concrete is fully matured. The relationship is as follows:

$$\frac{S}{S_u} = \frac{k_r\,(t_e - t_{or})}{1 + k_r\,(t_e - t_{or})}$$ (7)

where

  t_e = equivalent age at the reference temperature, days,
  t_{or} = age at the start of strength development at the reference temperature, and
  k_r = rate constant at the reference temperature.

Equation (7) has been termed the "rate constant model" because the shape of the relative strength gain relationship is uniquely defined by the value of the rate constant at the reference temperature and because the equivalent age is dependent on the rate-constant-versus-temperature relationship.

Equivalent age represents the age at the reference curing temperature which would result in the same fraction of the limiting strength as would result from curing at other temperatures. As was explained in Ref. [5], the equivalent age may be determined from the temperature history as follows:

$$t_e = \sum \left(\frac{k_T}{k_r} \right) \Delta t = \sum \gamma\, \Delta t$$ (8)

where

  k_T = value of rate constant at the temperature T during the time interval Δt, and
  k_r = value of rate constant at the reference temperature, T_r, (23 C was used at the reference temperature in this study)

---

[b] "Rate Constant Functions for Strength Development of Concrete," paper accepted for publication in the ACI Materials Journal (1990).

The ratio, γ, of the rate constants in Eq. (8) converts a curing interval at any temperature to an equivalent interval at the reference temperature. This age conversion actor has been called the "affinity ratio" [18]. The expressions for the affinity ratio, corresponding to the three rate constant functions (Eqs. (4), (5), and (6)) are as follows:

Linear function:
(9a)

Arrhenius function:
(9b)

Exponential function
(9c)

$$\gamma = \frac{T - T_o}{23 - T_o} \qquad \gamma = e^{-Q\left(\frac{1}{T+273} - \frac{1}{296}\right)} \qquad \gamma = e^{B(T - 23)}$$

where
  T = actual curing temperature, C, and the others constants are as defined previously.

In summary, to estimate the relative strength gain of a given concrete mixture, the actual temperature history of the concrete is converted to an equivalent age at the reference temperature. Then, Eq. (7) is used to estimate the relative strength. Thus three characteristics of the concrete must be known: the rate constant at the reference temperature, the age at the reference temperature when strength development begins, and the temperature dependence of the rate constant (as defined by T_o, Q, or B).

4 Test results

4.1 Isothermal strength development

The strength-age data for each mixture were used to determine the parameters of the strength-age relationship, Eq. (3). Table 1 gives the results for the concrete mixtures. The complete data are available [13].

As expected, the limiting strength, S_u, decreased with increasing curing temperature. However, the magnitude of the decrease varied widely. The water-cement ratios were controlled in the usual manner by accounting for the aggregate moisture contents determined from small samples. It is possible that the actual water-cement ratios for the various batches of the same mixture were not equal. This could partly explain the wide variations in the effect of temperature on limiting strengths. The quantitative effect of curing temperature on the limiting strength is an important practical problem deserving further study. Also as expected, the age at the start of strength development, t_o, decreased with increasing curing temperature. This age did not seem to be greatly affected by the water-cement ratio or the cementitious materials, except for the mixtures with a retarding admixture.

Table 1. Strength development parameters for the concrete mixtures

Cement	W/C = 0.45				W/C = 0.60			
	Temp (C)	S_u (MPa)	k_T (1/day)	t_o (day)	Temp (C)	S_u (MPa)	k_T (1/day)	t_o (day)
I	10.0	57.9	0.202	0.80	10.0	31.0	0.212	0.95
	21.5	44.1	0.401	0.10	22.0	29.0	0.336	0.22
	40.5	31.7	2.673	0.21	40.5	24.8	1.482	0.10
II	10.0	53.8	0.205	0.95	9.5	46.9	0.153	0.87
	22.5	46.2	0.351	0.37	21.5	30.3	0.287	0.28
	40.5	40.7	1.641	0.16	40.5	27.6	0.911	0.21
III	10.0	57.2	0.523	0.64	10.0	37.2	0.508	0.74
	21.0	44.8	0.844	0.28	21.5	31.7	0.832	0.28
	42.0	38.6	3.313	0.14	41.5	24.1	3.204	0.18
I + 20% Fly Ash	10.0	52.4	0.231	0.91	10.5	37.9	0.176	0.89
	21.5	49.7	0.623	0.45	20.5	33.1	0.291	0.26
	41.0	42.1	0.868	0.07	41.0	31.0	0.642	0.10
I + 50 % Slag	10.0	66.2	0.131	1.07	10.0	55.9	0.057	0.99
	20.5	64.1	0.194	0.38	21.0	53.1	0.113	0.41
	43.0	50.3	0.902	0.22	41.5	33.8	0.599	0.19
I + Accelerator	10.0	62.1	0.381	0.75	10.0	38.6	0.249	0.86
	20.5	53.1	0.659	0.20	20.5	35.9	0.397	0.29
	43.0	43.4	2.689	0.16	41.5	28.3	2.004	0.20
I + Retarder	10.0	59.3	0.206	2.18	10.0	60.7	0.153	1.82
	20.5	56.6	0.367	1.12	21.0	53.8	0.286	0.97
	42.5	52.4	1.117	0.41	42.5	37.9	0.835	0.31

4.2 Temperature dependence of the rate constant

Figure 1(a) shows the calculated rate constants, k_T, for the concrete mixtures. It is seen that the rate constants depend on the cementitious materials. For the hyperbolic strength-age model, the rate constant equals the reciprocal of the age beyond t_o required to achieve 50% of the limiting strength. Thus higher values of the rate constant mean that it takes less time to gain 50% of the limiting strength.

Figure 1(a) also reveals that the temperature dependence of the rate constant was not the same for the various mixtures. Figure 1(b) shows the extremes that were observed. The rate constant for the low-W/C concrete made with Type I cement had the greatest variation with curing temperature, while that for the low-W/C concrete made with Type I cement and fly ash had the least variation. For each mixture, the functions given by Eqs. (4), (5), and (6) were fitted to the rate constant values for the various curing temperatures. In most cases, the Arrhenius (Eq. (5)) and the exponential (Eq. (6)) functions described the temperature dependence better than the linear function (Eq. (4)). Figure 1(b) shows the best-fit linear and exponential functions for the two extreme mixtures.

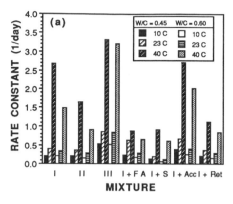 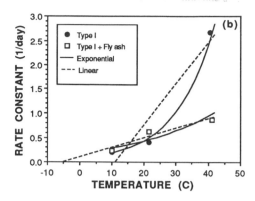

Fig. 1 Variation of the rate constant with curing temperature; (a) all concrete mixtures; (b) mixtures (W/C = 0.45) made with Type I cement and with Type I cement plus fly ash.

Table 2 summarizes the best-fit parameters for the three rate constant functions (Eqs. 9(a), (b), and (c)) for the concrete and mortar mixtures. Since the Arrhenius function is usually expressed using an activation energy, the Q-values in Table 2 were multiplied by the gas constant (R = 8.314 J/K-mole) to obtain corresponding activation energy (E) values.

Table 2 shows that the mixtures resulted in different T_0-values, and none of the values were -10 C, which is the value of the datum temperature in the traditional Nurse-Saul maturity function, Eq. (1).

The exponential function described the temperature dependence of the rate constant as well as the Arrhenius function. Because the exponential function is mathematically simpler than the Arrhenius function, the B-values of the various mixtures are examined further. According to Eq. (6), if the curing temperature increases by (1/B) degrees Celsius, the rate constant increases by a factor of 2.72 (base of natural logarithm). Thus higher B-values mean greater changes in the rate constant as the temperature changes. Figure 2(a) shows the B-values for the concrete mixtures. It is seen that the B-values depend on the cementitious materials and may also depend on the water-cement ratio (to be discussed further).

Published data on the temperature sensitivity of the rate constant have been expressed using values of activation energy. Because of the simplicity of the exponential function, it would be desirable to convert activation energies to the corresponding B-values. Figure 2(b) shows the B-values and the activation energies obtained for the concrete mixtures. There is excellent correlation between these parameters, and it was found that B = 0.00135 E, where E is expressed in units of kJ/mol.

Table 2   Best-fit parameters for the rate constant functions for concrete

| Cement | W/C = 0.45 | | | | W/C = 0.60 | | | |
	T_o (C)	Q (K)	E (kJ/mol)	B (1/C)	T_o (C)	Q (K)	E (kJ/mol)	B (1/C)
				CONCRETE				
I	11	7640	63.6	0.0862	9	5770	48.0	0.0652
II	9	6140	51.1	0.0694	6	5130	42.7	0.0578
III	7	5240	43.6	0.0587	7	5290	44.0	0.0595
I+Fly Ash	-5	3610	30.0	0.0400	0	3750	31.2	0.0419
I+Slag	8	5380	44.7	0.0600	10	6740	56.0	0.0755
I+Acc.	8	5360	44.6	0.0597	9	6040	50.2	0.0678
I+Ret.	5	4650	38.7	0.0518	5	4660	38.7	0.0519
				MORTAR				
I	11	7360	61.1	0.0824	7	5250	43.6	0.0585
II	9	6670	55.4	0.0749	5	4950	41.1	0.0554
III	6	4820	40.1	0.0540	6	5130	42.6	0.0573
I+Fly Ash	-2	3980	33.1	0.0442	3	4400	36.6	0.0492
I+Slag	7	5140	42.7	0.0573	9	6170	51.3	0.0691
I+Acc.	10	6500	54.1	0.0728	9	6270	52.1	0.0703
I+Ret.	6	5040	41.9	0.0562	2	4100	34.1	0.0456

It is concluded that the rate constant function, $k(T)$, can be accurately represented by the simple exponential function, Eq.(6). The exponential function leads to a mathematically simpler formula, Eq. 9(c), than the Arrhenius equation for computing equivalent age. The empirical relationship between B and E can be used to transform published activation energy values into B-values so that the exponential function can be used to represent the temperature dependence of the rate constant.

4.3 Effect of water-cement ratio
It has been suggested that the temperature dependence of the rate constant should be independent of the water-cement ratio [5,15]. However, Fig. 2(a) shows that this suggestion may be incorrect. To examine this further, Fig. 3(a) shows the B-values for the high-W/C mixtures plotted against the values for the low-W/C mixtures. The line of equality is also plotted. It is seen that for some of the mixtures, the B-values were not affected by water-cement ratio, but in other cases the low-W/C mixtures had higher B-values. The effect of W/C on the B-value deserves additional study because it determines whether the same maturity function can be used for different strength grades (different W/C) of concrete made with the same cementitious materials.

The water-cement ratio also affects the value of the rate constant at the reference temperature, k_r, as shown in Fig. 3(b). The k_r-values were computed at a reference temperature of 23 C using the best-fit exponential function for each of the mixtures [13]. It is seen that the k_r-values for the low-W/C mixtures were, on average, about 0.1 (1/day) larger than the values for the high-W/C mixtures.

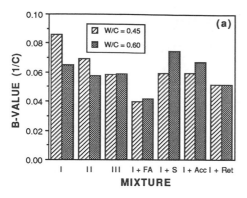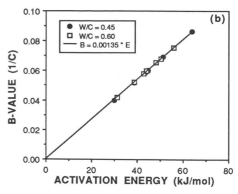

Fig. 2 (a) The temperature sensitivity factor (B-value) for concrete mixtures; (b) comparison of B-values with activation energies.

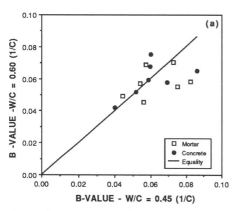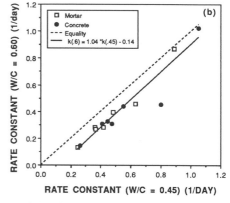

Fig. 3 Effects of water-cement ratio on (a) the B-values and (b) the rate constant at the reference temperature (23 C).

4.4 Comparison of concrete and mortar results

One of the objectives of the study was to determine whether the parameters needed to estimate relative strength gain of concrete under different curing temperatures could be obtained from tests of mortar specimens. Such an approach would simplify the laboratory pre-testing process required before using the maturity method in the field. Figure 4(a) compares the B-values obtained from the mortar tests with those obtained from the concrete tests. The straight line (slope = 0.996) is the best-fit line constrained to pass through the origin. On average, the B-values obtained from the two types of specimens are nearly equal as indicated by the slope of the line. However, for some mixtures, such as the one containing Type I cement and an accelerator, there are large deviations of the B-values from the line of equality.

The other parameters needed to estimate relative strength gain are the rate constant at the reference temperature, k_r, and the age at the reference temperature when strength gain begins, t_{or}. Figure 4(b) compares the k_r-values from the mortar tests with those from the concrete tests.

There is good correlation between the k_r-values, but those from the concrete tests are about 17% greater than those from the mortar tests.

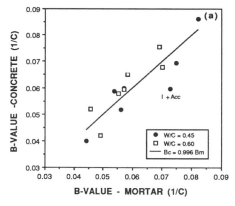 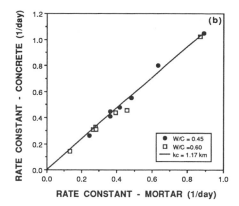

Fig. 4 Comparison of concrete with mortar test results: (a) B-values and (b) rate constant at the reference temperature (23 C).

Table 4 compares the values of t_o obtained from the mortar tests with those from the concrete tests for room temperature curing (close to the 23-C reference temperature). With the exception of the low-W/C mixtures made with Type I cement plus a retarding admixture, there is good agreement between values obtained from the mortar and concrete tests.

The next section discusses the significance of the differences in these parameters on strength estimation using the rate constant model.

Table 4 Values of t_o for room-temperature curing (day)

Cement	W/C = 0.45		W/C = 0.60	
	Concrete	Mortar	Concrete	Mortar
I	0.10	0.17	0.22	0.30
II	0.37	0.44	0.28	0.30
III	0.28	0.27	0.28	0.29
I + Fly Ash	0.45	0.50	0.26	0.34
I + Slag	0.38	0.45	0.41	0.44
I + Accelerator	0.20	0.20	0.29	0.30
I + Retarder	1.12	1.93	0.97	1.16

4.5 Accuracy of Rate Constant Model

It is proposed that the relative strength gain of concrete in a structure can be estimated from the in-place temperature history of the concrete and the strength gain parameters obtained from tests of isothermally cured mortar specimens. However, from the previous section it appears that mortar tests do not yield the same k_r-value and may not yield the same B-value as concrete tests. How much error in the estimated in-place relative strength will result if the mortar values are applied to concrete?

First the errors associated with k_r are discussed. The value of k_r affects the shape of the curve of relative strength versus equivalent age (Eq. (7)). Recall from Fig. 4(b) that the k_r-values from concrete tests were about 17% higher than the values from mortar tests. The mixtures with the highest and lowest k_r-values were those made with Type III cement and those made with Type I cement plus slag, respectively. Figure 5(a) compares the shapes of the relative strength gain curves for the concrete and mortar mixtures with the highest and lowest k_r-values. For the high k_r-values, the curves appear to be nearly identical; for the low k_r-values, the curves are different. Figure 5(b) shows the differences between the relative strength gain curves as a function of equivalent age. It is seen that differences of 17% in the k_r-values lead to errors of at most 0.04 in the estimated relative strength. This error would not be significant when estimating in-place strength gain.

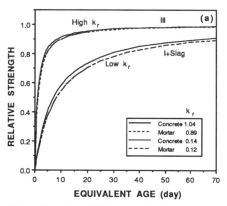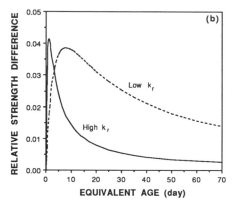

Fig. 5 (a) Effect of the rate constant k_r on relative strength gain; (b) difference in relative strength gain.

Errors in the B-value lead to errors in the calculated equivalent age (see Eqs. 8 and 9(c)). The greatest discrepancy between the B-values obtained from concrete and mortar tests occurred for the low-W/C mixtures made with Type I cement and an accelerator (see Table 2 and Fig. 4(a)). The concrete tests resulted in B = 0.0597 (1/C), and the mortar tests resulted in B = 0.0728. If the higher mortar B-value is used to calculate the equivalent age of the concrete, the equivalent age will be too low for curing temperatures below the reference temperature and too high for curing temperatures above the reference temperature [5].
To illustrate the significance of the errors when the relative strength gain of concrete is estimated using parameters obtained from mortar tests, consider the data for the low-W/C concrete made with Type I cement and an accelerator [13]. Figure 6(a) shows the relative strength values versus the equivalent age; in this case, equivalent age was calculated using the B-value from the concrete tests. The solid curve is the rate constant model (RCM) obtained using the parameters from concrete tests, i.e, t_{or} = 0.2 days and k_r = 0.802 (1/day). The average of the absolute differences between the curve and the data is 0.03. Figure 6(b) shows the same strength data, but in this case equivalent age was calculated using the B-value from the mortar tests. The solid curve is the rate

constant model for the parameters obtained from mortar tests, i.e., t_{or} = 0.2 days and k_r = 0.633 (1/day). The average of the absolute differences between the curve and the data is 0.05.

Thus it has been demonstrated that, even though the B-value and the k_r-value obtained from tests of mortar specimens differ from those obtained from tests of concrete specimens, they result in reasonably accurate estimates of the relative strength gain of concrete cured at different temperatures.

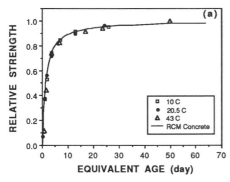 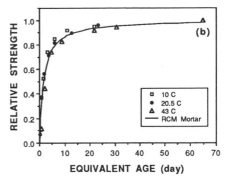

Fig. 6 Relative strength gain of concrete made with Type I cement plus an accelerator compared with prediction by the rate constant model using values of B and k_r obtained from: (a) concrete; (b) mortar.

5 Conclusions

- For the seven concrete and mortar mixtures used in the study, it was found that strength gain under isothermal curing conditions was modeled accurately using the three-parameter hyperbolic function [13].
- The variation of the rate constant with curing temperature was represented by a simple exponential function, which results in a simpler expression for computing equivalent age than the Arrhenius function.
- The B-value (or activation energy), which describes the temperature sensitivity of the rate constant was a function of the cementitious materials.
- Tests of mortar cube specimens resulted in B-values which were, on average, equal to the B-values resulting from tests of concrete cylinders. The rate constants were lower for the mortar tests.
- The effect of water-cement ratio on the B-value was not consistent and deserves further study. On the other hand, the rate constant was consistently higher for the low-W/C mixtures.
- Use of the rate constant model with B- and k_r-values obtained from tests of mortar appears to give accurate estimates of relative strength gain of concrete under isothermal curing conditions.

In summary, a procedure has been developed for implementing the maturity method which will result in reliable estimates of in-place relative strength development. It is proposed that mortar tests will provide the necessary information to define the relative strength gain of concrete. It is realized that the results were obtained under isothermal conditions. It is necessary to verify the methodology under variable temperature conditions.

6 References

1. Nurse, R. W., "Steam Curing of Concrete," Mag. of Conc. Res., V 1, No. 2, June 1949, pp. 79-88.
2. Saul, A. G. A., "Principles Underlying the Steam Curing of Concrete at Atmospheric Pressure," Mag. of Conc. Res., V 2, No. 6, March 1951, pp. 127-140.
3. McIntosh, J. D.,"Electrical Curing of Concrete," Mag. of Conc. Res., V 1, No. 1, Jan.1949, pp. 21-28.
4. Annual Book of ASTM Standards, Concrete and Aggregates, American Society for Testing and Materials, Section 4, V 04.02, Sept. 1987.
5. Carino, N. J., "The Maturity Method: Theory and Application," Journal of Cem., Conc., and Aggr., ASTM, V 6, No.2, Winter 1984, pp. 61-73.
6. Malhotra, V. M., "Maturity Concept and the Estimation of Concrete Strength," Information Circular IC 277, Dept. of Energy, Mines and Resources (Canada), Mines Branch, Nov. 1971, 43 pp.
7. Freiesleben Hansen, P. and Pedersen, J., "Maturity Computer for Controlled Curing and Hardening of Concrete," Nordisk Betong, V 1, 1977, pp. 19-34.
8. Lew, H. S. and Reichard, T. W., "Mechanical Properties of Concrete at Early Ages," J. Amer. Conc. Inst., V 75, No. 10, Oct. 1978, pp. 533-542.
9. Lew, H. S. and Reichard, T. W., "Prediction of Strength of Concrete from Maturity," SP-56, Amer. Conc. Inst., Detroit, 1978, pp.229-248.
10. Carino, N. J., Lew, H. S. and Volz, C. K., "Early Age Temperature Effects on Concrete Strength Prediction by the Maturity Method," J. Amer. Conc. Inst., V 80, No. 2, March-April 1982, pp. 92-101.
11. Kleiger, P., "Effects of Mixing and Curing Temperatures on Concrete Strength," J. Amer. Conc. Inst., V 54, No.12, June 1958, pp. 1063-82.
12. Carino, N. J., and Lew, H. S., "Temperature Effects on the Strength-Maturity Relations of Mortar," J. Amer. Conc. Inst., V 80, No. 3, May-June 1983, pp. 177-182.
13. "Standard Practice for Selecting Proportions of Normal, Heavyweight and Mass Concrete," ACI 211.1-81 (1985), Amer. Conc. Inst., Detroit.
14. Tank, R. C., "The Rate Constant Model for Strength Development of Concrete," Doctoral Dissertation, Polytechnic University of New York, June 1988, 209 pp.
15. Knudsen, T., "On Particle Size Distribution in Cement Hydration," Proceedings of the 7th International Congress on the Chemistry of Cement, V II, Paris, 1980, pp. I-170-175.
16. Filliben, J., "DATAPLOT - Introduction and Overview," Nat. Bur. of Stand., SP 667, U.S. Dept. of Com., Wash., D.C., June 1984, 112 pp.
17. Carino, N.J., "Maturity Functions for Concrete," Proceedings of the RILEM International Conference on Concrete at Early Ages, V I, Paris 1982, pp. 123-128.
18. Regourd, M., "Structure and Behavior of Slag Portland Cement Hydrates," Proceedings of the 7th International Congress on the Chemistry of Cement, V I, Paris, 1980, pp. III-2/11-2/26.

18 EXPERIMENTAL AND ANALYTICAL PLANNING TOOLS TO MINIMIZE THERMAL CRACKING OF YOUNG CONCRETE

F.S. ROSTASY and M. LAUBE
Technical University of Braunschweig, Germany

Abstract
The avoidance of thermal cracking in massive concrete structures is a long-standing goal of practice and science. Because the control of cracking by purely empirical methods is often not successful, complementary methods based on experiments and analysis are being developed. The report shows the necessary steps for the improvement of the methods for crack control and also the deficits of knowledge to be resolved by further research.

1 Introduction

In concrete structures of massive dimensions the occurrence of thermal stresses at young age is inevitable. These stresses are caused by the evolution of the heat of hydration and its subsequent transfer because the resultatent nonuniform temperature field leads to internal and external restraint. Quite often the thermal stresses produce cracks in the structural element, thereby endangering its durability and/or tightness if the crack width is excessive and severe environment prevails. The control of the widths of restraint cracks by reinforcement may lead to excessive amount of steel. Besides, in many instances such an approach to crack control is not sensible because it does not take into consideration the transient character of crack risk. Many practical cases prove that cracking occurs predominantly within the first fortnight after casting and/or striking of shutter. Consequently, it is this critical period of age during which the young concrete must be protected from stresses in excess of its developing tensile strength.

During the past decades practice and science jointly developed tools of planning and execution for the avoidance and control of cracks. These tools relate to structural design, composition of concrete, and execution of works. Fig. 1 shows these fields and the certainly incomplete naming of parameters which are influential for crack control. It is obvious that the fields and the parameters are interdependent.

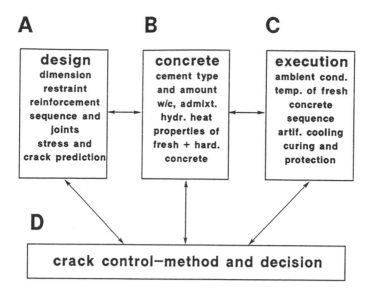

A

design
dimension
restraint
reinforcement
sequence and
joints
stress and
crack prediction

B

concrete
cement type
and amount
w/c, admixt.
hydr. heat
properties of
fresh + hard.
concrete

C

execution
ambient cond.
temp. of fresh
concrete
sequence
artif. cooling
curing and
protection

D

crack control—method and decision

Fig. 1. Parameters of crack control

The tools of crack control were for long time primarily based on experience. They proved their suitability in as many instances as they unexpectedly failed. Consequently, ways of improvement of the existing methods of prediction of the critical state of stresses are being sought today in many countries. In the next section the structure of promising improvement will be discussed.

2 Structure of Methodical Improvement

The structure of improvement of the methods for crack control is shown in Fig. 2. It represents a tool for planning and decision. Consequently it has to be linked to the fields A to C of Fig. 1, whose parameters must possibly be variied or corrected in course of the optimization process. The procedure described by Fig. 2 is generally accepted by practice and science. It consists of the following consecutive steps:

a) Experimental step

For a set of chosen parameters of Fig. 1 (type and amount of cement; W/C-ratio; temperature of fresh concrete; etc.) the adiabatic heat release will be measured. Choice of parameters may be based on pilot tests or on experience.

In conjunction with the adiabatic calorimetry, concrete specimens are cast and cured following temperature-time histories which are representative for specific locations in the structure (sealed conditions); some specimens should be cured at constant temperature. Strength testing at different ages reveals the actual degree of hydration of the chosen concrete composition.

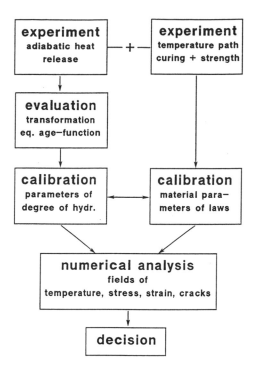

Fig. 2. Structure of improved procedures for crack control

b) Evaluation step

With a suitable maturity function the adiabatic heat release process is transformed into the maturity-equivalent isothermal heat release process (equivalent time transformation) for the formulation and calculation of the actual
temperature-dependent heat release in the structure.

c) Calibration step

With the data of adiabatic calorimetry and of strength testing the parameters
of the chosen function of the degree of hydration will be calibrated. With the
same data the parameters of the material laws as a function of the degree of
hydration can be determined.

d) Analytical step

Numerical analysis renders the fields of temperature, degree of hydration and
with the material laws as function of the degree of hydration the mechanical
material properties. Numerical analysis leads to the fields of stress, deformation
and eventual cracks.

e) Decision step

Results facilitate the decision with respect to acceptance or alteration of parameters. Optimization becomes possible. Additional tests may become necessary.

The described procedure can be fortified by on-site measurement of temperatures of a large-size concrete model specimen. This may be f.i. a realistic cut-out of a thick base slab with the same boundary condition as the real structure.

In spite of the inherent logic of the described procedure uncertainties still exist. These uncertainties relate to the material properties and their variability within the structure, to the stress and crack risk analysis in view of assumptions which cannot be verified and other problems. In the next section the above-mentioned steps will be dealt with.

3 Heat evolution, temperature, and degree of hydration

3.1 Rate of heat evolution and process time transformation
Presequite for the realistic estimation of thermal stresses in the hardening concrete body is the field of temperatures. This field is determined numerically on basis of the differential equation of heat conduction, Eq.(1), and for specific boundary conditions:

$$\frac{\lambda}{c\rho}\Delta_L T + \frac{q}{c\rho} - \frac{\partial T}{\partial t} = 0 \tag{1}$$

with

λ	coefficient of heat conduction	[W/mK]
c	specific heat of concrete	[kJ/kgK]
ρ	density of concrete	[kg/m^3]
Δ_L	Laplace operator	[-]
T = T(x,y,z,t)	temperature of concrete	[°C]
q = q(x,y,z,t,T)	heat evolution rate of concrete	[kJ/m^{3}h]

Eq. (1) shows that the rate q in a specific location also depends on temperature. As the temperature T relevant to q is a priori unknown, the solution of problems requires the coupling of the instationary temperature field with the heat evolution. The coupling equations are derived on basis of the theory of rates of chemical processes, [8,9]. Thereby, the unknown rate q is related to the known rate of a reference process.

According to this theory, the rate of a hydration process can be expressed by Eq. (2):

$$q = \frac{dQ}{dt} = r \cdot k(T(t)) \tag{2}$$

In this equation the process factor r describes parameters of reaction kinetics; it is independent on the process temperature T(t). The rate factor k is temperature-dependent. It is presupposed that the rate q of a reference hydration process $Q_1(T_1(t))$ has been determined a priori, e.g. by calorimetry. Then, the unknown rate q_2 of the process $Q_2(T_2(t))$ taking place under the temperature regime $T_2(t)$, e.g. in the structure, can be related to q_1. With Fig. 3 a relation between the time intervals dt_1 and dt_2 during which both processes release the same heat quantum dQ can be formulated. With

$$dQ_1(T_1(t_1)) = dQ = r_1 k(T_1(t_1))dt_1 \qquad (3)$$

and

$$dQ_2(T_2(t_2)) = dQ = r_2 k(T_2(t_2))dt_2 \qquad (4)$$

we arrive at the relation of intervals:

$$dt_1 = \frac{k(T_2(t_2))}{k(T_1(t_1))}dt_2 = \frac{1}{k'}dt_2 \qquad (5)$$

with $r_1 = r_2$ both processes refer to concretes of identical composition. k' is the time scale factor by which dt_2 is distorted in such a way that $dQ_1 = dQ_2$. By integration of Eq. (5) an implicite relation between the reaction times t_1 and t_2 at which both processes have released the identical total heat can be derived.

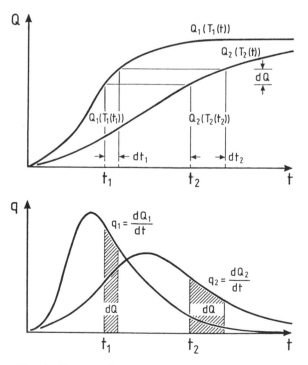

Fig. 3. Heat evolution and process transformation

With

$$Q_1(T_1(t_1)) = Q_2(T_2(t_2)) = Q \tag{6}$$

we obtain

$$\int_0^{t_1} k(T_1(t_1))dt = \int_0^{t_2} k(T_2(t_2))dt \tag{7}$$

The solution of Eq. (7) requires the formulation of the rate factor $k(T(t))$ which usually is called: maturity function. Evaluation of tests proves that the maturity function of the Arrhenius-type of reaction kinetics

$$k(T(t)) = p \cdot exp\left(-\frac{E}{R} \cdot \frac{1}{T_K(t)}\right) \tag{8}$$

is best suited, [2]. In this relation are:

E - activation energy [kJ/kmol]; R - universal gas constant, 8,315 [kJ/kmolK], and $T_K(t)$ - absolute process temperature [K]. Insertion of Eq. (7) into (6) renders:

$$\int_0^{t_1} exp\left(-\frac{E}{R}\frac{1}{T_{K1}(t)}\right)dt = \int_0^{t_2} exp\left(-\frac{E}{R}\frac{1}{T_{K2}(t)}\right)dt \tag{9}$$

Because of $Q_1(t_1) = Q_2(t_2)$ both processes have attained at the relevant times t_1 and t_2, resp., the same degree of reaction or degree of hydration.

Most of our knowledge about mechanical properties of concrete pertains to curing at 20 °C. It is therefore of interest to know the duration t_{eq} of a maturity-equivalent isothermal hardening condition $T(20°C; t_{eq})$ which leads to

$$Q(293, t_{eq}) = Q_2(T_{K2}(t_2)) \tag{10}$$

With $T_{K1}(t) = const = 293$ K the equivalent time t_{eq} can be expressed:

$$t_{eq} = \int_0^{t_2} exp\left(\frac{E}{R}(\frac{1}{293} - \frac{1}{T_{K2}(t)})\right)dt = \int_0^{t_2} \frac{k(T_2(t))}{k(20°C)}dt \tag{11}$$

if the function $T_{K2}(t)$ is known.

3.2 The adiabatic process and its transformation

For the formulation of the heat evolution q in the structure a defined reference process is needed. This process must be derived from calorimetry with the concrete composition envisaged for the pour of structure. By this approach all synergistic effects of cement, water, admixtures and additives are considered in the complex process of hydration. Due to many, especially experimental reasons adiabatic calorimetry is to be prefered. If $\Delta T_{ad}(t)$ is the increase of temperature above the temperature of fresh concrete T_o the total heat evolution amounts to

$$Q_{ad}(t) = c\rho \Delta T_{ad}(t) \tag{12}$$

with the rate

$$q_{ad}(t) = c\rho \frac{d\Delta T_{ad}(t)}{dt} \tag{13}$$

As the duration of test is about 7 to 10 days at the most, the maximum release must be approximated on basis of the ΔT_{ad}-curve and/or with the maximum heat release of the cement:

$$max\ Q = c\rho\ max\ \Delta T_{ad} = max\ Q_{cem} \cdot C \tag{14}$$

Computational reasons require the transformation of the adiabatic process to a maturity-equivalent isothermal process at T=20°C. This is performed with Eq. (13) by setting $t_2 = t_{ad}$. Fig. 4 proves the suitability of the Arrhenius-rate factor Eq. (8) for the transformation.

3.3 Degree of hydration and its transformation
From the adiabatic calorimetry the degree of hydration can be deduced

$$m(t_{ad}) = \frac{Q_{ad}(t_{ad})}{max\ Q} = \frac{\Delta T_{ad}(t_{ad})}{max\ \Delta T_{ad}} \tag{15}$$

For computational reasons $m(t_{ad})$ has to be described by an analytic expression. The formulation of Jonasson [6] permits a good depiction of the experimental behaviour:

$$m(t_{eq}) = exp\left(a(1n(1 + \frac{t_{eq}}{t_c}))^b\right) = \frac{Q(t_{eq})}{max\ Q} \tag{16}$$

with a,b,t_c[h] being material factors to be determined with the test data.

Because of Eq. (15) and (16) the rate of m corresponds to the rate of heat evolution q, the degree of hydration can be transformed into the maturity-equivalent process of the structure at different temperatures by the equivalent time procedure, Eq. (11). This leads to the rate of heat evolution

$$q(t) = max\ Q\frac{\partial m}{\partial t} = max\ Q\frac{dm}{dt_{eq}} \cdot \frac{dt_{eq}}{dt} \tag{17}$$

which is evaluated with Eq. (11) and (16) and which serves as master reference process for the incremental determination of heat evolution q(x,y,z,t,T) in the structure.

3.4 Fields of temperature and degree of hydration
The time-dependent field of concrete temperature in the structure is determined numerically with the FEM. Several procedures are described in the literature (e.g. [3,4,6,8]). The field of degree of hydration is simultaneously calculated. It is also needed for the generation of material laws.

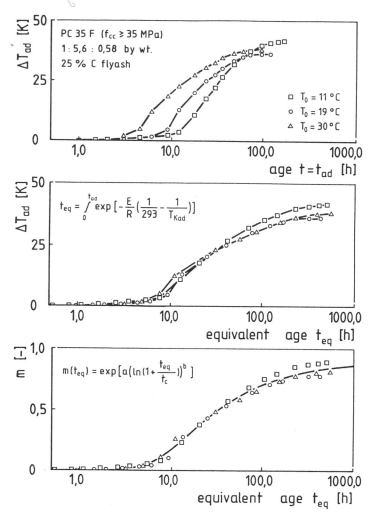

Fig. 4. Verification of Arrhenius-rate factor and degree of hydration

The formulation of the boundary conditions requires certain assumptions which cannot easily be verified. In some instances a model specimen could be cast on-site with the planned concrete composition and on the ground as for the future structure: f.i. a geometrically and thermally similar cut-out. Thereby, the field of temperature can be measured and the real boundary condition can be specified.

From the calculated fields of temperature the temperature-time function for the temperature controlled curing of specimens are derived. Calibration of the degree of hydration becomes possible.

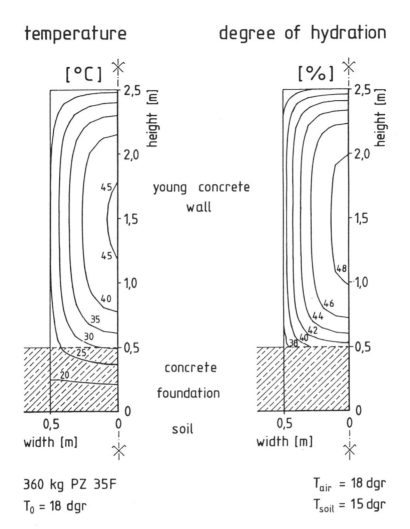

temperature   degree of hydration

[°C]   [%]

young concrete wall

concrete foundation

soil

360 kg PZ 35F
T_0 = 18 dgr

T_{air} = 18 dgr
T_{soil} = 15 dgr

Fig. 5. Fields of temperature and degree of hydration due to evolution of heat of hydration, 30 hrs after casting. Example: wall of 1 m width on foundation

4 Mechanical Properties

4.1 Strength and modulus of elasticity
The mechanical properties of young concrete are markedly dependent on age and the hardening environment. A vast amount of test data has been gathered on this topic. Reference is made to the sources, [1,3,10] which present an overview on experimental work and on modelling of mechanical behaviour. Usually the mechanical properties are described as function of equivalent time t_{eq}. Because the latter depicts the hydration progress in a nonlinear fashion, the mechanical properties should rather be

described in terms of the degree of hydration m which linearizes the hydration process. The experimental function of m has to be determined by adiabatic calorimetry, s. sec. 3.2, [7,12].

Fig. 6 shows normalized relations of the axial tensile strength f_t, compressive cylinder strength f_c and of tensile modulus E_t dependent on m [7]. These curves represent mean values. The degree of hydration must be determined for each specific set of constituents and composition of concrete.

Experiments have shown that a high curing temperature and its steep rise in the first two days of equivalent age may lead to an irrecoverable loss of strength at later ages in comparison to a curing at 20°C, [2]. This effect can be taken into account by reduction of the normalized property of Fig. 6. The reduction factor for the compressive strength is approximately:

$$\gamma_c(t, max\ T) \approx 1 - \frac{T_{max}^{-20}}{80} \frac{t}{t_{max}} \leq 1 \qquad (18)$$

with:

T_{max} - maximum temperature at the specific location;
t_{max} - real time to attain T_{max}.

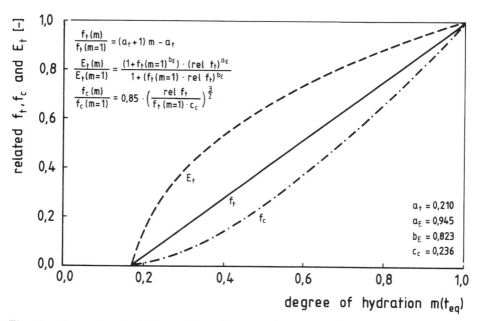

Fig. 6. Normalized relations of tensile strength, compressive strength, modulus of elasticity vs. degree of hydration

The functions of Fig. 6 must be calibrated by tests. Sealed specimens with the concrete composition already being investigated in the calorimeter are to be cured isothermally at 20°C and higher. To determine the specific strength gain and loss by lower or higher temperatures in the structure, it is advised to cure also specimens at temperature paths $T_c(t)$ corresponding to specific ones in the structure. The compressive and tensile splitting strength as well as the modulus E_t are tested at several ages 2, 7, and 28 d.

4.2 Tensile stress - displacement relation

The complete stress - displacement relation of concrete with its strain - softening descending branch and its implementation in FEM calculus have been dealt with intensively in literature [13,15]. Because most of our data pertain to old concrete, the behaviour of young concrete had to be studied experimentally.

Its ascending branch can be linearized up to $\sigma_t = f_t$ with the knowledge of $f_t(m)$ and $E_t(m)$, Fig. 4. The strain rate is of little influence [7]. The descending branch formulated as stress - crack opening displacement ($\delta = $ COD) relation depends markedly on the degree of hydration and on the COD-rate. The following relation proved to be a satisfactory description of experimental behaviour [7]:

$$\frac{\sigma_t}{f_t} = \frac{a\left(\frac{\delta}{\delta_p} + 1\right)}{a - 1 + \left(\frac{\delta}{\delta_p} + 1\right)^a} \tag{19}$$

with
$a = f(m, d\delta/dt)$;
δ_p - displacement at $\sigma_t = f_t$, $\delta_p = g(f_t, E_t, m)$.

As the compressive thermal stresses usually are considerably lower than the momentaneous compressive strength, Hooke's law with $E_c \approx E_t$ is an acceptable approximation of stress-strain behaviour.

4.3 Viscoelastic behaviour

The development of constitutive relations for the viscoelastic behaviour of young concrete is a complex matter. The noncontradictory formulation of the mechanical behaviour is aggravated by the lack of suitable test data. Therefore, relaxation is usually described by inversion of creep compliance, though thermal restraint is a relaxation phenomenon. In [1] an overview on the presently available constitutive relations is given. In order to enhance our knowledge relaxation and strain-rate controlled tensile tests were performed [7]. In the relaxation test, the specimen was subjected to a temperature path curing corresponding to a realistic history in a massive structure. Thus, for each instant the degree of hydration and equivalent age are known. At defined ages t_i and t_{eqi}, resp., initial mechanical strains corresponding to $\alpha_i = \sigma_i/f_{ti}$ ($\alpha_i = 0{,}5\text{-}0{,}9$) were forced upon the specimen and maintained constant. The stress σ_i was attained with a high strain rate in order to minimize initial viscous strain components.

Fig. 7 shows typical test results. The tensile relaxation function can be expressed on basis of the tests as residual stress:

$$\psi = \frac{\sigma(t, t_i)}{\sigma(t_i, t_i)} = \frac{\sigma}{\sigma_i} \tag{20}$$

and

$$\psi = \psi(t - t_i; m(t_{eqi})) = exp\left(-P_1(m_i)\frac{(t - t_i)}{t_b}^{P_2(m_i)}\right) \tag{21}$$

with

t-t_i	time under stress; t, t_i real time (age);
t_{eqi}	equivalent time (age) at strain step;
$t_b = 1$ [h]	
$m_i = m(t_{eqi})$	degree of hydration at t_{eqi};
P_1, P_2	linear function of m_i.

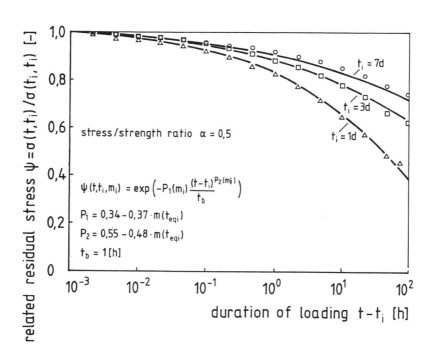

Fig. 7. Tensile relaxation-experiments and theory

In Fig. 7 also the theoretical lines acc. to Eq. (20) and (21) are drawn, observation is well described by them. Evaluation of other relaxation tests from literature proved that their results also could be described by Eq. (20) and (21) with only slightly differing values of P_1 and P_2.

It is justified to use Eq. (20) and (21) for compressive relaxation too, and for the inversion to creep compliances irrespective of stress sign. For massive structural elements drying and shrinkage are not of interest when dealing with the early thermal restraint. Thus, these effects can be neglected during the first weeks of age. Consequently, Eq. (20) and (21) pertain to basic relaxation.

5 Thermal stress analysis and cracking

Two types of thermal restraint occur simultaneously and inseparably. Variation of free thermal strains across the section of the structural member leads to internal restraint, i.e. plane-strain thermal eigenstresses. Variation of the free thermal strains within the entire structural member may lead to external restraint if the free thermal displacements along boundaries are impeded. Thus, each practical case has to be modelled individually, with distinction between the uncracked and cracked state of member, [3,4].

Basis for any formulation of the stress-displacement field is the uniaxial stress response to a time-dependent effective strain $\epsilon_{ef}(t)$ with the on-set of restraint at $t=t_0$:

$$\sigma(t, t_0) = \int_{t_0}^{t} \frac{\partial \epsilon_{ef}(t_i)}{\partial t_i} E(t_i) \psi(t, t_i, m_i) dt_i \qquad (22)$$

In the case of total axial restraint of the structural member the effective strain corresponds to the mean free thermal strain

$$\epsilon_{ef}(t \geq t_0) = -\alpha_T \Delta T_{ef}(t) = -\epsilon_{0T}(t) \qquad (23)$$

with
$\alpha_T = \alpha_T(t_i)$ coefficient of linear thermal expansion, $[\text{K}^{-1}]$
$\Delta T_{ef}(t)$ effective temperature difference after lapse of
                       dormant time t_d of concrete (m $\approx$ 0,17 at t_d), $[\text{K}]$.

If the structural member is attached to a resilient neighbour element the restraint will reduced. This fact can be expressed by a restraint factor $k_r \leq 1$ (1,0 for total restraint) which also depends on $E(t_i)$, [5].

The application of Eq. (22) is principally restricted to the uncracked state ($\epsilon_t \leq f_t/E_t$). Its use to predict the thermal eigenstresses by taking into account strain softening, Eq. (19), was shown in [5].

The suitability of Eq. (22) to predict the observed behaviour can be proved in various ways. In [1,3] axially and totally restrained specimens were subjected to a specified temperature path until failure. In [7] tensile tests with widely variable but constant strain-rate were performed. Thereby, the strain history $d\epsilon/dt = \epsilon \cdot (t - t_0)$ is known and the stress response will be measured. Fig. 8 shows the test results of these rate-controlled slow relaxation tests and their satisfactory prediction by Eq. (22). For comparison also the results of the stress-strain tests with a high strain-rate to minimize viscous strain components are depicted. The effects of relaxation, further maturing during loading, age at on-set, etc. are markedly disclosed.

The strict solution of the time-dependent state of restraint stresses for a practical problem is numerically possible with the FEM but cumbersome. For crack prediction a simple though reliable approach is desirable. For the everyday case of the wall restrained by the elder base a proposal is presented in [11]: from the field $T(j,t)$ the effective thermal strain field is known. It can be divided into the free mean axial strain and into the free curvature of the wall. With equilibrium and compatibility relations

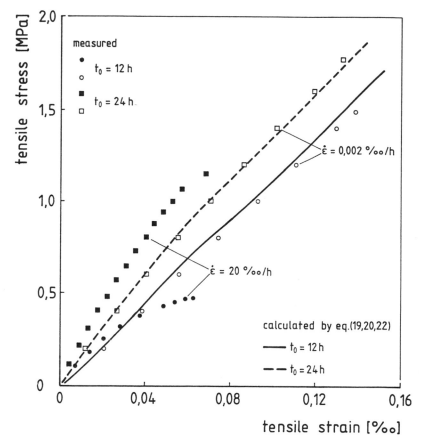

Fig. 8. Test results of slow rate-controlled tensile tests and prediction

the restraint actions and stresses can be determined if the plane strain condition of bending theory is presupposed. Time-dependence of material behaviour can be taken into account incrementally with the effective modulus of elasticity:

$$E_{tef}(t_i) = \frac{E_t(t_i)}{1 + \varphi(t, t_i, m_i)} \approx E_t(t_i)\psi(t, t_i, m_i) \tag{24}$$

Eq. (22) was also applied to determine the thermal eigenstresses for a practical problem. Fig. 9 shows in its upper half the temperature paths at various locations of a circular base slab of 4 m thickness. The lower half of the figure depicts the eigenstress time functions in the same locations. Cracking of the surface-near zone commences at an age of about 100 hrs. Cracks will open due the warping restraint of the slab. For regions with $\sigma_t \geq f_t$, cracking is expectable though considerations regarding crack stability and variability of material properties become yet necessary. Not enough known is about the real material properties in the structure and their scatter.

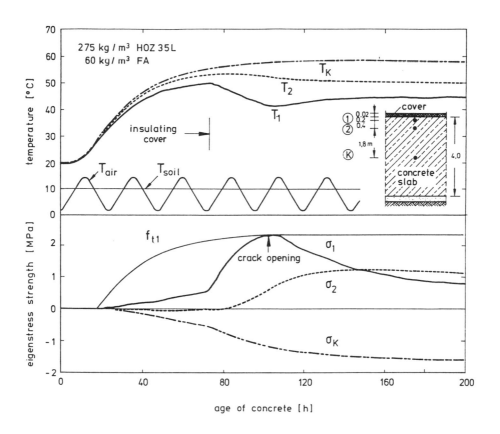

Fig. 9 Temperatures and thermal eigenstresses in a base slab (example)

221

MOUNT PLEASANT LIBRARY
TEL. 051 207 3581 Ext. 3701 .

6 Concluding Remarks

Our planning and execution tools for the control of thermal cracking must be improved. It must be possible to study well in advance the effects of envisaged compositions, materials, and practical procedures with respect to cracking. In some fields success is at hand, in others deficits prevail. Improved experimental and analytical tools are available. Practice should use them though they cost money, money which is well invested.

References

[1] Breugel van, K.: Relaxation of young concrete. Delft University of Technology, Dept. of Struct. Concrete, Res. Rep. 5-80-D8, 1980.

[2] Byfors, J.: Plain concrete at early ages. Swedish Cement and Concrete Research Institute, Fo 3:80, Stockholm 1980.

[3] Emborg, M.: Thermal stresses in concrete structures at early ages. Doct. Thesis, Lulea University of Technology, 1989: 73 D.

[4] Hamfler, H.: Berechnung von Temperatur-, Feuchte- und Verschiebungsfeldern in erhärtenden Betonbauteilen nach der Methode der finiten Elemente. DAfStb-Heft Nr. 395, 1988.

[5] Henning, W.: Zwangrißbildung und Bewehrung von Stahlbetonwänden auf steifen Unterbauten. Diss. TU Braunschweig, 1987.

[6] Jonasson, J.E.: Slipform construction - calculations for assessing protection against early freezing. CBI Research, Vol. 4, Stockholm, 1984.

[7] Laube, M.: Werkstoffgesetze für jungen Beton. Heft Nr. 87, Inst. für Baustoffe, Massivbau und Brandschutz, Technische Universität Braunschweig, 1989.

[8] Marx, W.: Berechnung von Temperatur und Spannung in Massenbeton infolge Hydratation. Diss. Universität Stuttgart, Heft 64 des Inst. für Wasserbau, 1986.

[9] Reinhardt, H.W., Blaauwendraad, I. and Jongendijk, J.: Temperature development in concrete structures taking account of state dependent properties. RILEM-Intern. Conf. on concrete of early ages. Paris 1982, Vol. 1, pp. 211 - 218.

[10] RILEM - International conference on concrete of early ages. Paris 6th to 8th April 1982. Edts. Anciens ENPC.

[11] Rostásy, F.S., Henning, W.: Zwang in Stahlbetonwänden auf Fundamenten. Beton- und Stahlbetonbau 84 (1989), H. 8, S. 207/214; H. 9, S. 232/237.

[12] Rostásy, F.S., Laube, M.: Verformungsverhalten und Eigenspannungsrißbildung von jungem Beton. Forschungs-Bericht, Inst. für Baustoffe, Massivbau und Brandschutz, TU Braunschweig, 1988.

[13] Shah, S.P., Swartz, St.E., Edts.: Fracture of Concrete and Rock. Springer-Verlag New York, Berlin, Heidelberg, 1989.

[14] Springenschmid, R., Breitenbücher, R., Balardini, P.: Vergleich zwischen Berechnungen und Messungen von Zwangspannungen im jungen Beton. Beton- und Stahlbetonbau 83 (1988), H. 4, S. 93/97.

[15] Wittmann, F., Edt.: Fracture Mechanics of Concrete. (Developments in Civil Engineering, 7); Elsevier Science Publishers B.V., Amsterdam, Oxford, New York, 1983.

19 THE CONCRETE HARDENING CONTROL SYSTEM: CHCS

W.R. de SITTER and J.P.G. RAMLER
Hollandsche Beton Groep (HBG) n.v., Eindhoven University of
Technology, The Netherlands

Abstract
Development of the strength and E-modulus of young concrete is linked with the hydration of cement. Hydration is accompanied by the generation of heat. The state-of-the-art on these phenomena has been gathered in a linked set of computerized algorithms; the CHCS Environment. Results on adiabatic tests of mixes have been compiled in a data-base which is linked to the calculation modules. The CHCS system is consulted in the work design stage to compare pouring and stripping procedures, the effects of artificial cooling or heating, the use of different mixes and types of cement. A special adiabatic test has been developed to define and measure heat generation and strength development of different proposed mixes for incorporation in the data base.

On site temperature measurements are carried out to monitor the hardening process and to use as input for the system for strength predictions and fine tuning of cooling/heating-, formwork stripping- and prestress procedures. Measurements on site are also used to check and to calibrate the CHCS system. The system has been used on more than 15 projects.

The accent of the paper is not on new scientific findings but on the application of the state-of-the-art in an integrated working environment in the day to day operations of a contractor.

1 Introduction

The development of the strength of concrete is linked to the degree of hydration of cement. During hydration heat is produced. The total energy that can be released in the form of heat by a certain type of cement -including other binders such as fly-ash- is a material property. The degree of hydration Hg is defined as the amount of heat which has been generated at a certain time expressed in percent (%) of the total amount of heat which would be generated after complete hydration of all the cement particles.

The development of properties such as compressive strength, tensile strength and moduli of elasticity for compression and tension as functions of Hg can be determined experimentally. If sufficient experimental data are available, the degree of hydration Hg can be used as a measure for the properties of the concrete at a certain stage during the hardening process.

The state-of-the-art of the know-how on these subjects has been used as a basis for a set of computer-programs. Experimental data are collected in a data-base. The data-base and the computer programs constitute the CHCS-Environment. In this way available know-how on the hardening process of concrete can be consulted in the CHCS-Environment.

The CHCS is based on research carried out at the Delft University of Technology by K. van Breugel and others. The system has been developed by the R & D department of Hollandsche Beton Groep nv (HBG). HBG is indebted to van Breugel for his cooperation and the stimulating discussions with his group.

2 What is the concrete hardening control system CHCS?

CHCS is a system to make know-how on the hardening process of concrete accessible and available for use in a contractors day to day operations. The system com prises three elements:

I) **The CHCS environment** consisting of calculation modules and data-base. The system is consulted using a keyboard and screen in the CHCS environment.

II) **Tests in the laboratory** consisting of adiabatic tests to determine the heat generation of a particular mix and isotherma tests to determine the properties of concrete at various degrees of hydration. Compressive strength and tensile splitting strength are relevant properties in this context.

III) **Verification on site** to check the calculated temperature development at various levels in the concrete. As well as tests to verify the calculated development of strength in the pieces which have been cast at the same time as the structure. These test pieces are stored under water in a tank. Thermocouples are cast in the concrete of the structure and coupled to a unit which steers a heating/cooling system of the water in the tank. In this way the test pieces can be subjected within $\pm 0.5°C$ to the same temperature regime as the concrete at a particular location in the structure.

Verification is needed to enhance the reliability of the system. The results are part of the data base for future reference.   In this way CHCS is an integrated system which provides answers in two phases of a construction project.

1. In the planning stage and during design of work methods.
2. On site during construction.

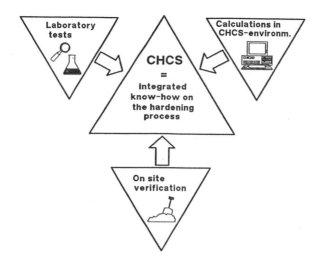

Fig. 1. CHCS overview

In the **planning stage** the system is used to:

- Compare the merits of different mixes with respect to development of strength in view of applying prestress and/or removal of formwork. The mix to be used in order to ensure that formwork can be reused within a specific number of days.

- Determine the amount and time of cooling, necessary to avoid cracks due to differences in temperature.

- Determine the amount and time of heating necessary to achieve the desired early strength but avoiding temperature cracking.

- Compare the relative merits of changes in the mix, applying insulation and/or heating.

- Compare the costs associated with different routes to achieve the desired effect. Changes in the mix and insulation versus heating, effects of increasing or decreasing cycle times of formwork, cooling versus casting massive structures in layers, ... etc.

Since the system provides answers rapidly, there is no practical restriction on the number of alternatives to be considered.

On site during construction, the results of the system are verified by measurement of the temperatures in the concrete and the development of strength. The system is also used for fine tuning of the construction procedures. For example:

- Low temperatures are expected. Is it necessary to change the mix or take other measures such as heating?

- The ready mix supplier wishes to switch to a different type of cement. Is there an influence on the required development of strength?

- Cracks have been observed. Can these be attributed to temperature stresses?

Answers can be provided practically on the spot within a couple of hours to a couple of days depending on whether the system already "knows" the structure (in terms of geometry, mix design, insulation, seasonal temperature variations, daily temperature variations, wind velocity ... etc).

The data-base and the computer programs constitute the CHCS-Environment. In this way available know-how on the hardening process of concrete can be consulted in the CHCS-Environment.

3 The CHCS environment

From the outset the following rules for development of soft-ware were adhered to:

1. The system must run on a current type of PC and execute calculations within minutes. Starting with an AT type of machine, this rule leads to a number of restrictions on the complexity of the problems to be handled; for instance only one dimensional (1D) heat flow. The increasing capacity of PC's will lead to a relaxation of the restrictions; at this time the inclusion of 2D heat flow is considered.

2. The time between the start of the development and the first practical applications had to be short. Therefore standard soft- ware packages were used.

   Lotus 1 2 3$^{(R)}$ provided ready made, menu driven input and output screens. Cardbox Plus$^{(R)}$ was used as a data base which is linked to the Lotus worksheets.

3. The system must provide sufficiently reliable answers within an acceptable time; calculations should not take more than a couple of minutes on a current PC AT type. "Sufficient reliable" for practical use on site is different from "scientifically correct". Therefore the user must be prepared to accept a fair amount of approximation.

The structure of CHCS is shown in the following flowchart.

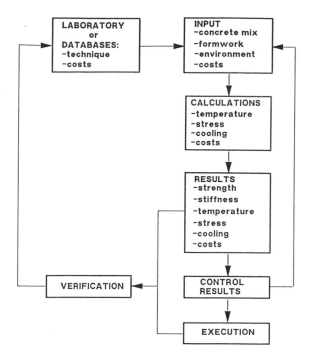

Fig. 2. CHCS structure

The input data and the calculation results have been incorporated into the LOTUS$^{(R)}$ software package. The mathematical calculations have been written in Turbo Pascal. The databases have been set up using the CARDBOX Plus$^{(R)}$ software package.

As far as input data are concerned, there is a distinction between temperature and stress calculations. A stress calculation can be carried out only following a temperature calculation. Consequently, there are various calculation models available, such as a temperature calculation involving no artificial cooling, a temperature calculation taking artificial cooling into account, and a stress calculation.

The results of these calculations can be shown numerically as well as graphically. After checking these results, it is possible to analyse other variants until the results are considered satisfactory and the work can be undertaken.

Input data for calculating the hydration temperatures are entered in a LOTUS$^{(R)}$ display menu. CHCS is equipped with a single-dimensional temperature calculation program. The following input data may be analysed.

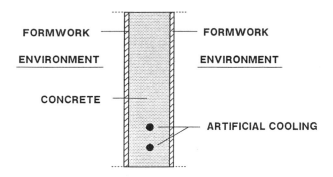

Fig. 3. Input data for temperature computations

Concrete:
- element thickness
- weight per m³ of cement
- coefficient of heat conduction
- specific heat
- density
- heat generation of cement
- maximum degree of hydration
- initial temperature of concrete
- dormant stage
- time constant adiabatic
- strength coefficients

Formwork:
- thickness
- coefficient of heat conduction
- time of formwork removal

Surroundings:
- average temperature of surroundings
- amplitude temperature of surroundings
- elapsed time until maximum temperature of surroundings is reached
- wind velocity
- temperatures may include heating effects

Cooling:
- extent of cooling
- starting time of cooling
- finishing time of cooling

Formwork as well as surroundings may be different on either side of the concrete element.

Execution within the CHCS environment is controlled by a batch program which steers the sequence of modules. The batch program calls the LOTUS$^{(R)}$ menu driven input screens. From the input it follows whether Cardbox Plus$^{(R)}$ data base must be consulted or the user supplies data on the mix by keyboard in one of the worksheets. Lotus$^{(R)}$ worksheets are stored in data files. The batch program calls the required Turbo Pascal calculation programs which use the data files. The results of calculations are stored in output data files. The Lotus$^{(R)}$ output screens are activated to use these output files. Output can be requested on screen and/or printer, in graphic and/or tabular form.

Input and output files may be stored for future reference and/or verification.

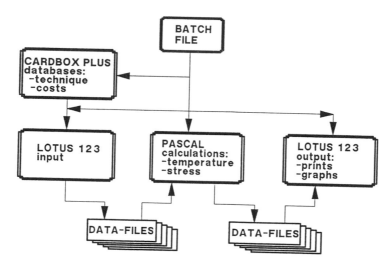

Fig. 4. Flow chart of CHCS-programs

4 The Prins test, adiabatic curve and strength development

4.1 Adiabatic temperature gradient

The temperature gradient of hydrating concrete mix depends on a great number of factors, such as the granular size, type and quantity of the cement, aggregates, additives, water/cement ratio, ambient factors, etc. Yet it is not possible to incorporate all these factors in a model in such a way that the temperature gradient can be forecast. This problem is overcome by submitting the concrete mix to an adiabatic test. A process may be called adiabatic when there is no interchange of heat with the surrounding. Using this adiabatic test, the adiabatic of a concrete mix can be determined. This will then form the basis for the temperature calculations.

In order to determine the adiabatic, the CHCS uses the Prins method, so called after a staff member of HBW, Hollandsche Beton en Waterbouw bv. Immediately after the mixture has been made, a cube is placed in a thermally insulated water tank. Thermocouples attached to a recorder are inserted in both the water tank and the cube.

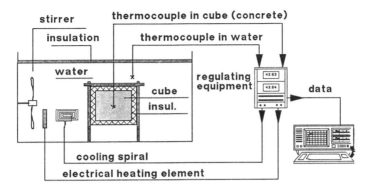

Fig. 5. Test arrangement for Prins method

By means of a ΔT-control the temperature of the water follows the temperature of the concrete, which results in a virtually adiabatic test.

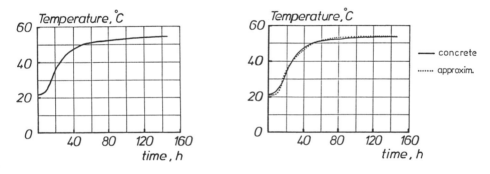

Fig. 6. Ascertaining the adiabatic by the Prins method

Fig. 7. Schematisation of the adiabatic

The adiabatic obtained from the adiabatic test is then schematised as an exponential function.

$$\Delta T = \Delta T_{max} \cdot \left(1 - e^{-(t-t^*)/t_\infty}\right) \tag{1}$$

in which

Δ T   = increase of temperature at time t [°C]

ΔT_{max} = maximum increase of temperature under
            adiabatic conditions [°C]

t   = momentary time [hour]

t^* = dormant stage [hour]

t_∞ = time constant [hour]

$$\Delta T_{max} = \frac{W \cdot C}{c \cdot \rho} \cdot Hg_{max} \tag{2}$$

in which

W   = theoretical heat generation (at 100 % hydration)
        of the cement [kJ/kg]

C   = cement content in concrete [kg/m³]

c   = specific heat of the concrete [kJ/kg°C]

ρ = density of the concrete [kg/m³]

Hg_{max} = maximum degree of hydration [%]

c and ρ are assumed to be constant during hydration

An alternative analytic approximation, which follows more closely the adiabatic S-curve, has been developed by D. Kiefer (TU- Darmstadt):

$$\Delta T = \Delta T_{max} \cdot exp(-a/(t + ts)^b)$$

in which
a = parameter depending on the initial temperature
b = parameter depending on the type of cement
ts= time shift

4.2 Strength development using the degree of hydration
The degree of hydration at a certain point in time t_i can be determined as follows:

$$H_{g_i} = \frac{\Delta T_1 + \Delta T_2 + \ldots\ldots\ldots + \Delta T_i}{\Delta T_{max}} \cdot 100\% \tag{3}$$

The figure below shows how the degree of hydration is determined.

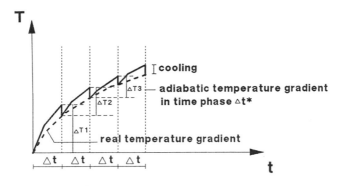

Fig. 8. Determining the degree of hydration

Concurrently with the execution of the adiabatic test, cubes are manufactured which are compressed at different points in time.

Using the schematised adiabatic (see Fig. 7) an isothermal computation is carried out at a temperature of 20°C. In this way a ratio is established of the degree of hydration to the compressive strength. That relation proves to be virtually linear.

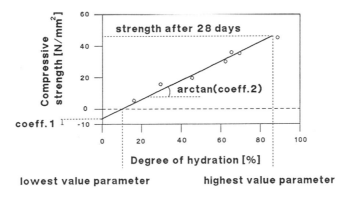

Fig. 9. Determining strength coefficients

The following coefficients then have to be input in the CHCS:

- compressive-coefficient 1   : negative compressive strength
                                after extending regression line to Hg = 0 %
- compressive-coefficient 2   : regression line gradient
- lowest parameter   : degree of hydration at 0 compressive strength
- highest parameter   : degree of hydration at strength after 28 days.

In order to determine the tensile strength the same method may be used.

The modulus of elasticity is computed with the use of the relation:

$$E = (1800 - 4f_{cc})\,(10 \cdot f_{cc})^{0.5} \tag{4}$$

5 Process temperature gradient

In reality, the hydration process of hardening concrete will not be adiabatic. During hydration, heat will be transferred to the surroundings. In the following description it is assumed that heat is lost because of the lower temperature of the surroundings. It is, however, possible that during the hydration process an additional source of heat is involved, in the case of steamcuring for example.

To calculate the real temperature gradient, the CHCS can draw upon a one-dimensional temperature computation program. It divides the thickness of the concrete slab into a number of elements of 0.05, 0.10 or 0.20 m. The thickness of the elements depends on the thickness of the slab.

The program computes how much adiabatic heat is being released from each element during each time phase of one quarter of an hour or one hour. This adiabatic increase in temperature depends on the initial temperature at the outset of each time phase. As heat is lost to the surroundings, the temperature at the outset of a time phase will be lower than the adiabatic temperature at that moment. At a lower starting temperature the binding reaction will happen less quickly, which results in a lower increase of temperature. This delaying effect is taken into account by shifting the time axis as follows:

$$\Delta t_{new} = \Delta t_{old} \cdot g^{(T_{real} - T_{adiabatic})}/10 \tag{5}$$

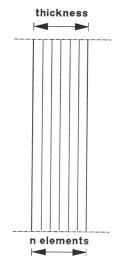

Fig. 10. Division into elements

The g-factor represents the influence of the temperature on the reaction speed. The value of g fluctuates between 1.3 and 3.5. The value of g has provisionally been assumed as being 2. Comparison with actual measurements has shown that this seems to be a correct assumption. When more data become available, it will be possible to adjust the value of g accordingly.

The algorithm described above can be represented graphically as follows.

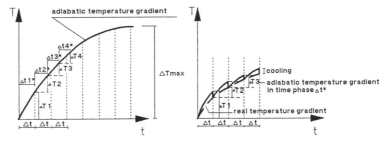

Fig. 11. Process temperature gradient

Heat exchange with the surroundings may take place in three ways:

- conduction

- convection

- radiation

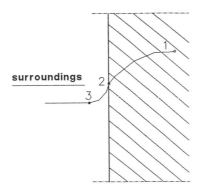

Fig. 12. Heat exchange with surrounding

Heat exchange through conduction takes place between points 1 and 2. Transfer of heat between 2 and 3 is caused by conduction and convection. In addition, heat may be lost or gained through radiation exchange with the surroundings. Due to solar radiation, the surface temperature of the concrete slab may increase considerably.

The phenomena described above have been laid down in mathematical equations (Appendix A).

6 Case histories

To illustrate the use of the system in practice, two examples are given. The first example refers to sections of a reinforced concrete tunnel which were fabricated in a dry dock and subsequently floated and submersed at the site. The temperature development due to hydration was calculated on the basis of the adiabatic curve of the mix, environmental conditions on site, the properties of the formwork and the striking time of forms.

Willemsspoortunnel (Rotterdam)

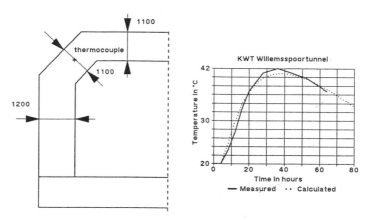

Fig. 13. Tunnel section and comparison of predicted and measured temperatures

In Fig. 13 a comparison is given of the calculated and the measured temperatures. This example and many others like it show that predictions of temperature development are sufficiently accurate for practical purposes.

The second example refers to the planning stage of a submersed tunnel under the Noord. In this case the fabrication was executed in two stages. In the first stage the bottom slab was cast. After hardening of the bottom slab the walls and the roof were cast in the second stage.

The development of the temperatures and of the tensile strength in the wall were predicted for two alternatives; without cooling as well as with cooling pipes in the wall. Taking into account the temperatures and the development of E-moduli in the bottom slab, the walls and the roof, the eigenstress in the slab, the walls and the roof have been calculated. Fig. 15 shows the stresses in the underside and the upperside of the wall for the case without cooling.

It appears that the tensile stresses will surpass the predicted development of the tensile strength. For the case of a particular arrangement of cooling pipes it is shown in Fig. 16 that the tensile stresses will remain well below the tensile strength.

In this case a rather crude relaxation model was used and the non- linear behaviour after cracking was not taken into account. This explains why in Fig. 15 the calculated stresses surpass the tensile strength, a situation which will not occur in reality. Since we were anyhow aiming for a solution with stresses below the tensile strength, such inconsistencies are for the time being accepted from a practical point of view.

In cases where we want to avoid cracking we are not interested in an exact represen- tation of post cracking behaviour.

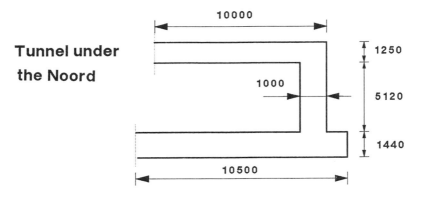

Fig. 14. Dimensions of tunnel section

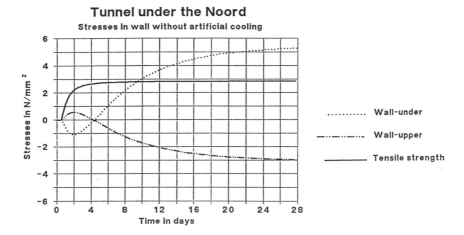

Fig. 15. Stresses in the case without cooling

237

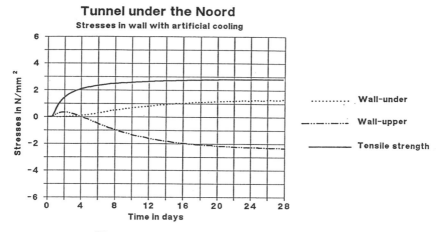

Fig. 16. Stresses if artificial cooling is applied

7 How reliable is CHCS?

The proof of the pudding is in the eating. So far we may conclude from our experience the following:

a) In the laboratory a curve is determined which gives the relation between development of strength and the degree of hydration Hg. The agreement between the strength derived from the experimental curve (Hg) and site tests on cubes subjected to a temperature regime leading to an identical value of Hg is good and sufficiently reliable.

It is safe to use temperature measurements as a basis for decisions on removal of formwork and application of prestress. As such CHCS is a valuable tool for Quality Assurance.

b) The agreement between calculated temperatures and the measurements are very good; in the order of 1 to 2°C. When the concrete is subjected to artificial cooling, larger deviations have been observed. It is not yet clear whether these differences must be attributed to the calculation model or to differences in the actual cooling characteristics on site with respect to the assumed cooling procedure in calculations. It has been observed that the generation of heat due to solar radiation can have a marked influence on temperatures in a layer of 0.1 to 0.2 m of concrete near the surface.

c) The calculated temperature development is the basis for the calculation of E moduli, temperature stresses and tensile strength. If the temperature stresses are larger than the tensile strength, then cracking may be expected. Or rather,

238

if the tensile strain exceeds the ultimate tensile strain of the concrete. Direct verification of temperature stresses and temperature strains has not been carried out. So far we can only cite one case were cracking was predicted six days after pouring. Cracks were observed seven days after pouring. In most cases measures were taken to prevent cracking. In this respect CHCS has proved to be successful. It is not known, however, if the measures are over cautious.

8 Future development

In the future further development is needed in the following areas:

a) Relaxation of (tensile) stresses is accounted for in CHCS in a very crude, rough manner. A better approximation of relaxation will lead to more refined predictions with respect to cracking.

b) Cracking is predicted on the basis of stress calculations, including redistribution of stresses between different parts of the structure. The inclusion of a strain criterion for the prediction of cracking in combination with a non-linear stress strain relationship could lead to better predictions. A tri-linear stress/strain relationship is considered using compression-tensile-softening-crack curves.

c) For cooling calculations a 2D model would be a major improvement. With the increased capacities of P.C. hardware this will become feasible without increasing calculation times to unacceptable levels.

Of course a number of other areas of interest may be indicated but the three listed items seem suitable candidates for first priority.

References

[1] Prins, P.C., Bau von Flusskraftwerken in den Niederlanden, Beton-Informationen, 3/4 (1988).

[2] de Sitter, W.R., Van materiaaltechnologie tot procesbeheersing, Cement 1988, nr. 12

[3] Maatjes, E., Berlage, A.C.J., Beheersing van het verhardingsproces, Kennis van betontechnologie en uitvoeringstechniek gecombineerd in een programmapakket. Cement 1989, nr. 3.

[4] Horden, W.C., Maatjes, E., Berlage, A.C.J., A computerized concrete hardening control system and its application in tunnel construction, Conference on Immersed Tunnel Techniques, Institution of Civil Engineers, 11-13 april 1989, Manchester

Appendix A

The temperature gradient as a consequence of the hydration process is a particular case of the fluctuating heat flow which can be expressed with the use of the Fourier equation:

$$\frac{\delta T}{\delta t} = a \cdot [\frac{\delta^2 T}{\delta x^2} + \frac{\delta^2 T}{\delta y^2} + \frac{\delta^2 T}{\delta z^2} + \frac{1}{\lambda} \cdot w(x,y,z,t)] \qquad (A.1)$$

in which
T = temperature [°C]
t = time [hour]
a = temperature compensating coefficient [m²/hr]
λ = heat conduction coefficient [J.hr/m°C]
w = heat generation per volume and time unit [J/m³ hr]
x,y,z = coordinates [m]

$$a = \frac{\lambda}{\rho \cdot c} \qquad (A.2)$$

in which
ρ = density [kg/m³]
c = specific heat [J/kg°C]

As described before, the CHCS incorporates an one-dimensional temperature program. In this instance, the Fourier equation can be simplified to:

$$\frac{\delta T}{\delta t} = a \cdot \frac{\delta^2 T}{\delta x^2} + \frac{1}{\rho \cdot c} \cdot w(x,t) \qquad (A.3)$$

It is assumed that a, ρ, λ and c are constants and, therefore, independent of time, place and temperature.

The pre-condition regarding the concrete surface is:

$$(\frac{dT}{dx})_{x=0} = \frac{T_{surf} - T_{surr}}{\lambda / \alpha} \qquad (A.4)$$

in which :
T_{surf} = temperature of concrete surface [°C]
T_{surr} = temperature of surroundings [°C]
α = heat exchange coefficient [w/m² °C]

The effect of formwork is taken into account by using a resultant heat exchange coefficient of α_r instead of α.

$$\frac{1}{\alpha_r} = \frac{1}{\alpha} + \frac{d_k}{\lambda_k} \qquad (A.5)$$

in which :
α_r = resultant heat exchange coefficient [w/m² °C]
α = heat exchange coefficient formwork-surroundings [w/m² C]
d_k = formwork thickness [m]
λ_k = heat conduction coefficient of formwork [w/m °C]

240

The solution to the differential equations A.3 may be approximated using a differential calculation. The concrete slab is divided into a number of elements Δx (see figure 11).
The solution to this differential equation is determined via time-steps Δt. Equation A.3 can then be restated as:

$$\frac{\Delta T}{\Delta t} = a \cdot \frac{\Delta^2 T}{\Delta X^2} + \frac{1}{\rho \cdot c} \cdot \frac{\Delta W}{\Delta t} \qquad \text{A.6)}$$

Assuming that : n = element number
                 k = time phase number,

the following diagrams can be drawn:

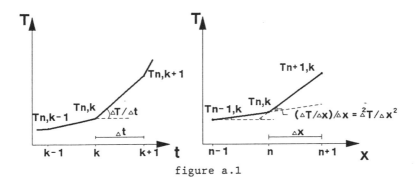

figure a.1

From figure a.1 it follows:

$$\frac{\Delta T}{\Delta t} = \frac{T_{n,k+1} - T_{n,k}}{\Delta t} \qquad \text{(A.7)}$$

From figure a.2 it follows:

$$\frac{\Delta^2 T}{\Delta X^2} = \frac{\Delta \left(\dfrac{\Delta T}{\Delta X} \right)}{\Delta X} = \frac{\dfrac{T_{n+1,k} - T_{n,k}}{\Delta X} - \dfrac{T_{n,k} - T_{n-1,k}}{\Delta X}}{\Delta X}$$

$$\frac{\Delta^2 T}{\Delta X^2} = \frac{T_{n+1,k} - 2 \cdot T_{n,k} + T_{n-1,k}}{\Delta X^2} \qquad \text{(A.8)}$$

241

Replacing equations A.7 and A.8 in equation A.6 results in:

$$\frac{T_{n,k+1} - T_{n,k}}{\Delta t} = a \cdot \frac{T_{n+1,k} - 2 \cdot T_{n,k} + T_{n-1,k}}{\Delta X^2} + \frac{1}{\rho \cdot c} + \frac{\Delta W}{\Delta t}$$

Rewritten as:

$$T_{n,k+1} = T_{n,k} + 2 \cdot a \frac{\Delta t}{\Delta x^2} \cdot \left(\frac{T_{n+1,k} + T_{n-1,k}}{2} - T_{n,k} \right) + \frac{\Delta W}{\rho \cdot c} \qquad (A.9)$$

This approximation of the differential equation A.1 is valid only on the condition that:

$$2 \, a \cdot \frac{\Delta t}{\Delta x^2} \leq 1 \qquad (A.10)$$

In the part of the CHCS program where temperatures are calculated equation A.9 is used.

20 TESTING TEMPERATURE GRADIENTS OF YOUNG CONCRETE

G. SCHICKERT
Bundesanstalt für Materialforschung und -prüfung (BAM),
Federal Institute for Materials Research and Testing,
Berlin, Germany

Abstract
Young concrete is particularly sensitive to temperature gradients. Uncontrolled cracking, especially within the layer just under the surface, may be caused by temperature induced stress exceeding the strength of the young concrete. The purpose of this article is to point out that the heat resistance of the formwork and the geometry of the cross section of the concrete element as well strongly influence the heat distribution caused by hydration after casting.

1 Introduction

When the formwork has been removed for a certain period the surface temperature of the concrete usually differs greatly from the ambient air. Then the layers just beneath the surface adjust to the ambient temperatures within a relative short time. In contrast the temperature field of the center of the concrete volume, however, at the very beginning is scarcely influenced. Therefore considerable temperature gradients may develop between core concrete and the respective surface. And this may lead to inherent stresses within the inner concrete structure which the young concrete - in respect to its still poor and just developing strength and deformation characteristics - cannot withstand. Up to now manufactures have usually paid no additional attention to such temperature induced stresses. Actually it is unknown to which extent creeping diminishes such stresses and how severe they really are compared to other inherent stresses for instance due to shrinkage [1].

In a special case, however, problems during the production of fiber reinforced cement boards doubtless were caused by remarkably high temperature gradients. The 2.5 x 1.25 m² plates were stacked 1 m high immediately after manufacturing and steam hardening. In the cores of the stacks very high temperatures existed while the outer areas of course showed lower temperatures. These temperature gradients revealed to increase as the different areas got drier and drier. By virtue of this relationship a numerical simulation of the heat and moisture transfer was established and the production process could thus be optimized [2,3]. In agreement with the measured values the results of the calculations revealed that even 30 days after manufacturing the stacks of the fiber reinforced cement boards had not achieved a uniform temperature distribution. Fig. 1 shows a typical combination of temperature and moisture. At this time the corners have higher temperature and less moisture values compared to the core.

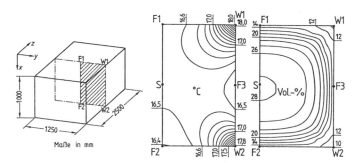

Fig. 1. Isotherms and moisture distribution in vol.-% within a vertical cross-section of a stack of fiber reinforced cement boards [3]

Similar temperature gradients or temperature induced stresses respectively, occur very often and mostly under different circumstances. As for young concrete a regulation expressly calls attention to the fact that steel formwork compared with a wooden one may more strongly affect the heat flow of the hydration process causing greater temperature gradients. Also heat insulation measures in connection with curing are mentioned to improve concrete durability [4].

In the following such questions are explained in more detail. With respect to durability microcracking expecially within the skin of concrete members must be avoided. Otherwise the resistance to corrosion, moisture, frost and to the residue of harmful substances would be diminished. Young concrete, however, is not only very sensitive to microcracking caused by shrinkage but also to stresses caused by certain temperature gradients. This will be illustrated on the basis of temperature distributions calculated with numerical methods and of thermograms which are taken with an infrared camera.

2 Heat flow

The temperature induced deformation response of young concrete caused by setting heat has already been studied. Wall like sections were chosen as specimens [5].

Still greater temperature gradients, however, as in such cases of concrete walls can be expected for those structural elements which do not have a uniform thickness but cross sections with a special shape. This is true for instance for a member with an

I-shaped cross section. Here and in similar cases the ratio of the amount of the surface area to the corresponding volume is greater than of a rectangular-shaped cross section. Consequently this influences the heat flow. In addition the heat flow is governed by the time dependant development of the setting heat and parameters like moisture content, velocity of the surrounding air, solar radiation etc. There is already some research work on such parameters, not for young concrete but for concrete specimens 2 months and 5 years old [6,7].

Fig. 2 shows to which remarkable extent the shape of the cross section in principle influences temperature distribution. At first equal setting heat develops in all small elements of the volume and the heat flow normally finds its way to the periphery of the concrete member. With respect to this in the case of Fig. 2 the calculation is based on a stationary heat flow. It reveals for any distinct instant that temperatures within the cross section are influenced not only by its shape but moreover by the different thickness of its flanges. This makes clear that regarding the setting heat the highest temperatures in principle occur in such areas of the cross section where the ratio of the relevant volume to the corresponding outer surface is highest. Hence follows that the outer corners of cross sections which have an extreme value of this volume/surface ratio show the greatest difference in temperature compared to the core concrete. And this makes clear, too, that outer surfaces must show areas with temperature differentials in spite of stationary heat flow. Thus if one follows the contours of Fig. 2 the periphery indeed shows different temperatures zones.

These results have been achieved with well proved computer programming which is also the basis for the following statements of this report [2]. It is obvious to investigate not only one but several cross sections and to include in this research other parameters like reinforcing bars and different types of cement. The accompanying tests, however, are just at the beginning. Thus only first results can be reported.

Fig. 2. Relative temperature distribution within an I-shaped concrete cross section during hydration

3 Experiments

The tests have been performed in a big enviromental chamber of the BAM which has a volume of 55 m³. In this way uncontrolled influences of fluctuation in room temperature or radiation etc. are excluded. The concrete specimens have a height of 1 m and as basis a cross section of 30 x 60 cm². The concrete mixture is the same in all tests with 291 kg/m³ cement and a water/cement ratio of 0.56. The siliceous aggregates are graded up to 16 mm. For the first series the specimens are not reinforced and a blast furnace cement (HOZ 35 L) has been chosen that means a cement with relatively slow hydration. For comparislon purposes a fast portland cement will be taken later on. The cross sections will be rectangular, I-shaped, L-shaped and circular.

The test series started with a special cross section which has cutouts of different sizes at opposite corners. Fig. 3 gives an impression of the formwork and shows the concreting and the vibratory compaction. By this photo one becomes aware of the fact that the two cutouts act as a very good heat insulation. Of course this will govern the heat flow to a great extent.

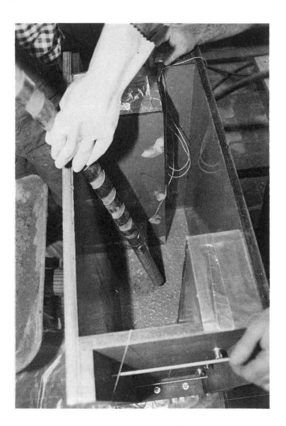

Fig. 3. Concreting a specimen with cutouts at opposite corners

4 Test results

At first it is Fig. 4 that outlines up to what differences in temperature can be expected for such special shaped cross sections. Shown, however, is only that in principle developing temperature regime due to setting heat, that means the graph is plotted without true relation to the really existing temperature gradient expressed in Kelvin. So far it is only a relative figure. For instance in the case of very cold ambient air the temperature gradient between an outer and a central zone will be very high and for room temperature of course smaller. In addition Fig. 5 illustrates the same result three- dimensionally [8].

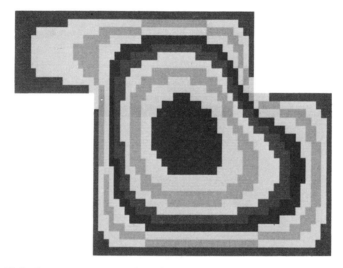

Fig. 4. Relative temperature distribution within a specially shaped concrete cross section during hydration

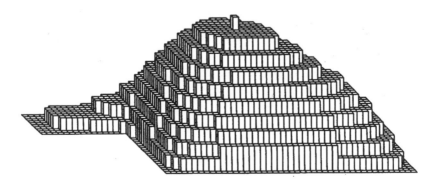

Fig. 5. Three-dimensional illustration of the relative temperature pattern of Fig. 4

While Fig. 4 or Fig. 5, respectively, refer to a demolded concrete member, Fig. 6 shows the situation with the connected formwork as already explained by Fig. 3 and this implies highly insulated areas in connection to the two cutouts. And this time the difference between the highest and the lowest temperatures has been measured with thermocouples and then for graphic reasons put to 100 %. The maximum temperature the blast furnace cement achieved was about 36°C after 22 hours (Fig. 7).

It is worth mentioning that only one of the four measuring points showed a clear difference in temperature. This point was near to the less insulated surface and therefore registered the smallest values.

In other words this temperature distribution of Fig. 6 is explained to a great extent by the extreme insulation of the cutouts. The heat flow nearly stops at these areas contrary to the other surfaces.

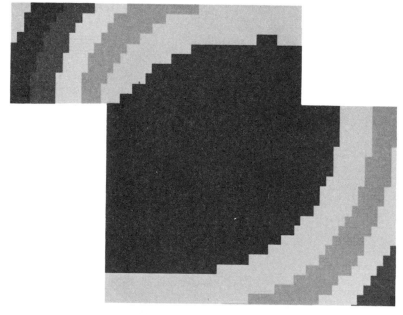

Fig. 6. Temperature distribution during hydration within a concrete cross section insultated by wooden formwork with additional very good insulation at the cutouts (down left and right above), central area 36°C, steps 1 K

This situation is confirmed by a thermogram which was taken with an infrared camera immediately after demolding the specimen. For this instant the differences in temperature are relatively small. Therefore only a powerful image processing provides the informations wanted with averaging - in this case on the basis of 20 images - to reduce the noise level (Fig. 8).

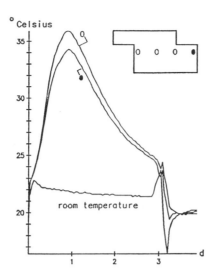

Fig. 7. Measuring points within the cross section and temperature time curves with sudden drop of room temperature after 3 days

In the other case, however, where the heat insulation of the formwork all around is uniform and moreover small the stationary temperature distribution due to setting heat would result in a temperature field pattern as in principle Fig. 9 shows. After a certain time also the temperature regime of Fig. 6 would change to this pattern. The real difference in temperature expressed in Kelvin, however, of course in Fig. 9 is smaller than in Fig. 6 where at the instant of culminating temperature more than 4 K had been measured at the measuring points graphed in Fig. 7.

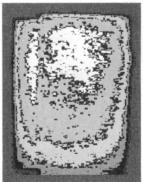

Fig. 8. Infrared or thermal image thermogram immediately after casting. The surface shown corresponds to the contour at the foot of Fig. 6

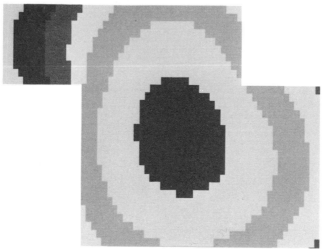

Fig. 9. Like Fig. 6 but poor insulation all around

It is interesting to compare the contour at the foot of Fig. 9 with the thermogram in Fig. 10.

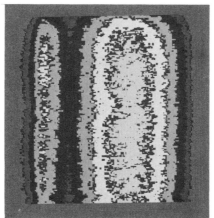

Fig. 10. Thermogram of the situation according to Fig. 9

In addition different computer procedures still can improve the visibility of the iso-therms within such an infrared image (Fig. 11). In this manner it is possible to determine quite exactly the time dependant development of the temperature fields on the concrete surface. And this, however, would allow fo draw conclusions on the temperature distribution within the cross section of the concrete member by help of such numerical methods which result in graphs like Fig. 6 or 7. Following this line also the determination of maturity would be possible and therefore in principle estimated values even of the strength of young concrete could be achieved.

Thus it can be seen that the infrared camera in combination with proper image processing at least verifies the numerical result of the calculation method. To make this more obvious or to give a clear idea of the temperature distribution at the concrete surface of the investigated specimen, respectively, Fig. 12 finally shows a thermogram of the same situation as in Fig. 10 but in an oblique view.

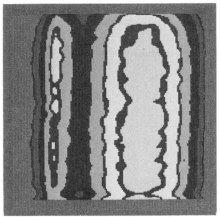

Fig. 11. Thermogram of Fig. 10 after image processing

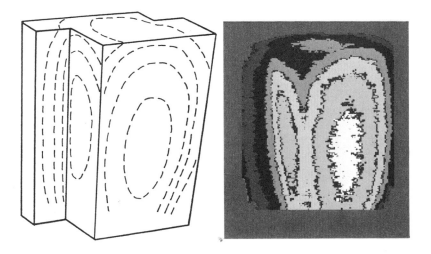

Fig. 12. Thermal image (oblique view from above) belonging to Fig. 10

5 Conclusion

In principle the setting heat creates a uniformly distributed temperature within each element of the concrete volume. Influenced by the surrounding conditions this heat normally flows to the outer surfaces. It could be shown, however, that this regime does not at all result in a uniform temperature distribution within the concrete. On the contrary the resulting temperature gradients are strongly influenced by the effect of the formwork as heat insulation and especially by the shape of the cross section. Temperature induced stresses may develop which possibly cause the microcracking of young concrete and consequently diminish the protection against corrosion and the durability of structural concrete members.

6 Acknowledgement

The numerical calculations and computer graphics, respectively, were performed by Dr.-Ing. M. Weber. His efforts are gratefully appreciated.

References

[1] Wischers, G.: Die mathematische Erfassung der Spannungen infolge Schwindens. Betontechnische Berichte 1960, Beton Verlag, Düsseldorf, p. 73-82

[2] Rudolphi, R. and R. Müller: Bauphysikalische Temperaturberechnungen in FORTRAN. Band 1: Zwei- bzw. dreidimensionale stationäre Probleme des Wärmeschutzes. Teubner- Verlag, Stuttgart, 1985, 227 p.

[3] BAM -Jahresbericht 1986, Report Lab. 2.44 "Numerical Methods of Building Physics"(R. Rudolphi, R. Müller).

[4] Richtlinie zur Nachbehandlung von Beton. Deutscher Ausschuß für Stahlbeton, Berlin, 1984.

[5] Rostasy, F.: Eigenspannungsrißbildung und Verformungsverhalten von jungem Beton. Forschungsbericht TU Braunschweig, 1989, 237 p.

[6] Springenschmid, R., U. Wagner-Grey and M. Schwarzkopf: Temperaturspannungen in Beton bei sommerlicher Erwärmung. Der Bauingenieur 53(1978), p. 265-267

[7] Shkoukani, H. and G. Walraven: Time dependent behaviour of concrete at elevated temperatures. Annual Journal "Darmstadt Concrete", Vol. 3(1988), p. 167-178

[8] Weber, M.: VG3D; Zeichenprogramm für die Vektorgraphische Darstellung dreidimensionaler Strukturen. BAM- Forschungsbericht Nr. 150, Berlin 1988, 83 p.

21 IN-PLACE TESTING FOR QUALITY ASSURANCE AND EARLY LOADING OPERATIONS WITH PULLOUT TESTING AND MATURITY MEASUREMENTS

C.G. PETERSEN
Germann Instruments A/S, Copenhagen, Denmark

Abstract
Pullout testing systems and maturity measurements for quality assurance and early loading operations are briefly reviewed. Correlation data and variability of pullout testing is stated. Present applications are illustrated by a number of test cases based on the statistical evaluation procedure in the Danish Concrete code DS 411 from 1984 and the Danish Standard for Evaluation of Test Results DS 423.1 from 1985 with particular reference to pullout testing in-situ.

1. THE NEED FOR IN-PLACE STRENGTH TESTING DURING CONSTRUCTION

The traditional standard cylinder or cube test measures the potential compressive strength of a given mix after 28 days, when the concrete under ideal laboratory conditions is cast, compacted and cured in water or moist at 20°C.

The strength of the structure itself is different, depending on the effects of transportation of the concrete, pumping and casting, compaction and curing as well as the maturity developed on-site.

In an attempt to achieve a more realistic evaluation of the in-situ strength, cylinders or cubes have traditionally been made on-site and cured alongside the structure. The results of such tests are often significant different from the structures and have higher variation, because it is difficult and often impossible to ensure identical casting, compaction, curing and maturity. The results may be directly misleading if the test specimens are mistreated or cured differently than the in-place concrete.

Test results from such traditional testing are matched with a required 28-day strength, which usually is obtained by multiplying the design strength by a factor of 1.5 partly motivated by the wellknown fact that the in-place strength may be lower than the test specimens.

Developments the past 20 years in the world of in-situ testing have
resulted in considerable progress to overcome the inherent deficiencies in using standard test specimens for the strength evaluation.

Although in-situ testing with the new techniques attempts to give
a direct estimate of the in-place strength and although such testing
is much more economic and quick to perform than traditional testing,
it is mainly from two special needs the practical applications have
originated, namely

>   - to quality control the critical cover layer
>   for durability to make sure it is well cast,
>   compacted and cured and

>   - a wish to time early loading of a hardening
>   structure safely.

As far as the first need is concerned, it is wellknown that the quati-
ty of the cover layer protecting the reinforcement is one of several
critical factors to ensure a durable structure. If this part is
free from unintended porosities, micro- and macro-cracks, the structure will more easily resist the attacks from e.g. chlorides, moist,
oxygen, carbondioxide, acids, sulphates and freeze-thaw it was designed for. The traditional standard cylinder or cube test generates no
such information. This deficiency may be overcome by testing the critical cover layer after 28 maturity days and compare the results with
those of the lab-crete. If they are of the same order and within certain variations, it has been proved that the real-crete has the same
quality and homogeneousnes as the lab-crete. Otherwise, the reasons
for lacking compliance may be found by petrographical analyses and
immidiate corrective measures taken. From recent experience with modern high strength concretes designed to resist long term chloride
chloride attacks, a 50% loss of strength of the cover layer was eviden-
ced caused mainly by unsufficient compaction and curing conditions.

Timing of removal of forms, shores and other early loading oprations
involve the possibility of cutting down on a construction schedule
and lead to substantial benefits in terms of more economic use of
form materials, saved interest and earlier rentals. Such benefits far
exceeds the extra costs usually required by using better concretes
than specified, protection of the structure, controlled heating and
in-situ testing. Even if parts of an in-situ cast structure, like
massive columns quickly develop maturity, a number of flexural and
prestessed members do not necessarily develop strength before they
are required to accept large percentages of their design loads, especially in cold weather conditions. Also out of safety reasons, such
early loading operations require in-situ testing. If such testing
indicates adequate strength, e.g. 24 hours after casting, it is evident
that the 28-days standard specimen test is only of academic interest.

Both aspects - ensuring durability and achieving early loading -
usually requires the use of quality concretes with a potential higher
strength than needed from only a 28-day design strength point-of-view.
Increasingly the practice has been to use mixes with low w/c-ratios,
which gives 50% to 200% higher strength than required from a purely
static calculation. This trend makes the 28-day cylinder or cube test
even more obsolete as the only means of quality control.

2. IN-SITU TEST SYSTEMS

To evaluate the strength in-place a number of different test systems
are available today. They comprise hammer testing, ultrasonic, resi-

stance to penetration, break-off, pull-off, internal fracture, push-out cylinders, drilled cores, pullouts and maturity.

The methods differ in terms reliability, reproducibility and accuracy of the strength estimate, simplicity, speed of testing, degree of needed planning and the required training of the testing personel as well as the degree of destruction and the costs involved.

The methods are described in British Standard BS 1881 or in ACI report 228.1R-89.

Comparisons between the systems calibrations to standard specimen compression tests have during the years been made by a number of researchers around the world, among who Poulsen (1975), Bellander (1979), Malhotra and Carette (1980), Bungey (1982) and Keiller (1982) are some of the most prominent.

3. PULLOUT TESTING TIMED BY MATURITY MEASUREMENTS

Pullout testing has during the years attracted special attention because of the excellent correlation obtained between pullout force and standard specimen compression tests. The system is furthermore flexible, simple and quick to use and gives only a minimal destruction to the structure.

The system is today standardized by ASTM (ASTM C-900), British Standard (BS 1881, part 7), International Standard (ISO/DIS 8046), Swedish Standard (SS 137238) and by Danish Standard (DS 423.31).

In the 1970´s pullouts were carried out partly using design by Richard (1972) and partly by Kierkegaard-Hansen (1975). Both these systems required special designed inserts to be embedded in the fresh concrete.

Later development by Petersen (1980) allows an insert to be installed at random in the hardening or fully hardened structure. All the experience described and data given in this paper refers to the use of the latter two systems, named LOK-TEST and CAPO-TEST respectively.

Simultaneously with the development of LOK-TEST and CAPO-TEST, Hansen (1981) designed a simple and quick-to-use maturity meter called the COMA-Meter to time the pullouts. This meter is described as well in the following.

Finally a number of testing cases will illustrate the applications of the systems based on the Danish Concrete Code DS 411 from 1984 and the Danish Standard for Evaluation of Test Results DS 423.1 from 1985.

4. LOK-TEST AND CAPO-TEST

4.1 THE TESTING PRINCIPLE

The testing principle is to measure the force by which a 25mm disc or ring placed in a depth of 25mm is pulled out of the concrete through a 55mm inner diameter counterpressure placed on the testing surface, and relate the pullforce to standard cylinder or cube compressive strength by means of a correlation curve.

The LOK-TEST utilizes a disc embedded in the fresh concrete, the CAPO-TEST a ring to be expanded in a hole undercut as illustrated below.

255

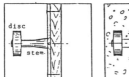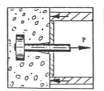

Fig.1 LOK-TEST testing procedure

Fig.2 CAPO-TEST testing procedure

4.2 TEST EQUIPMENT

The LOK-TEST disc is theaded to a stem, both parts are threadlocked and coated. The unit is delivered in various configurations depending on the purpose of testing. The L-42 insert with L-44 steel plate is installed on a removable plug drilled out of the form. This allows the plug to be removed and the insert to be tested prior to form removal. The L-40 insert (shown above in figure 1) is installed on a watertight masonite plate nailed to the form. When the form is removed, the plate breaks in the screw-hole or follows the form. Also floating inserts to be embedded in top surfaces are available.

The test equipment consists of a manually operated hydraulic precision equipment, which ensures a constant and uniform loading rate during pullout. The weight of the equipment is 2.5 kg and it is delivered in a small briefcase with all accessories.

Testing with LOK-TEST usually only takes place exactly to failure. Then the test equipment is unloaded and disconnected from the disc, leaving only a slightly raised 55mm ring visible on the surface as shovn in figure 3. The surface needs usually not to be repaired.

With CAPO-TEST an 18mm hole is drilled perpendicular to a plane surface outside reinforcement disturbance with a watercooled diamond bit. A router undercuts a 25mm hole 10mm deep in a depth of 25mm from the surface. The folded ring is inserted in the hole and expanded on a special tool in the undercut hole. The same hydraulic equipment as used for LOK-TEST is attached to the tool and activated by hand. Pullout takes place until the CAPO-TEST failure occurs and the cone is fully dislodged, as shown in figure 4. The cone hole may be repaired with a polymer modified mortar.

The complete test equipment for CAPO-TEST is contained in two small portable briefcases.

One LOK-TEST usually takes 3-5 minutes to perform, CAPO-TEST 10-15 minutes depending on how well trained the testing personel is.

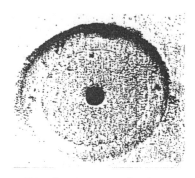 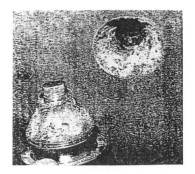

Fig.3 LOK-TEST failure   Fig.4 CAPO-TEST failure

4.3 CORRELATION BETWEEN PULLOUT FORCE AND STANDARD SPECIMEN TESTS

The failure mechanism in a pullout is complicated, and it has not
so far been possible to calculate convincingly a compressive strength
from a given pullout force. Consequently, it has been necessary to
compare pullout forces with the socalled uniaxial compressive strength
measured on standard cylinders or cubes, to obtain a valid correlation.

Such correlations have been performed during the past 15 years in a
large number of series in various countries all over the world.

Three different procedures have been used:

1. 150mm x 300mm (6"x12") standard cylinders have been installed with
   LOK-TEST inserts in the bottom resting against L-44 steel plates.
   At the time of testing the inserts are pulled exactly to failure,
   the cylinder capped and tested in compression. To prevent radial
   cracking during pullout, the cylinder bottom have been clamped
   into a steel ring, especially if the maximum aggregate size has
   been higher than 32mm or the strength measured was higher than
   40 MPa. To prevent radial cracking in-place, a minimum distance
   between inserts and edges or corners has to be 100mm (Petersen, 1984

   Comparative measurements have usually been performed at different
   maturities, usually after 1, 1½, 2½, 3, 5, 7 and 28 days at 20ºC
   for the particular concrete mix to be used. For each of the mentioned ages it has become practice to test 3 cylinders.

   From the data two curves are generated, one with the strength development in dependence of the M_{20} days (maturity days at 20ºC),
   and another comparing the pullout force in kN to cylinder compressive strength in MPa (or PSI).

   This procedure has been predominant in Canada and USA.

2. In Denmark cylinders were cast without inserts embedded. To test
   for pullout strength seperate 200mm cubes were cast, usually with
   two LOK-TEST inserts placed centrally in opposite vertical faces.
   The remaining vertical faces were tested with CAPO-TEST.

   Both sets of specimens have been cast, compacted and cured as identical as possible, and the testing took usually place at 28 M_{20} days
   A wide variaty of concrete parameters were used in the concrete mixes as later described.

257

3. In Sweden, Norway, the Netherlands, England and the Golf Countries
   the normal procedure has been to cast two sets of 150mm cubes, test
   one in compression and the other with LOK-TEST or CAPO-TEST centrally
   placed on opposite vertical faces. In some correlations 200mm cubes
   were used for pullout testing or a steel frame was used to secure
   the 150mm cube in before pullouts, especially if large aggregates
   were used or at higher strength ranges.

   Testing were carried out in some cases at various maturities and
   in other test series at 28 M_{20}-days.

In the test series conducted it was found that the use of lightweight
aggregates influenced the position of the correlation significantly.
For all other parameters investigated the correlation was found to be
stable as indicated in the figures 5 and 6. The parameters investigated were: w/c-ratio, type of cement, curing conditions (water cured,
water and air cured and mistreated), maturity and age, source/form/
size of aggregates (up to 38mm maximum aggregate size), fibers, air-
entrainment, flyash and microsilica content.

The findings from 24 such major correlation series are reported on the
following page from Krenchel & Petersen (1984).

The coefficient of correlation ranged from 0.91 to 0.99, typically
it was 0.95.

Krenchel (1982), Bellander (1983) and Bungey (1983) found similar
correlations for LOK-TEST and CAPO-TEST compared to uniaxial compressive strength.

Later correlations made after 1984 have shown the same general connexion
between pullout force and uniaxial compressive strength as indicated
in the figures 5 and 6.

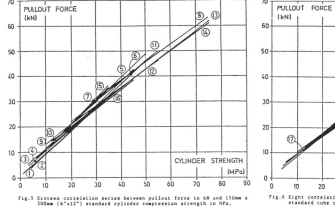

Fig.5 Sixteen correlation series between pullout force in kN and 150mm x
300mm (6"x12") standard cylinder compression strength in MPa.
(3041 pullout- and 2067 cylinder tests)

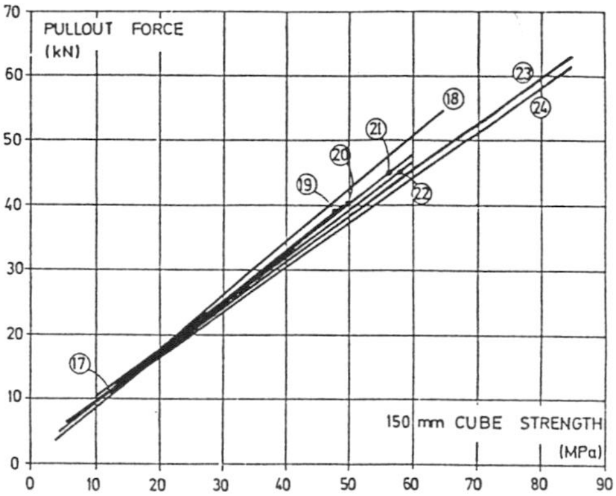

Fig.6 Eight correlation series between pullout force in kN and 150mm (6")
standard cube compressive strength in MPa.
(1087 pullout- and 860 cube tests)

A statistical analyses of the data gives the correlations shown in figures 7 and 8. The deviation between those recommended correlations and
any of the ones indicated in figures 5 and 6 is within 10%.

The precision of the pullout test is shown in figure 7 based on Danish
data for 16mm and 32mm maximum aggregate size and in figure 8 for 18mm
and 38mm maximum particel size calculated from Swedish data.

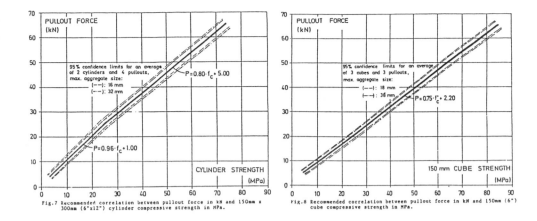

Fig.7 Recommended correlation between pullout force in kN and 150mm x 300mm (6"x12") cylinder compressive strength in MPa.

Fig.8 Recommended correlation between pullout force in kN and 150mm (6") cube compressive strength in MPa.

It should be mentioned that if correlations are carried out on the structure by comparing pullout force with compression tests of drilled out cores, deviations from the above found correlations to standard specimen compression test may be found, depending on the quality of the cover layer compared to the quality of the concrete at the depth where the cores are taken. Also the curing conditions (dry or wet) of the cores before testing is important as well as the diameter size and the height-diameter relationship. One example of such a calibration is given in Rockström and Molin (1989). They compared CAPO-TEST to 100mm x 100mm cores on a number of Swedish bridges. The cores were air cured 3 days in the laboratory before testing.

4.4 VARIATION OF PULLOUT TESTING

The variation of pullouts have been investigated and reported by Bickley and Fasullo (1981), Bickley (1982) and Krenchel and Petersen (1984). The results are summarized below for laboratory testing as well as testing in-situ with the standard deviation "s", coefficient of variation "v" and the number of tests "n" stated.

Laboratory testing:

Table 1. Variability of concrete specimens cast in the laboratory

Specimens	PULLOUT TEST			REF. COMPRESSION TEST		
	s (kN)	v (%)	n —	s (kN)	v (%)	n —
1) Pullout vs. standard cylinder	1.9	7.5	957	1.7	6.4	994
2) Pullout vs. standard cylinder	2.8	9.9	2084	1.6	4.2	1073
3) Pullout vs. standard cube	2.5	6.8	1087	2.4	6.2	860
4) Pullout vs. cores (shotcrete)	1.3	4.5	125	4.4	8.9	36

1) Cylinders used for both pullout force and compression strength determination. Pullout test positioned in cylinder buttom.

2) Pullouts positioned centrally on two opposite vertical faces of 200mm cubes. Standard cylinders cast seperately without embedded inserts for pullout.

3) Pullouts positioned centrally in one vertical face of 150mm cubes, if tested at higher strength levels 200mm cubes were used. Seperate 150mm cubes without pullout inserts were used for compressive strength determination.

4) Pullout force measured on top-part of horisontal shot panels in the laboratory. 80mm x 160mm cores were used for compression tests.

The cylinders and the 150mm cubes were always cast in three layers, each compacted on vibration table. The 200mm cubes were also cast in three layers, but compacted using hand-rodding.

In-situ testing:

Table 2. Variability of concrete elements on-site

| Type of element | PULLOUT TEST | | |
	s (kN)	v (%)	n -
Beams and columns	2.7	7.8	325
Slabs, buttom part	3.1	9.7	4190
Walls and foundations	3.2	10.0	753
Slabs, top part	3.5	12.5	274
Shotcrete, walls	4.0	13.4	150
Dubious structures	4.5	14.7	1001

The catagory "dubious structures" mentioned in table 2 comprises for example construction elements having been subjected to fire, acids, alkali-silica reactions, freezing at an early age, temperature cracking or bad cast, consolidated and cured elements.

4.5 FAILURE MECHANISM INVESTIGATIONS

To explain the excellent correlations found between pullout force and uniaxial compressive strength, the failure mechnism has during the years been investigated by a number of researchers, of who three will be mentioned in the following.

Jensen and Bræstrup (1976) investigated first the failure mechanism by means of Coulumbs criterion for sliding failure. They concluded, that the pullout force in a LOK-TEST is directly proportional to the compressive strength of concrete.

Ottosen (1981) used a non-linear finite element analyses to describe the failure mechanism. His conclusion is,"that large compressive forces run from the disc in a rather narrow band towards the counterpressure, and this constitutes the load-carrying mechanism. Moreover, the failure in a LOK-TEST is caused by the crushing of the concrete and not by cracking. Therefore, the force required to extract the embedded steel disc is directly dependent on the compressive strength of the concrete in question," Ottosen (1981, pp. 602).

Krenchel and Shah (1985) investigated experimentally the crack formatio: in a LOK-TEST. The findings are further referred in Krenchel and Bickley (1987). The conclusion is, see figure 9.

"The internal rupture during this type of test is a multi-stage process where three different stages with different fracture mechanisms can be clearly separated. In the first stage, at a load level of about 30-40% of the ultimate load, tensile cracks are formed starting from the notch formed by the upper edge of the pullout disc. These cracks are running out in the concrete with a very open angle (cone angle between 100° and 135°). The total length of this first crack is typically some 15 to 20mm from the edge of the disc. As a result of this first stage cracking,the material between the top face of the pullout disc and the bottom face of the counter pressure ring is now free, so that straining

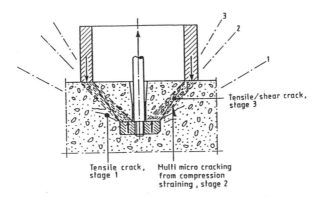

Fig.9. Internal cracking stages during a pullout

in the material is now concentrated and all load is taken up in the truncated zone between the two plane faces. In the <u>second stage</u> of internal rupture a multitude of stable microcracks are formed in the above mentioned truncated zone, the main direction of these cracks running from the top of the disc to the bottom of the ring forming a cone angle of approximately 84°. The formation of this second cracking pattern is very much parallel to the formation of more and more vertical microcracks inside a concrete cylinder or prism during ordinary uniaxial compressive tests. Development of the acoustic emission activity during this second stage of the test also follows an exponential function quite parallel to the AE-development in ordinary uniaxial compressive tests. If more and more oil is pumped into the pullout jack, even after the load has stabilized at the peak point, then the <u>third stage</u> of internal rupture occurs by the formation of a tensile/shear crack all the way around, running from the outside edge of the disc to the inside edge of the counter pressure ring and forming the final pullout cone with a cone angle of about 62°", Krenchel and Bickley (1987, pp. 165-166).

Krenchel concludes (1987, pp. 167): "As the micro cracking of stage number two is responsible for and directly related to the ultimate load in this testing procedure, it seems quite logical that such close correlations with the concrete <u>compressive</u> strength is always obtained."

4.6 RECOMMENDED CORRELATION PROCEDURE

For quality control of the cover layer, it has become practice in Scandinavia to use the recommended general correlations as illustated in figure 7 and figure 8, for LOK-TEST as well as for CAPO-TEST.

Also the relationships are used for timing of early loading of in-situ cast structures in England, in Canada, USA and the Benelux.

However, it is still a good idea to correlate ones own particular concrete mix in a pre-testing program prior to starting up a major project, and compare the results with the data in this paper. The following rules should be observed when performing such a correlation:

a. The correlation ought to be conducted over a span of at least 35 MPa strength range. Otherwise the slope of the curve will not be determined with sufficient precision.

b. One correlation should consist of minimum 7 clusters of corresponding observations, equally distributed in the strength range decided upon.

c. Each cluster should consist of the following minimum number of specimens, alternatively:

       c-1. One set of three 150mm x 300mm cylinders with LOK-TEST L-42 inserts/L-44 steel plates in the bottom or

       c-2. One set of two 150mm x 300mm cylinders and two 200mm cubes with two LOK-TEST L-42 inserts in each, placed centrally on two opposite vertical faces. The remaining two faces may be used for CAPO-TEST or

       c-3. One set of three 150mm cubes with LOK-TEST L-42 inserts placed centrally in one vertical face or

       c-4. One set of two 150mm cubes and two 200mm cubes with two LOK-TEST L-42 inserts in each, placed centrally on two opposite vertical faces. The remaining two faces may be used for CAPO-TEST.

d. The specimens should be cast, compacted (preferably on vibration table) and cured in water at $20^{\circ}C$, identically.

e. The seven sets of specimens may be tested at the following maturities: 1, $1\frac{1}{2}$, $2\frac{1}{2}$, 4, 7, 14 and 28 M_{20} days, one set at each maturity.

f. If the specimens for compression tests contain a LOK-TEST L-42 insert, the testing is performed by first pulling the insert exactly to failure and no further, record the pullout force and then compress the specimen with the face containing the insert resting against one of the test machine compression plates, e.g. after capping (of cylinders).

g. From the test data two curves are generated,

       - one comparing pullout force to compressive strength using liniear regression analyses and calculation of the coefficient of variation and the coefficient of correlation,

       - another showing the strength development in relation to maturity days (M_{20}-days).

The first relationship is compared with the general recommended correlations in this paper (figure 7 or figure 8). The second is used for timing of pullouts with maturity before early loading operations. If e.g. the required strength is supposed to be present in the structure after 1.5 M_{20}-days, the COMA-Meter described in the following has to show 1.5 M_{20}-days before the pullout testing takes place. By using this procedure, not only the potential strength strength of the structure is checked, but also the actual strength before critical early loading operations is measured in-place.

5. COMA-Meter

The COMA-Meter (COncrete MAturity-Meter) was developed to measure accurate, simple and quick the maturity of in-place concrete without troublesome daily temperature registrations or sensible electronic maturity computers involved.

The meter follows the Arrhenius equation for maturity

$$M_{20} = \int_o^t exp\left(\frac{E}{R}\left(\frac{1}{293} - \frac{1}{T(t)}\right)\right) dt \qquad (1)$$

with an accuracy as indicated in table 3. E is the activation energy chosen to 40 kJ/mol, R the gas constant (kJ/mol ^oK) and T the temperature (^oK).

Table 3. The accuracy of the COMA-Meter compared with maturity after the Arrhenius equation (M_{20} after one day).

Temp. °C	Arrhenius (M_{20})	COMA-Meter (M_{20})
-5	0.07	0.17
0	0.15	0.26
5	0.29	0.37
10	0.50	0.50
15	0.75	0.71
20	1.00	1.00
25	1.26	1.17
30	1.57	1.54
35	1.95	2.00
40	2.41	2.57
50	3.58	4.13

The COMA-Meter, shown in figure 10, operates by breaking a closed capillary tube filled with a special liquid at zero, when the concrete has been cast in-place. The tube is placed on a scale which is inserted a container. The meter is pressed into the concrete and the temperature inside the meter will stabilize with the concretes after 10 minutes. The liquid in the capillary will start evaporating and the level will on the scale indicate the number of days the concrete is old at 20°C. The maturity may be read by unscrewing the scale from the container at any time. Afterwards the scale is reinserted and the integration of time and temperature will continue.

Fig.10. The COMA-Meter

Two measuring ranges are available, either 0 to 5.5 M_{20} days or 0 to 14 M_{20} days.

The COMA-Meter is described in detail in Hansen (1981).

6. TESTING CASES

Pullout testing is today mainly used for quality control of new
structures or elements, for timing of early loading operations or
for evaluation of the residual strength of old concrete structures.

In the following a number of testing cases will illustrate the applications within the first two catagories. Evaluation of the results are
based on the statistical evaluation procedure outlined in the Danish
Concrete Code DS 411 from 1984 and the Danish Standard for Evaluation
of Test Results DS 423.1 from 1985. A brief summary of the evaluation
procedure is given in appendix 1.

Only "variable control" will be mentioned leaving out "alternative
control",since pullout testing in practice almost always is performed
to the maximum peak-load and not only to a required level. This is
partly caused by the testing personels curiosity "to learn what the
concrete strength really is", and partly that the CAPO-TEST cones
always have to be pulled out. However, if LOK-TEST is only used, the
simple "alternative control"-procedure may as well be used.

6.1 QUALITY ASSURANCE BY CONTROL IN TOTAL

A slab in an office building suffered from severe cracking in a 10 m^3
pour. The slab was one week old.

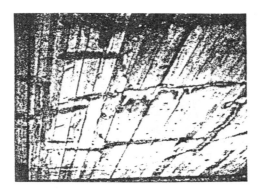

Fig.11. Fully cracked slab from below

The consulting engineer ordered the slab to be tested in-situ to
make sure the specified 28-day strength $f_c=20$ MPa was present in-
place.

The pour consisted of concrete from three truckloads. It was decided
to test each truckload with three CAPO-TEST from the top of the slab.

The maturity of the concrete was not measured by the contractor. The
daily temperatures was procured from the Institute of Meteorology and
the maturity was estimated to be 35 M_{20} days at the time of testing on
the assumption of a slightly higher temperature two days after casting.
Since the specified strength 20 MPa was supposed to be present 28 M_{20}
days after casting, the measured in-place strength was corrected
according to the maturity curve in appendix 2 for the cement type
used, rapid cement.

The testing was carried out as illustrated in figure 12 after the reinforcement was located and the rough surface ground smooth and plane.

Fig.12. Completed CAPO-TEST

The test results are exhibited in table 4.

Table 4. Test data from in-situ testing with CAPO-TEST on cracked slab

CAPO-TEST No.	CAPO-Strength (35 M_{20}) (kN)	Cyl.strength (35 M_{20}) (MPa)	Corr.factor ɤ	Cyl.strength (28 M_{20}) (MPa)	Observation Average (MPa)
1	21,9	21,8	0,95	20,8	
2	18,3	18,0	0,95	17,1	18,2
3	17,8	17.5	0,95	16.7	
4	17,2	16,9	0,95	16,1	
5	16,7	16,4	0,95	15,6	17,5
6	21.9	21.8	0,95	20.8	
7	22,4	22,3	0,95	21,2	
8	23,9	23,9	0,95	22,8	21,7
9	22,4	22.3	0,95	21,2	

As no of the observation were lower than 80% of the 20 MPa, the slab was accepted to be in compliance with DS 411 and DS 423.23.

The testing lasted 2½ hours. The consulting engineer received the test report the following morning.

The ready mix plant supplying the concrete reported two days later the cylinder results. The average was 22.5 MPa with almost no variation involved.

6.2 QUALITY ASSURANCE BY SAMPLE TESTING

In the specifications for a rainwater reservoir it was stated that the in-place concrete had to be tested with LOK-TEST to make sure the construction could resist attacks from chlorides specially during the winter period, where chlorides would be present in the water from de-icing salts.

The reservoir contained 34 truckloads of concrete in the bottom slab, the walls and the top slab. Eight of the truckloads were specified to be tested at random.

The testing was carried out at 15 M_{20} days (test no. 1-6) and at 35 M_{20} days (test no.7-8) indicated by the COMA-Meter.

Fig.13.Testing of bottom slab with LOK-TEST and COMA-Meter.

Fig.14. Testing of wall with LOK-TEST

The results are indicated in table 5.

Table 5. Test data from in-situ testing with LOK-TEST of rainwater reservoir

LOK-TEST No.	LOK-Strength (kN)	Cyl.strength (MPa)	Maturity (M_{20})	Corr.factor γ	Cyl.strength (28 M_{20}) (MPa)
1	35,2	37,8	15	1,05	39,8
2	38,8	42,3	15	1,05	44,5
3	27,9	28,6	15	1,05	30,1
4	36,2	39,0	15	1,05	41,1
5	34,3	36,6	15	1,05	38,5
6	35,2	37,8	15	1,05	39,8
7	36,3	39,1	32	0,96	37,5
8	38,3	41,6	32	0,96	39,9

$$\bar{f}_c = 38,9$$

266

It should be noticed that test number 3 was a floating L-49 LOK-TEST inserts with a major air accumulation below the test surface caused by wrong installation by the contractor. He could have chosen to conduct a CAPO-TEST instead, but wanted to stick to the low result as it had no influence on the acceptability of the concrete.

The specified strength in-place was 35 MPa. With an undocumented coeffecient of variation of v=0.16 and the number of tests ‾n=8, the k_n factor is 1.27. This means that the mean strength in-place at least has to be

$$\bar{f}_c \geq 0.8 \cdot k_n \cdot f_c = 0.8 \cdot 1.27 \cdot 35 = 35.6 \text{ MPa}$$

As the measured mean strength was 38.9 MPa the structure was accepted to be in compliance with DS 411 and DS 423.1.

The testing took one hour and the contractor received the testing report the following morning.

The laboratory cylinder results averaged 39.6 MPa.

6.3 EARLY LOADING BY CONTROL IN TOTAL

A bridge box girder had to be tensioned in segments. The following example is from one of the segments being cast and tensioned in cold weather conditions.

One segment contained 10 truckloads. In the truckload close to the anchor zone four LOK-TEST inserts were installed, while the remaining contained one insert in each. Type L-40 and L-42 inserts were used.

COMA-Meters were installed through the formwork or from the top of the slab, one close to each LOK-TEST inserts.

The specified strength before tensioning was f_c=30 MPa. As every truckload was tested, the type of control was "control in total" and consequently all observations had to be higher than 80% of 30 MPa, before tensioning could take place.

The segment was covered and insulated after casting. Heating from inside of the box girder was applied. The outdore temperature was -5°C with very strong windy conditions.

Three extra installed LOK-TEST inserts were tested after 3.0 M_{20} days, averaging 28.1 MPa.

Fig.15. LOK-TEST inserts type L-40 installed
close to the anchor zone ready for test.

Fig. 16. Flying form system

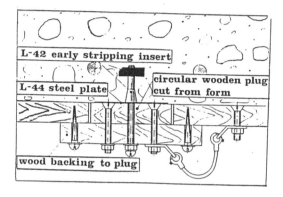

Fig. 17. Principal installment of
L-42 inserts with L-44 steel plate
through slab forms.

Fig.18. LOK-TEST L-42 insert with
L-44 steelplate installed in slab
formwork together with one COMA-
Meter (seen from casting side).

As the COMA-Meters close to the LOK-TEST inserts all showed minimum 3.0 M_{20} days, it was decided to test the 14 installed LOK-TEST inserts.

Table 6. Test data from in-situ testing with LOK-TEST of segmental box girder.

LOK-TEST. No.	LOK-Strength (kN)	Cyl. strength (MPa)	Observation (MPa)
1	29.7	30.9	30.9
2	30.3	31.6	31.6
3	28.1	28.9	28.9
4	27.0	27.5	27.5
5	27.0	27.5	27.5
6	26.8	27.3	27.3
7	27.5	28.1	28.1
8	27.2	27.7	27.7
9	28.3	29.1	29.1
10	31.8	33.5	
11	30.6	32.0	
12	28.7	29.6	30.0
13	27.8	28.5	
14	26.2	26.5	

Since all observations in-place were higher than the required strength 24 MPa, the consulting engineer accepted the segment to be tensioned, a operation which took place right afterwards.

The testing took 65 minutes.

At the time of testing with pullouts, the contractor tested six cylinders cast on-site and cured on the bridge slab under wet blankets covered with insulation mats and a pile of sand. The average strength of the cylinders was 37.5 MPa. The maturity of the cylinders was not measured.

6.4 EARLY LOADING BY SAMPLE TESTING

Each floor in a 33 storey office building contained 520 m^3 of concrete. Every pour consisted of a 75 m^3 casting. The data reported in the following is taken from one such typical pour.

Ten LOK-TEST inserts type L-42 were installed in the flying form system equally distributed througout the casting and placed in the middle between the supporting columns.

Two COMA-Meters were mounted in the formwork, one at the beginning of the pour and one at the end.

As one truckload contained 4.5 m^3 of concrete, every truckload would not be tested. Consequently the type of control was control by sample testing.

The specified strength was 28 MPa (28-days). The required average strength before form removal was 21 MPa. The contractor decided to use a 35 MPa concrete to speed up the early age strength gain.

The in-place strength of the pour had to average as a minimum:

$$\bar{f}_c \geq 0.8 \cdot k_n \cdot f_c = 0.8 \cdot 1.37 \cdot 21 = 23.0 \text{ MPa}$$

with a k_n factor of 1.37 calculated from appendix 1 based on an un-documented coefficient of variation $v=0.19$ and the number of test $n=10$.

From earlier experience with the concrete mix chosen it was known, that the strength was supposed to be present in the slab after 1.8 M_{20} days if the real-crete had a potential strength as intended and the pumping, the casting, the compaction and the curing was carried out as normally.

The testing with pullouts took place after the COMA-Meter placed at the end of the pour showed 1.8 M_{20} days. The result are given in table 7.

Table 7. Test data from in-situ testing with LOK-TEST of slab prior to early form removal

LOK-TEST No.	LOK-Strength (kN)	Cyl. Strength (MPa)	Maturity (M_{20})
1	25.2	25.2	2.0
2	24.4	24.4	-
3	22.2	22.1	-
4	25.8	25.8	-
5	21.0	20.8	-
6	25.2	25.2	-
7	26.5	26.9	-
8	27.0	27.5	-
9	24.5	24.5	-
10	25.8	25.8	1.8

$$f_c = 24.8 \text{ MPa}$$

As it will be seen, the average strength in-place at the time of testing was 24.8 MPa which surpassed the required strength of 23.0 MPa. The flying form system was removed after acceptance of the test results and the calculations by the consulting engineer. Immidiately afterwards the slab was supported by shores to prevent deflection.

The pullout testing lasted 1½ hour and since the inserts cast-in only were pulled exactly to failure, no repair was needed. The preliminary cutting of portholes in the formwork took 3 hours.

7. DISCUSSION

Usually it is the owner of the structure who decides to use pullout testing for quality assurance. Similar it is the contractor who is interested in performing early loading operations. Conflicts of interest are not unusual in such situations.

To overcome such conflicts it is a very good idea to arrange a short one day course where the theoretical background, the systems operation, the equipments, the practical testing, the sources of error, the flow of information, the decision authority and the costs/benefits are illustrated and discussed. All the parties in the contract should participate at management level. It is also a very good idea prior to a project involving pullout testing to train the testing personel in correct installment of inserts and meters, in the practical testing on-site and in maintenance of the equipment. This may be done on a one to two day course, preferably at the site.

270

The world of concrete testing is very conservative. Deeply rooted habits and ideas are not changed over-nigth, not to speak of building codes, standards of testing and evaluation procedures.

But it is the authors experience that if the users understand the basic purpose of the systems illustrated, the functioning of the test equipments and the proper required maintenance, and after they have realized how reliable, simple and quick the systems are, the usual reaction is to ask why this type of testing has not substituted cylinders or cubes long time ago.

The successor exist, she is young - only 20 years - experienced, quick and reliable, and then the structure itself is tested without guesswork or assumptions involved.

8. REFERENCES

Bellander, U. (1979): "Quality control of concrete structures", RILEM, Part 1, 2 & Proceedings, Swedish Cement and Concrete Research Institute, Stockholm, June 17-21 1979.

Bellander, U. (1983): "Quality Control of Concrete Structures", Nordisk Betong, No.3-4, 1983.

Bickley, J.A. & Fasullo, S. (1981): "Analyses of Pullout Test Data from Construction Sites", Transportation Research Board, Washington, D.C., 1981.

Bickley, J.A. (1982): "The Variability of Pullout Tests and In-Place Concrete Strength", Concrete International, Vol.4, No.4, April, 1982.

Bungey, J.H. (1982): "Testing of Concrete in Structures", Surrey University Press, Glasgow, 1982.

Bungey, J.A. (1983): "An Appraisal of Pullout Methods for Testing Concrete", Proceedings of the International Conference on Non-Destructive Testing, Engineering Technics Press, London, 1983.

Hansen, A.J. (1981): "COMA-Meter - The Mini Maturity Meter", Nordisk Betong, No.4, 1981.

Jensen, B.C. & Bræstrup, M.W. (1976): " LOK-TEST Determine the Compressive Strength of Concrete", Nordisk Betong, No.2, 1976.

Keiller, A.P. (1982): "Preliminary Investigation of Test Methods for the Assessment of Strength of In-Situ Concrete", Technical Report No. 42.551, Cement and Concrete Association, Wexham Springs, 1982.

Kierkegaard-Hansen, P. (1975): "LOK-Strength", Nordisk Betong, No.3, 1975.

Krenchel, H. (1982): "LOK-Styrkeprøvning og CAPO-Styrkeprøvning af Betons Trykstyrke", Department of Structural Engineering, Technical University of Denmark, Serie I, No.71, 1982.

Krenchel H. & Petersen, C.G. (1984): "In-Situ Pullout Testing with LOK-TEST, Ten Years Experience", Proceedings, International Conference on In-Situ/NDT Testing of Concrete, Canmet, ACI, CSCE & NBS, Ottawa, October 1984.

Krenchel, H. & Shah, S.P. (1985): "Fracture Analyses of the Pullout Test", RILEM, Materials and Structures, Research and Testing, No. 108, November-December, 1985.

Krenchel, H. & Bickley, J.A. (1987): "Pullout Testing of Concrete", Nordic Concrete Research, The Nordic Concrete Federation, No.6, 1987.

Malhotra, V.M. & Carette, G. (1980): "Comparison of Pullout Strength of Concrete with Compression Strength of Cylinders and Cores, Pulse Velocity and Rebound Number", ACI Journal, Proceedings V.77, No.3, May-June, 1980.

Ottosen, N.S. (1981): "Nonlinear Finite Element Analyses of Pullout Test", Proceedings, ASCE, V.107, ST4, 1981.

Petersen, C.G. (1980): "CAPO-TEST", Nordisk Betong, No.5-6, 1980.

Petersen, C.G. (1984): "LOK-TEST and CAPO-TEST Development and Their Applications", Proceedings, Institution of Civil Engineers, Part I, 76, May, 1984.

Richard, W. (1972): "Pullout Strength Tests of Concrete", Research Paper, ACI, Annual Meeting, Dallas, Texas, USA, 1972.

Poulsen, P.E. (1975): "Vurdering af Betons Styrke med Konstruktions-prøvning, LOK-Prøvning, Udborede Kerner, Støbte Cylindre, Ultralyd-Betonhammer Kombineret, Betonhammer og Ultralyd", Danmarks Ingeniør-akademi, Bygningsafdelingen, RAM-Rapport Nr. 75:64, Del 1-2, November 1975.

Rockström, J. & Molin, C. (1989): "Begränsad Studie av Sambandet mellam Hållfasthet mätt met Studshammera, CAPO-TEST resp. Utborrade Betongcylindrar", Statens Provningsanstalt, Byggnadsteknik, Arbetsrapport Nr.17, Stockholm, 1989.

APPENDIX 1

EVALUATION OF THE STRENGTH OF CONCRETE IN-PLACE BY MEANS OF THE
DANISH CONCRETE CODE DS 411 AND THE DANISH STANDARD FOR EVALUATION
OF TEST RESULTS DS 423.1 WITH LOK-TEST AND CAPO-TEST

It is described how the in-situ concrete strength may be evaluated.
The review is based on the requirements in the Danish Concrete Code
DS 411 from 1984 and the Standard for Evaluation of Testresults DS
423.1 from 1985 together with practical experience.

1. APPLICATIONS
1.1 PULLOUT TESTING WITH LOK-TEST
   Pullout testing with LOK-TEST may be used for

- Quality assurance of in-situ strength
  according to the requirements outlined
  in the concrete specification for a
  given control section.

- Timing of early loading for concrete
  elements like early form stripping,
  tensioning or cutting of strands.

1.2 PULLOUT TESTING WITH CAPO-TEST
   Pullout testing with CAPO-TEST is used for

- Supplementary testing needed if the concrete
  a control section has been rejected by LOK-
  TEST

- Testing of a control section where no LOK-
  TEST inserts have been installed

- Supplementary testing if LOK-TEST inserts
  have been installed erroneously

- Testing of a control section where it is
  complicated or troublesome to install LOK-
  TEST inserts, like in slipformed surfaces,
  trowelled top surfaces, if the concrete is
  dry or if gunite or vacuum-concrete has to
  be tested

2. CHOISE OF CONTROL SECTION
2.1 WHO DECIDES THE SIZE OF A CONTROL SECTION?
   When the purpose is to quality assure a structure, the size of the
   control section is decided by the design engineer.
   For early loading of elements, the sixe of a control section is
   chosen by the contractor together with the supervisor. A control
   section may typically be one pour.
   When decided, the contractor may always sub-divide a control section, if he finds it necessary. On the other hand, the contractor
   is not allowed to add up several single control sections into one
   and evaluate them together.
2.2 BATCHES
   A control section will consist of a number of batches. For a control section of concrete one batch is the uniform amount of con-

crete either from one mixing charge at the mixing plant or from one
truckload being supplied to the site.

2.3 THE SIZE OF A CONTROL SECTION

The delimitation of a control section is decided from the following
criterions:
- The material property of the concrete and other characteristics
  is aimed to be uniform in the scetion.
- A control section must not contain more than 200 batches.
- A control section must be cast in one continous working operation. No change in the production or transition to other material
  constituents, e.g. admixtures, is allowed.
- The size of the control section has to be evaluated in relation to
  the consequence of rejection. However, the contractor may subdivede
  a control section in this case.

3. INSTALLMENT OF LOK-TEST/CAPO-TEST INSERTS

The following rules are applied for positioning of pullout inserts:
- At least 6 tests must be conducted in a control section, randomely
  distributed.
- Inserts placed in one batch constitutes one test.
- One test may consist of one LOK-TEST/CAPO-TEST insert, but it is
  recommended to use minimum 2 inserts per test.
- Whatever the number of minimum inserts is chosen, all inserts in
  one test are placed at the same horisontal level.
- Inserts are placed with a 100 mm mimimum distance to edges or cor-
  ners and at least 30 mm from reinforcement.
- The mutual distance between inserts in one test must be maximum 300
  mm and minimum 200 mm.
- Foreign bodies are not allowed in the failure zone.

4. OBSERVATIONS

One observation is defined as the average of the pullout forces of
one tests single test results. One test may be one LOK-TEST or CAPO-
TEST, but it is recommended that a minimum of two are carried out.
If several tests are performed within each placed batch, the average
constitute the observation.

5. MINIMUM SAMPLE SIZE

If the number of observations are less than the number of batches in
the control section the number of observations are called the sample
size of the control section.
The required sample size depends on the number of batches in a control
section as follows

No of batches	Minimum sample size	No of batches	Minimum sample size
1-3	control in total	88-96	14
4-15	3	97-105	15
16-21	4	106-114	16
22-27	5	115-123	17
28-33	6	124-132	18
34-40	7	133-141	19
41-47	8	142-150	20
48-54	9	151-160	21
55-62	10	161-170	22
63-70	11	171-180	23
71-78	12	181-190	24
79-87	13	191-200	25

6. MATURITY AT THE TIME OF PULLOUT TESTING

6.1 REQUIREMENT
The specified strength of the concrete in a control section has to be met 28 M_{20} days after casting if the purpose is to quality assure a structure.
For timing of early loading of a structure, the required number of maturity days before pullout testing is executed is found by pre-testing of the concrete mix in the laboratory.

6.2 MEASUREMENT OF THE MATURITY
The maturity of the concrete is measured in a depth of 30-40 mm from the surface and close to the pullout test, the recommended distance being 250 mm. If the concrete of one placed batch has the same maurity it is allowed to place one maturity meter or sensor in the casting.

6.3 CORRECTION OF PULLOUT RESULTS FOR QUALITY ASSURANCE
If the concrete maturity deviates maximum plus minus 20% from the required 28 M_{20} days it is allowed to correct the pullout results to the required 28 M_{20} days strength, if a documented relationship exists between the strength and maturity for the particular cement type used.
Such documented relationships for Aalborg Portlands three different types of cement are outlined in appendix 2.

7. NORMAL REQUIREMENTS
The required concrete strength in a control section is stated in the specification for the concrete structure in question. The minimum required strength depends of the socalled "class of protection" as given in Danish Standard DS 411.

7.1 CHARACTERISTIC STRENGTH
The characteristic strength f_c by testing cylinders (DS 423.23) has to meet the specifications.
This requirement may be superseded if durability demands a required w/c-ratio leading to a higher strength than required by Danish Standard DS 411. It is not unusual that higher requirements are stated in the concrete specification than outlined by the minimum requirements in DS 411. It may be an economic advantage to make use of the higher strength required because of an aggressive environment.
A similar circumstance may exist if the purpose is to time early loading operations and a concrete with high early strength gain is selected exceeding the specified 28-day cylinder strength.

If the static calculations demands a higher strength than required by the minimal requirements in DS 411, this should be stated clearly in the specification for the concrete structure.

7.2 TWO-SIDED CONTROL
Where a homogeneous concrete strength is required, the specification may contain a required upper and lower limit whithin of which range the strength values have to be situated. Two-sided control of the concrete strength will usually require the application of "alternative control" illustrated later, since "variable control" is to complicated for this application.

7.3 IN-SITU TESTING
The Danish Standard DS 411 states, that a strength requirement is met if the structure is tested in-situ, if 80% of the cylinder strengt is achieved.

7. CONVERSION EQUATIONS

The following equations may be regarded as general relationships for concrete with a maximum aggregate size less than 38 mm. The relationships applies for normal concrete, only not for the use of lightweight aggregates.

7.1 PULLOUT FORCE TO 150mm x 300mm CYLINDER COMPRESSIVE STRENGTH

The conversion equation between pullout force P in kN and 150mm x 300mm cylinder compressive strength f_c in MPa is recommended as:

$$P = 0.96 \cdot f_c + 1.00 \qquad \text{for } 2 \text{ kN} \leq P \leq 25 \text{ kN}$$

$$P = 0.80 \cdot f_c + 5.00 \qquad \text{for } 25 \text{ kN} \leq P \leq 60 \text{ kN}$$

7.2 PULLOUT FORCE TO 150mm CUBE COMPRESSIVE STRENGTH

The conversion equation between pullout force P in kN and 150mm cube compressive strength f_c' is recommended as:

$$P = 0.75 \cdot f_c' + 2.20 \qquad \text{for } 5 \text{ kN} \leq P \leq 60 \text{ kN}$$

8. EVALATION BY CONTROL IN TOTAL

8.1 DEFINITION

Control in total is said to be performed, when every single batch in a control section is tested. The observation is average of the pullout forces of LOK-TESTS or CAPO-TESTS made within the batch placed.

8.2 ACCEPTANCE CRITERION

The concrete of a placed batch is accepted if the observation is minimum 80% of the specified strength as measured on standard cylinders or cubes. A control section is only accepted if all batches are accepted accordingly.

8.3 WHEN TO CHOOSE CONTROL IN TOTAL

Control in total has to chosen if the number of batches in a control section is 3 or less.

The contractor may choose control in total instead of control by sample testing as later described.

The design engineer may choose control in total in a control section where a great degree of safety is required. In this case it should be clearly stated in the specification.

9. EVALUATION BY CONTROL BY SAMPLE TESTING

9.1 DEFINITION

Control by sample testing is said to be performed when the number of batches tested is less than the number of batches placed in the control section, but larger than or equal to the number of minimum sample size as mentioned in clause 5.

The following rules are applied:

- The sample size n (number of observations) and the average $\bar{f}_c$ of the n observations is calculated using the conversion equations in clause 7.

- From the following expression the un-documented coefficient of variation v is calculated:

$$v = 0.22 \qquad \text{for } 5 \leq f_c \leq 10 \text{ MPa}$$
$$v = 0.23 - 0.002 \cdot f_c \qquad \text{for } 15 \leq f_c \leq 35 \text{ MPa}$$
$$v = 0.14 \qquad \text{for } 40 \leq f_c \leq 50 \text{ MPa}$$

or found from the following table:

f_c (MPa)	5	10	15	20	25	30	35	40	45	50
v (%)	0.22	0.22	0.20	0.19	0.18	0.17	0.16	0.14	0.14	0.14

- The factor k_n is calculated from:

$$k_n = \exp\left(\left(2.28 + \frac{1}{\sqrt{n}}\right) \cdot v - 0.1875\right)$$

or found from the following table:

k_n factor for:	n				
v (%)	3	6	9	15	30
0.06	0.98	0.97	0.97	0.97	0.96
0.08	1.04	1.03	1.02	1.02	1.01
0.10	1.10	1.08	1.08	1.07	1.06
0.12	1.17	1.14	1.13	1.12	1.11
0.14	1.24	1.21	1.20	1.18	1.17
0.15	1.27	1.24	1.23	1.21	1.20
0.16	1.31	1.27	1.26	1.24	1.23
0.18	1.39	1.34	1.33	1.31	1.29
0.20	1.47	1.42	1.40	1.38	1.36
0.20	1.55	1.50	1.47	1.45	1.43

9.2 ACCEPTANCE CRITERION

The control section is accepted if

$$\bar{f}_c \geq 0.8 \cdot k_n \cdot f_c$$

otherwise the section is rejected.

10. EVALUATION BY ALTERNATIVE CONTROL
10.1 DEFINITION

Alternative control is said to be performed if the testing with pullout testing is generating a yes/no-statement for the observed properties or characteristics.

Alternative control may be used if

- The pullout insert is only loaded to a certain pullforce. The observation is a "yes" if the insert is not pulled to failure, otherwise it is a "no".

- If two-sided control is needed, where the pullout forces have to be placed within a declared and accepted interval, e.g. from 40 kN to 50 kN.

10.2 ACCEPTANCE CRITERION BY ALTERNATIVE CONTROL

By sample testing using alternative control a yes/no statement is given for each observation. The number of no-statements in relation to the sample size decides if the section is accepted or rejected.

The decision rule appears from the following table, where the sample size has to be larger than or equal to the one assigned in clause 5.

Sample size	Max.No. of defects	Sample size	Max.No. of defects
3-12	0	65-79	6
13-19	1	80-94	7
20-29	2	95-109	8
30-39	3	110-124	9
40-49	4	125-145	10
50-64	5		

If the control section is tested by control in total, no observations are allowed to be defect.

11. EKSTREME VALUES
11.1 REASONS FOR EXTREME VALUES
The following reasons may be causing extreme pullout test values:

- To week concrete caused by mistakes in the production or in the curing procedure
- Incorrect installed LOK-TEST or CAPO-TEST inserts
- A failure cone of the pullout test which is not acceptable
- Foreign bodies in the failure cone

Detailed instruction in the evaluation of the correctness of a pullout test is given in the instruction manual supplied with the equipment.

11.2 SUPPLEMENTARY TESTING
An extreme low pullout force is allowed to be excluded in the evaluation if a mistake in the testing procedure is documented. Supplementary testing with CAPO-TEST may replace such test results.

11.3 PETROGRAPHIC ANALYSES
If the supplementary testing with CAPO-TEST consistantly indicates to low pullout forces, the reason has to be established by petrographical analyses of the concrete in question.

12. REPORT
The report has to contain all the informations required in Danish Standard DS 423.31 covering pullout testing, e.g. all relevant concrete information, position of observations, type of evaluation used, test equipment identification, testing institution and testing personel together with relevant standards and procedures.

APPENDIX 2

Maturity functions for Aalborg Portlands three types of cement

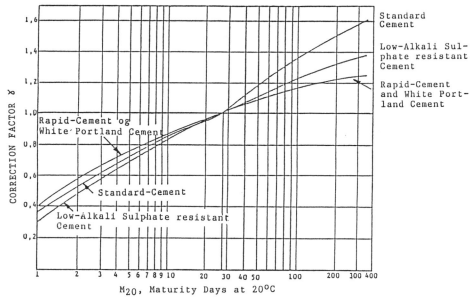

The correction factor γ as a function of the concrete
maturity (M_{20}, maturity days at 20°C).

Reference: CtO, Aalborg Portland, Beton-Teknik 01/22/85   p.3

22 MONITORING CONCRETE QUALITY OF A NUCLEAR CONTAINMENT STRUCTURE

P.K. MUKHERJEE
Civil Research Department, Research Division,
Ontario Hydro, Canada

Abstract
Extensive testing for the quality of the materials and concrete was carried out prior to and during the construction of a nuclear containment structure. The specification required use of low-heat cement and a concrete strength of 35 MPa at 90 days and, in some instances, 30 MPa at 28 days. This paper includes some of the test data on the materials and concrete. The properties of the in-place hardened concrete were determined by making test specimens after the concrete was pumped and vibrated into the formwork. The paper also includes some data on long-term creep and shrinkage tests carried out under laboratory conditions. The results of these various tests will be used as a baseline data for future monitoring of the performance of the structure.

1 Introduction

Concrete construction practices in the Nuclear Industry are well recognized for their excellence. However, authorities [1,2] have emphasized, time and again, the need for a well-organized inspection and testing program during the construction phase of a plant. This paper describes some of the quality control and assurance tests that have been carried out on the concrete and the related materials during the construction of a nuclear power plant at Darlington, Ontario, Canada. Also, the in-place quality of the concrete has been monitored to evaluate its long-term durability.

For brevity, only the results of tests carried out between the period of mid-1983 and the beginning of 1986 have been discussed here. During that period one of the major pressure containment structures, namely the Vacuum Building, was built. Therefore, the concrete and the related materials are representative of this building.

A brief description of this building is given below.

2 Vacuum building

The Vacuum Building is one of the important safety features of Ontario Hydro's CANDU (Canada Deuterium Uranium) reactors [3]. The building is a large cylindrical concrete structure consisting of a main chamber, a water storage tank, and an upper chamber (Fig. 1). The building is connected to the reactor building via pressure relief ducts. The main chamber is made of a concrete perimeter wall, a domed roof and a floor slab on a rock base. The main chamber is kept at a negative pressure (90 to 95 kPa) at all times. The perimeter wall and the dome are post-tensioned. The water tank is supported by a cylindrical shaft and a network of internal columns and beams.

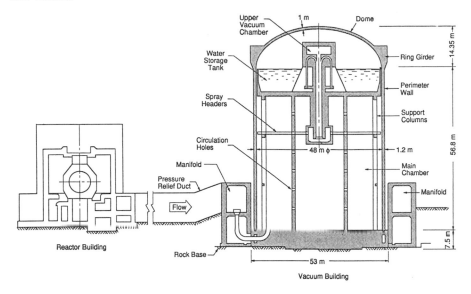

Fig. 1. Schematic view of vacuum bldg. Connected with reactor bldg.

The function of the vacuum building is to receive, by suction, and contain all the gases and vapours that may be produced due to the unlikely event of a loss of coolant accident in the reactor. A dousing system is automatically activated due to the pressure differential created between the upper chamber and the main chamber. The large volume of water in the tank is released via spray headers and condenses the gases that enter the main chamber.

3 Materials

The following materials were used in the concrete.

Cement: Low-heat cement, meeting the specification requirements of the Canadian Standard, CAN3-A5, Type 40.

Samples of the cement were obtained periodically from the job site, as well as from the cement plant. These were tested for their chemical analysis and physical properties.

Table 1 gives a summary of these properties. A total of 41 samples were tested. Fig. 2 shows the running average of five tests for compressive strengths of 50-mm mortar cubes at various ages. Prompt action was taken by the manufacturer to correct the cement quality, whenever a trend for low strength was noted. This is evident in the plot for 28-day and 91-day strength test results.

Table 1. Statistical analysis of cement tests results for chemical analysis and physical properties

Chemical Analysis (%)	No. of Tests	Average	Standard Deviation	Range Maximum	Minimum
loss on ignition	19	1.02	0.1	1.24	0.89
silicon dioxide	19	23.67	0.28	24.13	23.14
aluminum oxide	19	3.85	0.12	4.06	3.62
ferric oxide	19	3.5	0.13	3.73	3.25
calcium oxide	19	62.6	0.62	64.36	61.34
magnesium oxide	19	1.97	0.34	3.37	1.74
potassium oxide	19	0.57	0.1	0.83	0.43
sodium oxide	19	0.21	0.02	0.29	0.16
sulphur trioxide	19	2.28	0.14	2.58	2.05
titanium dioxide	19	0.2	0.02	0.24	0.18
phosphorus pentoxide	19	0.13	0.1	0.54	0.07
insoluble residue	19	0.18	0.1	0.43	0.06
total alkalies (as Na2O)	19	0.59	0.06	0.77	0.47
Calculated Compound Composition (%)					
tricalcium silicate	19	35.18	4.38	47.5	28
dicalcium silicate	19	41.34	3.99	48	30.8
tricalcium aluminate	19	5.05	0.33	5.5	4.2
tetracalcium alumino–ferrite	19	10.68	0.39	11.4	9.89

Physical Properties		No. of Tests	Average	Standard Deviation	Range Maximum	Minimum
pass 75um sieve (%)		22	96	1	98	93.7
Blaine air perm (m^2/kg)		19	360	25	401	329
autoclave soundness expansion (%)		20	0.03	0.01	0.054	0.005
heat of hydration – 7 days (kJ/kg)		19	233	20	279	207
Time of Set –	initial (min)	41	144	30	243	95
(Vicat method)	final (min)	41	260	42	364	185
false set –	@23 degC	41	50	0	50	48
11 min. penetration	@10 degC	40	46	10	50	9
compressive strength –	3 day	41	12	1	14.3	8.7
(MPa)	7 day	41	17	2	22.5	12.6
	28 day	41	29	3	33.3	23.2

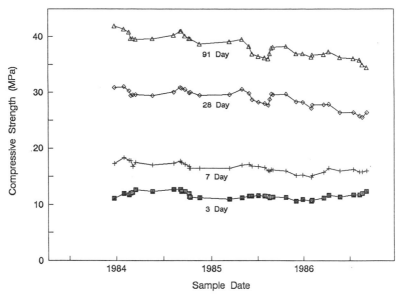

Fig. 2. Running average of five compressive strength tests at various ages (50 mm Mortar cubes)

Aggregates: Graded crushed gravel of maximum size 40 mm and washed natural sand from pit run materials, meeting the requirements of Canadian Standard CAN3-A23.1 and that of the project specification.

The aggregate pit was periodically inspected for visual changes in the quality of the excavated materials. In the event the materials were unacceptable, an alternative source of materials was also monitored for use when needed. Samples of the processed materials, stockpiled at the concrete mixing plant, were obtained periodically for testing. The frequency of testing for various properties is shown in Table 2.

Representative test results are included in Tables 3 and 4. General petrographic types of the coarse aggregate are shown in Table 5. Further, prior to acceptance, the aggregates were tested to determine their performance in concrete subjected to various curing and exposure conditions. These tests were carried out to evaluate the potential cement alkali-aggregate reactivity (ASTM C 227 Mortar Bar Test and CAN3-A23.2 Concrete Prism Tests at 23 C and 38 C) and that of the resistance to rapid freezing and thawing cycles in water (ASTM C 666, Procedure A). The concrete mix proportions for these tests were the same as Mix 1 (described later) except that Type 10 cement was used.

Table 2. Frequency of Testing of Aggregates at the mixing plant and in the laboratory

Properties	Test Method	Frequency
Gradation	CAN3-A23.2	Daily
Moisture Content	ASTM C 366	Daily
Passing 80 μm Sieve	CAN3-A23.2	Weekly
Organic Impurities	CAN3-A23.2	Daily
Flat and Elongated Particles	CAN3-A23.2	6 Months
Relative Density	CAN3-A23.2	6 Months
L.A. Abrasion	CAN3-A23.2	6 Months
Potential Reactivity	ASTM C 227	12 Months
Soundness	CAN3-A23.2	6 Months
Petrographic	CAN3-A23.2	6 Months
Clay Lumps	CAN3-A23.2	6 Months
Low Density	CAN3-A23.2	6 Months
Freeze-Thaw Durability	ASTM C 666(A)	12 Months

Table 3. Representative test results of coarse aggregate (Tested according to CAN3-A23.2 Test Method. Also, for guidance see Notes and Specification limits given below.

Sieve Size mm	Amount Passing %	Relative Density (SSD)*	Absorp-tion %	Soundness (in MgSO4) 5 cycles loss %	Abra-sion Wear %	PN (**)	Particles Pass 80 μm %
40	99.5						
28	76.7						
20	49.7	Max: 2.69	0.52	7.6	27.8	110	1.7
14	29.1	Min: 2.69	0.34	2.4	24.9	104	0
10	10.3						
5	1.1	Avg: 2.69	0.43	3.9	26.3	107	0.61
Specification Limits							
Maximum:		-	3.0	12	30	140	1.0
Minimum:		2.55	-	-	-	-	-

Notes:

1. Clay lumps, low density and friable materials: maximum of 0.4 % weighted according to grading.

2. A maximum of 5 % of total of all deleterious particles (including pass 80 μm) is allowed in the specification.

3. The specification required that the total of flat and elongated particles not to exceed 30 % by mass of the full sample. The particle width to thickness ratio > 3 is termed as flat while length to width ratio > 3 is termed as elongated. The total amount was less than 5 % in all samples tested.

* SSD: Saturated Surface Dry

** See Table 5 for Details of Representative Petrographic Analysis and Petrographic Number (PN).

Table 4. Representative test results of fine aggregates (Tested according to CAN3-A23.2 Test Methods. Also, for guidance see Note and Specification limits given below)

Sieve Size mm	Amount Passing %	Relative Density (SSD*)		Absorption %	Soundness in MgSO4 5 cycles loss %	Particles Pass 80 μm %
10	100					
5	97.3	Max:	2.68	1.79	15.47	3.2
2.5	82.9					
1.25	63.3	Min:	2.51	0.95	11.9	1.63
0.63	42.4					
0.315	21.9					
0.160	7.6	Avg:	2.65	1.13	13.64	2.46
FM(**)	2.85					
Specification Limits						
Maximum:			-	4.0	16.0	3.0
Minimum:			2.55	-	-	-

Notes:

1. Deleterious particles, i.e., clay lumps, low density and friable particles: maximum of 0.81 % was recorded.

2. A maximum of 4 % of total of all deleterious particles (including pass 80 μm) is allowed in the specification.

3. The organic impurities were checked by appropriate CSA test method and were within acceptable limits.

* SSD: Saturated Surface Dry

** The fineness modulus (FM) values were mostly between 2.7 and 2.9, a few were recorded up to a maximum value of 3.1. The specification requirement is between 2.3 and 3.1.

Table 5. Representative test results of petrographic analysis for coarse aggregate

PROJECT Darlington GS - Quality Assurance Program		REQUESTED BY G.M. Kidd	SAMPLE IDENTIFICATION S— 7454 P— 3699
SUPPLIER Highland Creek Sand and Gravel			DATE August 3, 1984
ORIGIN Fleetwood Pit, Bethany (stockpiled at CBM Plant, Darlington GS)			CHARGE NO. 740642-162-450
FORM Composite crushed gravel sample from 40,20 and 14mm stockpiles	DATE SAMPLED May 9, 1984	SAMPLED BY J. MacPherson and staff	FILE NO. 842.21 DAR(P) REPORT NO. SB84-4-H

PETRO NO. MULTI FACTOR *	TYPE NO. *	AGGREGATE TYPE		SIEVE —40 %	PETRO NO.	40—28 %	PETRO NO.	28—20 %	PETRO NO.	20—14 %	PETRO NO.	14—10 %	PETRO NO.	10—5 %	PETRO NO.
			% RETAINED / NO.OF PIECES			36 / 200		22 / 200		21 / 200		14 / 200		7 / 200	
X1	1	CARBONATES	(HARD)			82.5	82.5	77.5	77.5	81.0	81.0	81.0	81.0	80.0	80.0
	20	CARBONATES	(SLIGHTLY WEATHERED)			5.0	5.0	12.0	12.0	8.5	8.5	10.0	10.0	8.0	8.0
	2	CARBONATES	(HARD, SANDY)					.5	.5	.5	.5			.5	.5
	21	CARBONATES	(MEDIUM HARD, SANDY)												
	23	CARBONATES CRYSTALLINE	(HARD)												
	8	GRANITE – DIORITE	(HARD)			1.5	1.5			1.5	1.5	3.0	3.0	3.5	3.5
	4	GNEISS	(HARD)			8.0	8.0	7.0	7.0	5.5	5.5	4.0	4.0	4.5	4.5
	7	VOLCANIC	(HARD, SL. WEATHERED)												
	5	QUARTZITE	(COARSE GRAINED)							.5	.5				
	3	SANDSTONE	(HARD, MED.HARD)												
	6	GREYWACKE – ARKOSE	(HARD)			.5	.5								
	9	TRAP	(HARD)												
		GABBRO – ULTRABASICS	(HARD)			2.0	2.0								
X3	42	CARBONATES	(DEEPLY WEATHERED)			.5	1.5	2.5	7.5	2.5	7.5	1.5	4.5	2.0	6.0
	41	CARBONATES	(SOFT, SANDY)					.5	1.5						
	40	CARBONATES	(SOFT, OR SL. SHALY)												
	24	CARBONATES CRYSTALLINE	(SLIGHTLY WEATHERED)												
	27	GRANITE – DIORITE	(BRITTLE)												
	25	GNEISS	(BRITTLE)											.5	1.5
	28	VOLCANIC	(SOFT)												
	47	QUARTZITE	(FINE GRAINED)												
	52	ENCRUSTATIONS													
		CHERT 25—75%	(UNLEACHED)												
		CHERT >75%	(UNLEACHED)												
X6	43	CARBONATES	(CLAYEY OR SHALY)									.5	3.0	1.0	6.0
	44	CARBONATES	(OCHREOUS)												
	49	CARBONATES CRYSTALLINE	(SOFT)												
	51	GRANITE – DIORITE	(FRIABLE)												
	50	GNEISS	(FRIABLE)												
	48	VOLCANIC	(VERY SOFT, POROUS)												
	46	SANDSTONE	(SOFT, FRIABLE)												
	55	SCHIST	(SOFT)												
		ARGILLITE													
		CHERT 25—75%	(LEACHED)												
		CHERT >75%	(LEACHED)												
		SILTSTONE													
X10	62	CLAY													
	61	SHALE													
	60	OCHRE													
	63	VOLCANIC	(DECOMPOSED)												
		CARBONATES	(DECOMPOSED)												
		TOTALS				100.0	101.0	100.0	106.0	100.0	105.0	100.0	105.5	100.0	110.0

PETROGRAPHIC } NUMBER

	BY FRACTION	101	106	105	106	110
	TOTAL SAMPLE — WEIGHTED ACCORDING TO GRADING					104

*Based on Procedure for the Petrographic Analysis of Coarse Aggregates, Ontario Ministry of Transportation, Engineering Materials Office, Soils and Aggregates Section, No: LS-609, October, 1987.

The percentage length changes of the mortar bars and those of the concrete prisms are shown in Figures 3 and 4. The results of the freeze thaw test are given in Table 6.

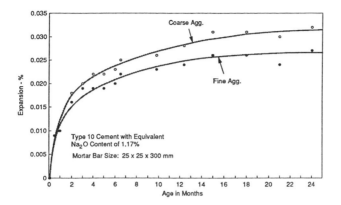

Fig. 3. Mortar bar expansion test according to ASTM C 227 method

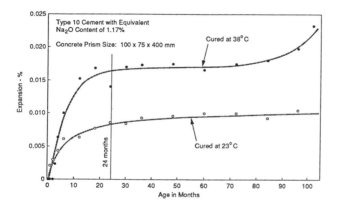

Fig. 4. Concrete prism expansion test according to Canadian Standard test method CAN3 A23.2 AT 23°C and 38°C

Table 6. Results of freezing and thawing tests on concrete (ASTM-C 666, Procedure A)

Reference and Test: Batch-Specimen No.		No. of Cycles at end of Freeze/Thaw	Dynamic Modules GPa		Pulse Velocity m/s		Change in Mass %	Durability Factor
			Initial	Final	Initial	Final		
Reference:								
	1-1	300	35.95	35.09	4570	4570	-2.5	97.6
(*)	1-2	300	36.41	34.7	4570	4570	-1.62	95.3
	1-3	300	36.55	34.83	4570	4570	-1.41	95.3
	2-1	300	37.80	36.48	4620	4670	-1.46	96.5
	2-2	300	36.12	33.99	4570	4570	-1.61	94.1
Average:		300	36.57	35.02	4580	4580	-1.72	95.8
Test:	1-1	300	37.74	36.42	4670	4670	-1.40	96.5
(**)	1-2	300	37.16	35.85	4670	4670	-1.54	96.5
	1-3	300	37.80	36.48	4670	4670	-1.53	96.5
	2-1	300	36.52	35.66	4620	4620	-1.41	97.6
	2-2	300	37.07	36.20	4730	4730	-1.01	97.6
	2-3	300	36.63	35.77	4670	4670	-1.74	97.6
Average:		300	37.15	36.06	4670	4670	-1.54	97.1

(*) Reference concrete is made with laboratory standard materials of known good performance, using 290 kg of Type 10 (CAN3-A5) cement per m^3 of concrete with water to cement ratio of 0.5.

(**) Test concrete is made with job site aggregate (40 mm maximum size) using 290 kg of Type 10 (CAN3-A5) cement per m^3 of concrete with water to cement ratio of 0.5.

Admixtures and Water: A lignosulphonate based water-reducing admixture, meeting the requirements of Canadian Standard, CAN3-A266.2, Type WN. An air-entraining admixture, meeting the requirements of Canadian Standard, CAN3-A266.1 and Potable Water meeting the requirements of CAN3-A23.1.

The admixtures were tested each time a shipment was received at the mixing plant. Twenty-nine samples of each admixture were tested during the period 1984 to 1986. A summary of the test results of these admixtures is shown in Table 7.

Table 7. Uniformity test results of water reducing and air-entraining admixtures

Admixtures		Relative Density	Residue %	pH	Amount of Chloride (ion) mg/liter	Infrared Analysis: Compared to the Initial Sample
29 samples of water reducing admixture (Lignosulphanic acid based)	Max:	1.226	46.2	7.4	720	No change
	Min:	1.212	43.6	6.0	45	
	Average:	1.219	44.7	6.8	247	
29 samples of air-entraining admixture (Alkyl-naptha-Sulphonate based)	Max:	1.030	Not tested	12.7	345	No change
	Min:	1.025	-	12.1	211	
	Average:	1.027	-	12.3	258	

Notes:

1. The water reducing admixture met the requirements of Canadian Standard (CAN3-A266.2, Type WN).

2. The air-entraining admixture met the requirements of CAN3-A266.1.

3. The amount of water soluble chloride ion content in a representative concrete sample was found to be 0.03 % by weight of cement.

Water from a municipal water supply was used throughout the project. Therefore, no periodic tests were conducted.

4 Mix proportions

Concrete incorporating two sizes of coarse aggregate was used. Mix 1, with 40 mm maximum size aggregate was used in the perimeter wall. Mix 2, with 20 mm maximum size aggregate, was used in the dome, the ring girder and the internal supporting columns and beams. The mix proportions were as follows:

Materials (kg/m^3)	Mix 1	Mix 2
Cement:	290	310
Sand:	775	815
Stone - 40 mm:	590	nil
20 mm:	400	760
14 mm:	170	330
Water-Cement ratio:	0.45	0.45

Admixtures (mL/m^3)

Water-reducing:	725	775
Air-entraining:	100	230

5 Tests on concrete

Perimeter Wall: The 56.8 metre high and 1.2 metre thick perimeter wall was built by continuous slip-forming in ten days. The quality of the concrete was monitored on a regular basis during this period. A total of 110 tests was carried out, averaging about 10 to 12 tests in a 24-hour period. A test program comprised of determining slump, air content, temperature, density and making a set of 150-mm diameter cylinders for compressive strength testing at 7, 28, and in some cases for 56 and 90 days.

In addition, the cement content and the water-cement ratios were monitored at the mixing plant. These values were recorded after making proper adjustments for changes in the fresh concrete densities and the moisture content of the aggregates. The specified strength of the concrete was 35 MPa at 90 days. The average strength of the moist-cured (at 23°C) cylinders was 39 MPa at 90 days. The average and the range of values for the strength tests are shown in Fig. 5.

Additional concrete cylinders and prisms (400 x 75 x 100 mm) were made in the laboratory using concrete with the same mix proportions as Mix 1. These specimens were tested for creep and shrinkage properties (ASTM C 512 and ASTM C 157 Test Methods). The results of these tests are shown in Figures 6 and 7.

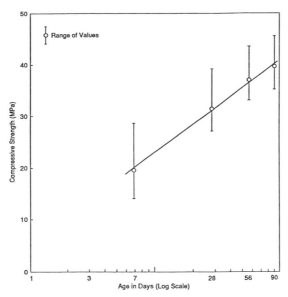

Fig. 5. Compressive strength vs. age of moist cured (23°C, 100 % RH) concrete cylinders made from mix 1. The 7 and 28 day values are average of 106 tests, while 56 and 90 days are for 17 tests. The water to cement ratio varied between 0.45 and 0.49. The cement content was between 285 and 294 kg/m^3 of concrete.

Ring Girder and the Dome: The ring girder at the top of the perimeter wall was placed in several sections over a period of two months and was followed by the placement of the dome sections during the next four months. The concrete was placed by pumping. The vertical lift of the 100-mm diameter pump line was about 56 metres, while the horizontal run at the base and at the top was about 30 metres each.

In order to monitor the changes that may have occurred in the quality of the concrete due to the pumping operation, a series of tests was conducted before and after pumping of the concrete. The concrete after pumping was obtained either from the delivery end of the pump line or from the formwork after the concrete was consolidated in-place. No significant difference was noted in the test results for the latter two cases. Therefore, these were not separately analyzed, but reported as "after pumping."

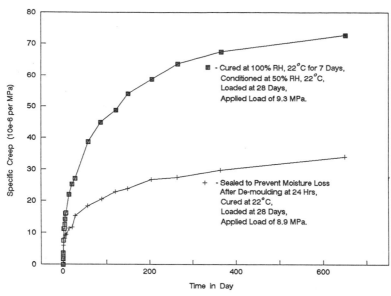

Fig. 6. Creep test results on concrete cylinders (150 x 300 mm). (ASTM C 512 Test Method)

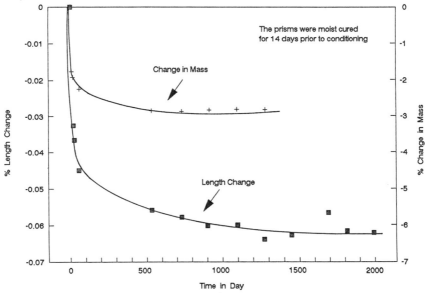

Fig. 7. Changes in length and mass of concrete prisms (400 x 75 mm) due to conditioning at 50 % rh, 22°C (ASTM C 157 Test Method)

A number of 150 mm diameter, 300 mm high test cylinders were made during the construction of the girder and the dome. Some of these were made before pumping and some were made after pumping. The cylinders were tested for compressive strengths at various ages, i.e., 7, 28, 56, and 90 days. Also, sets of cylinders, before and after pumping, were prepared and tested for air-void system analysis (ASTM C 457, Modified Point Count Method). The results of the strength tests and the air-void system analysis are shown in Fig. 8 and Table 8, respectively.

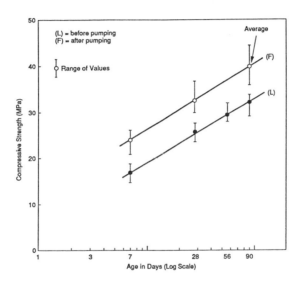

Fig. 8. Compressive strength vs. age of moist cured (23°C, 100 % rh) concrete cylinders made from mix 2. The 7 and 28 day values are average of 33 tests, while 56 and 90 days are for 11 and 29 tests, respectively. The water to cement ratio varied between 0.40 and 0.46. The cement content was between 293 and 324 kg/m³ of concrete.

MOUNT PLEASANT LIBRARY
TEL. 051 207 3581 Ext. 3701.

Table 8. Air void system of hardened concrete (Test according to ASTM C 457, modified point-count method)

Test Specimen (*)	Air Content (**) A (%)		Specific Surface α (mm^{-1})	Spacing Factor $\bar{L}$ (mm)	Paste Content P (%)	Number of Voids per 25 mm N
1A	7.0	(7.9)	30.8	0.11	23.9	13.5
1B	3.3	(3.7)	24.2	0.17	24.6	5.0
2A	6.6	(8.0)	32.8	0.12	23.1	13.5
2B	4.2	(5.0)	23.8	0.20	22.5	6.3
3A	6.5	(7.4)	34.5	0.10	23.1	14.0
3B	3.1	(3.8)	25.8	0.21	22.4	5.0
4A	3.3	(5.5)	43.5	0.12	22.2	9.0
4B	3.6	(4.6)	32.3	0.17	25.4	7.3

(*) Tests 1A to 4A were carried out on concrete before pumping. Tests 1B to 4B were carried out on concrete after pumping and consolidation in place.
(**) Air content values given in ( ) are measured on fresh concrete.

It may be noted that the concrete after pumping showed an increase in strength. This is mainly due to the loss of entrained air during pumping and handling operations. The loss of air was compensated for by increasing the initial air content of the concrete. Also, it may be noted that the air-void system was somewhat adversely affected due to the pumping operation; however, the values were still within acceptable ranges.

A set of companion cylinders, from those tested for the air-void system analysis, is being monitored under outdoor exposure conditions for changes in their mass and ultrasonic pulse velocity. The results of these tests would provide data for evaluation of the long-term durability of the concrete subjected to the natural freezing and thawing cycles under moist conditions. The test results are shown in Table 9. No significant changes in the values have been noted in four years of natural weathering.

Table 9. Changes in mass and ultrasonic pulse velocity (UPV) of concrete test specimens (150 mm dia. x 300 mm) due to outdoor exposure

Test (*) Specimen Cylinder No.	December 28/85 Mass (g)	December 28/85 UPV (m/s)	June/1987 Change in Mass (%)	June/1987 UPV (m/s)	June/1988 Change in Mass (%)	June/1988 UPV (m/s)	June/1989 Change in Mass (%)	June/1989 UPV (m/s)
1A	12805	4480	-0.12	4360	-0.34	4240	-0.23	4540
1B	13380	4690	-0.08	4630	-0.31	4630	-0.25	4690
2A	12685	4360	-0.08	4360	-0.32	4360	-0.16	4420
2B	12905	4630	0.02	4540	-0.19	4480	-0.19	4630
3A	12900	4540	-0.07	4510	-0.22	4480	-0.22	4590
3B	13222	4720	0.02	4660	-0.12	4590	-0.10	4720
4A	13183	4590	0.11	4510	-0.07	4360	-0.02	4660
4B	13295	4630	0.06	4590	-0.12	4450	-0.04	4660

(*) Same concrete as noted in Table 8 under Test Specimens.

7 Concluding remarks

For any large construction project and particularly for a nuclear containment structure, a well organized quality assurance program is absolutely essential. The on-going testing of the concrete and the related materials provided good control and it is expected that minimal variations in the properties of the in-place concrete was achieved.

On several occasions during the construction period, production adjustments were made promptly as deviations in the test results were noted.

The test results obtained during the construction phase will be used as baseline data for future monitoring of the performance of the structure.

8 Acknowledgements

The permission to publish the data was granted by the Director of the Research Division, Ontario Hydro, Toronto, Canada.

References

[1] Gallagher, E.J., Nuclear Standards Licensing, Enforcement and Construction Inspection Programs Session I, Chairman's Summary of five papers published in the Proceedings of the Conference on Construction of Power Generation Facilities, Edited by J.H. Willenbrock, Pennsylvania State University, Sept 16-18, 1981. American Society of Civil Engineers, 345 East 47th Street, New York, NY 10017

[2] Tuthill, L.H., Quality Attainment and Common Sense in Nuclear Concrete Construction, ACI-Concrete International, March 1979, Vol. 1, No. 3, pp 32-37

[3] Morison, W.G., (Ontario Hydro), Penn, W.J., (Ontario Hydro), Hassmann, K., (West Germany), Stevenson, J.D., (USA), Elisson, K., (Sweden), Containment Systems Capability. 5th International Meeting on Thermal, Nuclear Reactors Safety Proceedings, Vol. 1, Dec. 1984, pp 111 to 138 (held in Germany on Sept. 9 to 13, 1984), Report No. KFK-3880-1, Publisher: Kernforschungszentrum, Karlsruhe, GMBH, Germany.

23 DETECTION OF CRACKS IN CONCRETE BY ACOUSTIC EMISSION

C. SKLARCZYK, H. GRIES and E. WASCHKIES
Fraunhofer-Institut für zerstörungsfreie Prüfverfahren,
Saarbrücken, Germany

Abstract

Acoustic emission testing is a method to detect on-line the formation and the growth of cracks and crack-like defects in loaded components and buildings. A large concrete beam has been loaded up to rupture and the acoustic emission (AE) generated thereby has been measured and analysed. It could be shown that it is possible to reveal growing cracks by AE. Furthermore, non-growing cracks could also be detected by a weak but measurable AE due to crack friction. Thus it seems to be possible to develop AE-testing as a nondestructive method for on-line monitoring of concrete-buildings.

1 Introduction

The term acoustic emission (AE) refers to the sound waves emitted by a sudden release of elastic energy stored in loaded materials [1]. The main sources of AE are crack formation and propagation, crack surface friction [2], plastic deformation (mainly in metals) and phase transformation. The frequency range used for AE-method lies between 20 kHz and a few MHz. In contrast to conventional ultrasound testing, no external sound waves are put into the material under test, but the defect is revealed by the sound waves it produces itself. The greatest sensitivity of AE with regard to crack growth events is observed in brittle materials. Based on this background, it was obvious to check the suitability of AE-method for monitoring and surveillance of concrete buildings, especially in view of the formation of crack-like defects [3].

The object of this investigation was to find out whether AE-testing is a suitable method to detect cracks in concrete buildings. In detail following questions have been asked at the beginning of this work:

- What are the characteristical features of AE due to crack growth and friction between crack faces (or between matrix and reinforcement and tendons)?

- What is the minimum detectable crack size?

- Is it possible to detect non-growing cracks by AE generated by frictional effects?

2 Experimental procedure

2.1 Concrete specimen
To answer these questions, a T-shaped concrete beam with a length of 21 m, a height of 1 m and a breadth of 0.2 m has been loaded in many steps up to rupture (four-point-bending). Besides the usual steel reinforcements the beam was prestressed by 19 glass fibre tendons. The experiment was performed at the University of Gent (Belgium) and several institutions were involved in this test.

2.2 AE-Equipment
The mechanical vibrations of the specimen were transformed into electrical vibrations by the transducers. These signals were amplified 100 times and were then fed into the AE-system, which was a commercial equipment from Physical Acoustics Corporation (PAC) of type SPARTAN and possessed eight independent channels. AE-data were processed by a microcomputer of type 3000 from PAC and stored on floppy disk. Each AE-event is characterized by the signal parameters arrival time (necessary for location, peak amplitude, duration, rise time (time interval between beginning and peak of signal) and area under signal envelope (here denoted as "energy". Besides this AE-system, a transient recorder (Date 6000 from Data Precision) can be applied to receive and to plot the whole time signal.

Fig. 1 gives a scheme of the used AE-apparatus.

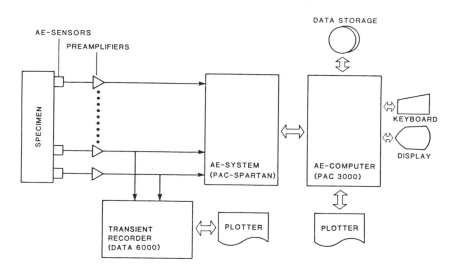

Fig. 1. Block Diagram of Acoustic Emission Equipment

The piezoelectrical transducers (PAC-A3) were sensitive in view of acceleration and had a frequency range of 20-100 kHz (resonance frequency 30 kHz). This frequency was very low compared to usual AE-trials, but it was necessary because of the high attenuation of the sound waves in concrete. Fig. 2 shows some frequency spectra of simulated AE-events produced by ultrasonic transducers. Mainly frequencies above 50 kHz are damped after distances between source and receiver of more than 1 m.

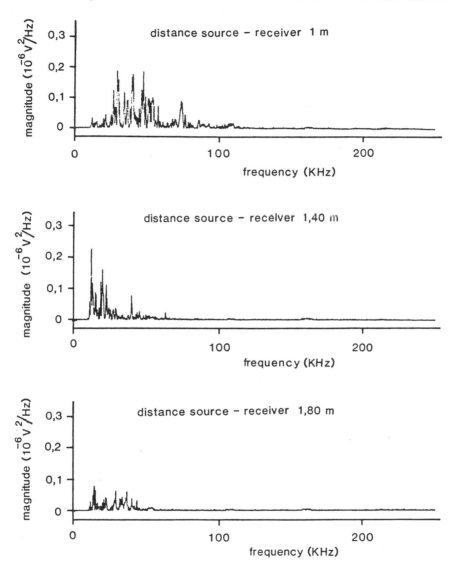

Fig. 2. Sound propagation in the large concrete specimen

The arrangement of the six AE-transducers is shown schematically in Fig. 3. It allowed a linear location of the AE-sources along to the main axis of the beam. Localization is based on the differences of arrival times at the different sensors. The effective sound velocity (group velocity) in the concrete beam under test is 2600 m/s. The accuracy of localization lies in the range of 0.2 m.

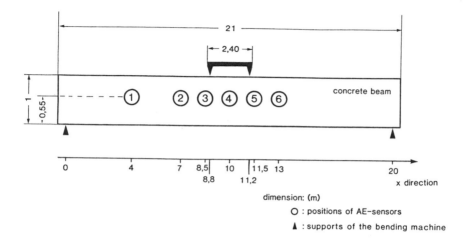

Fig. 3. Dimensions of the concrete beam and AE-sensor instrumentation

2.3 Loading program
The performed loading program (four-point-bending) was as follows:

1. overload which produced an array of large cracks with a depth of about 0.5 m in the area between the two inner loading points

2. cyclic loading: 5 - 25 2kN, 0.3 Hz, 250 cycles, no further crack formation and growth

3. monotonic increasing load: 0 - 35 2kN, about 1 h, some new cracks are formed, existing cracks grow

4. cyclic fatigue loading: 28000 cycles

   22.5 - 35 2kN, 0.35 Hz at beginning
   27.5 - 35 2kN, 0.52 Hz at the end

5. monotonic increasing loading up to the rupture: 0 - 133.5 2kN

The first overload was performed by error. Therefore no AE-data have been captured in this phase. The unit "2kN" means that the beam is loaded by two points for each of the two opposite sides.

3 Experimental results

3.1 Loading phase 1 and 2

After the first (unintended) damaging there were about eight cracks in the concrete with lengths of about 0.5 m each. Fig. 4 (bottom) shows that AE is active over the whole loading range but decreases with increasing load. The AE-sources are concentrated in the middle area of the beam (8 - 12 m in x-direction, see Fig. 4 top). The AE-energies have values lower than 300 (Fig. 4 middle, one energy counting impulse is equivalent to 10 Vμs).

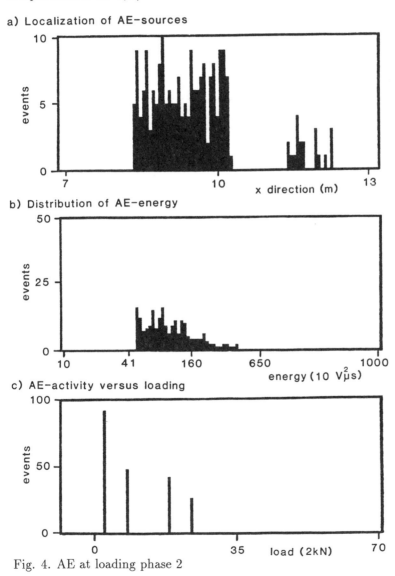

Fig. 4. AE at loading phase 2

3.2 Loading phase 3

During this experimental stage some new cracks are formed which could be seen visually on the specimen and by comparing the AE-location distributions (Fig. 4 and 5 top). AE is predominantly active at higher loads (Fig. 5 bottom). AE-energies extent from small values up to about 700 (Fig. 5 middle) and two maxima can be distinguished.

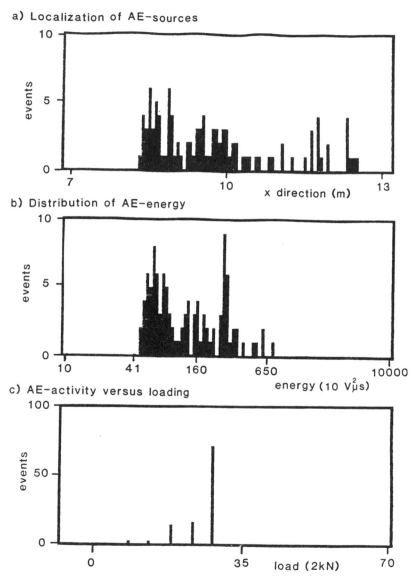

a) Localization of AE-sources

b) Distribution of AE-energy

c) AE-activity versus loading

Fig. 5. AE during loading phase 3

3.3 Loading phase 4

After this stage more than 20 cracks are to be observed in the concrete beam (Fig. 6). Figure 7 gives the AE-activity at the beginning of this phase. The location distribution becomes broader and broader (Fig. 7 top). Many AE-energies are higher than at the previous stages (Fig. 7 middle). At the end of this phase, AE-activity decreases and the AE-energies become smaller (Fig. 8 middle). Some indications of cracks near the position 11 m disappear compared to the beginning of the test (Fig. 7 and 8 top).

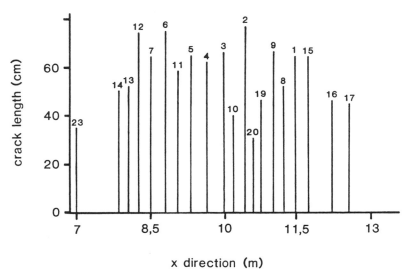

Fig. 6. Distribution of the cracks in concrete after phase 4

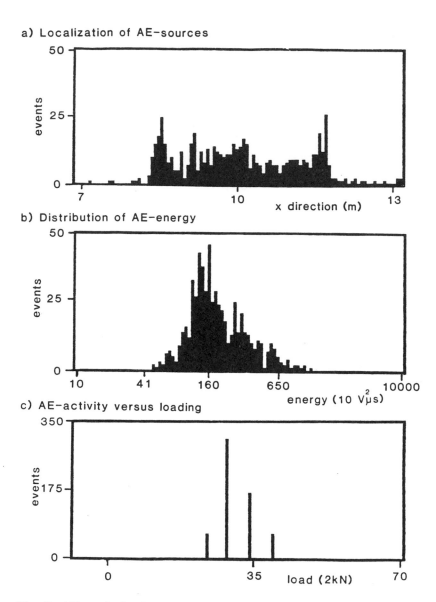

a) Localization of AE-sources

b) Distribution of AE-energy

c) AE-activity versus loading

Fig. 7. AE at the beginning of the loading phase 4

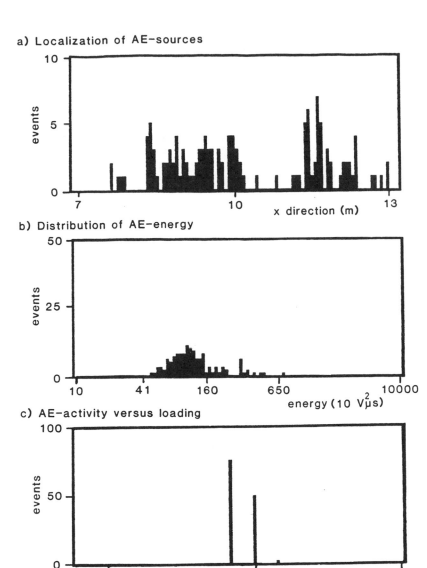

a) Localization of AE-sources

b) Distribution of AE-energy

c) AE-activity versus loading

Fig. 8. AE at the end of the loading phase 4

3.4 Loading phase 5
Now a lot of cracks is formed beyond the middle area, which can be seen visually and by AE-location distribution (Fig. 9 top). A lot of events with very high energies can be observed (Fig. 9 middle). Firstly AE-activity increases, but after having reached a maximum it decreases again (Fig. 9 bottom).

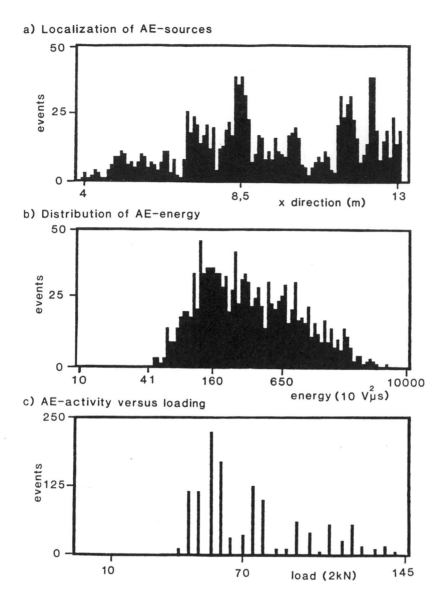

a) Localization of AE-sources

events

50

25

0

4   8,5   13
x direction (m)

b) Distribution of AE-energy

events

50

25

0

10   41   160   650   10000
energy (10 Vμs)^2

c) AE-activity versus loading

events

250

125

0

10   70   145
load (2kN)

Fig. 9. AE at the loading phase 5

306

4 Interpretation of the results

During all the experiments two types of AE-signals can be distinguished: signals with high energies and signals with low energies. The high energy signals arise at loading stages with crack formation and propagation, whereas the low energy signals occur during all experiments with existing cracks. Thus our assumption is as follows:

- High energy AE-signals (> 300 energy impulses) mainly originate by crack formation and growing.

- Low energy AE-signals (< 300 energy impulses) mainly originate by processes on existing cracks, like crack surface friction.

Due to the already mentioned unintended overload at test begin, the first crack formation in the concrete beam could not be recorded by AE. In loading phase 2, where no visible crack growth occurred, a low energy AE could be observed, which can be attributed to crack friction noise at opening of the crack faces (Fig. 4). Thus this experiment demonstrates that non-growing cracks can be detected by AE-method. For the most cracks a load of about 20 - 30 % of the load necessary to produce cracks was sufficient to generate frictional AE. At higher loads the cracks are completely open and no more frictional AE can be produced: AE drops down as shown in Fig. 4 bottom.

At the next loading phase (no. 3), the surfaces of the already existing cracks opened and some new crack growth occurred. Consequently, two AE-energy maxima, which can be attributed to friction events (low energies) and crack events (high energies), can be observed (Fig. 5 middle).

In loading phase 4 (fatigue cycling), both processes (crack growth and friction) are active, and a broad energy distribution can be observed (Fig. 7 middle). Main AE-activity is found at loads clearly below maximum. This behaviour is not understood up to now. At the end of fatigue cycling nearly no more high energetic AE is observed (Fig. 8 middle): obviously crack growth activity becomes very low. Some of the cracks near position 11 m cannot be detected any more. This may be due to the fact that the crack surfaces are smoothed and polished more and more; thus frictional AE is diminished.

At the loading phase up to the rupture again both processes can be observed (Fig. 9). At high loads AE-activity decreases. This is probably due to the fact that most cracks have reached the compressive (prestressed) zone in the top part of the beam and do not grow further.

5 Summary and conclusions

The above described experiments show that it is possible to detect growing and also non-growing cracks in concrete by AE-testing. Crack surface friction and crack growth processes generate AE-signals which can be separated approximately by their energies: frictional AE possess lower energies than crack growth AE. In these tests the borderline between the two AE-source types amounted to about 300 energy impulses. The minimum crack area to be detected by crack friction and crack growth cannot be given up to now. This shall be investigated in a future experiment. This research has shown that AE-testing is a promising NDT-method to predict and monitor the deterioration of concrete structures.

References

[1] Eisenblätter, J. Acoustic emission analysis: introduction, present status and future development. Acoustic Emission, DGM 1980, 1-16

[2] Sklarczyk, C., Waschkies, E. Detectability of defects in reactor pressure components by location and interpretation of AE-sources. Journal of Acoustic Emission 8 (1989), pp 93-96

[3] Ohtaki, T., Oh-Oka, T. AE technique evaluation of deteriorated viaduct, Progress in Acoustic Emission IV, Jap. Soc. for NDI (1988), 328-335

PART FIVE
DURABILITY-RELATED TESTING

24 IN-SITU PERMEABILITY TESTING OF CONCRETE

H.K. HILSDORF
University of Karlsruhe, Germany

Abstract
Various mechanisms of the transport of liquids, gasses, and ions in concrete are summarized. Test methods which are suitable for site applications to determine permeability and water absorption of concrete are described and discussed with respect to their sensitivity to technological and environmental conditions. It is shown that there is a lack of information relating permeability measurements to concrete durability.

1 The Problem

A possible approach to the testing of concrete during construction is the observation and characterization of the structure and the properties of the surface near regions or the "skin" of a structural concrete member. Such observations are of particular significance because the properties of the skin play a decisive role in the durability of a concrete structure: Durability of concrete is governed by its resistance to the ingress of aggressive media. Since the concrete skin is exposed to such media more frequently than the interior parts of a section the durability of the skin governs the durability of the entire section.

The microstructure and the properties of the concrete skin frequently differ substantially from the average properties of the section. This is due to a non-uniform distribution of the composition of the concrete, e.g. the skin may have a higher content of fines including cement paste and no large aggregate particles. Furthermore, the skin is particularly vulnerable to insufficient curing since it dries out rapidly after termination of curing. Thus hydration of the skin ceases whereas it continues

in the interior parts of the section so that the average compressive
strength of a sufficiently thick section continues to increase for quite
some time after curing.

Aggressive media such as various liquids, ions or gasses may penetrate
the concrete skin through different transportation mechanisms.

They can be described by different mathematical expressions which are
summarized in section 2. of this paper and are governed by different material characteristics.

The objective of testing the properties of the concrete skin during construction, therefore, is to estimate on the basis of in-situ measurements
the resistance of the concrete to the penetration of aggressive media. On
the basis of such measurements it should be possible to estimate whether
sufficient durability of the structure can be expected for specified exposure conditions and a specified design life. Only as a secondary result
some information may be obtained on the expected strength development.
The information gained through in-situ testing of the concrete skin may
be used for adjustments of the mix proportions and construction practices
such as compaction for subsequent parts of the structure. It may be particularly useful to determine the point in time at which curing of a particular structural member can be terminated.

In this paper the principles of some of the test methods to characterize
the properties of the concrete skin will be discussed. Because of the emphasis on in-situ testing, methods to evaluate concrete properties on
companison specimens will not be dealt with.

2 Transport mechanisms

Some knowledge of the mechanisms by which aggressive media penetrate the
concrete skin is essential in order to define those materials characteristics which govern the resistance of the concrete to such penetration
and to properly interprete the obtained test results. Also refer to [19].

The take-up of liquids, in particular water or aqueous solutions, may take place by permeation of the liquid under an external pressure, by diffusion of the liquid in the gaseous state or by capillary suction.

P e r m e a t i o n   o f   a   l i q u i d   through a porous body under a constant pressure gradient generally is described by Darcy's law:

$$Q = K_L \cdot \frac{h}{l} \cdot A \cdot t \qquad\qquad (1)$$

where

Q = Volume of liquid [m^3] flowing during time t [sec]

h/l = pressure gradient in terms of hydraulic head [m/m]

A = penetrated area [m^2]

K_L = coefficient of permeability of the liquid [m/sec]

T h e   a b s o r p t i o n   o f   a   l i q u i d , in particular of water, in the pore system of the concrete can be expressed in terms of the volume and diameter of capillary pores. According to [22], [23] the volume of water V_w [cm^3/m^2] absorbed per unit contact area can be expressed by

$$V_w = \frac{A_w}{\rho_w} \cdot t^n \qquad\qquad (2)$$

where

A_w = coefficient of water absorption [$\frac{g}{sec^n} \cdot \frac{1}{m^2}$]

ρ_w = density of water [g/cm^3]

t = suction time [sec]

$\qquad$ $0.2 < n < 0.5$

The coefficient of water absorption can be expressed by pore characteristics of the concrete acc. to eq. 3:

$$A_w = K \sqrt{2r_h} \cdot \frac{\varepsilon_{abs}}{a_{Tabs}} \qquad\qquad (3)$$

where

ε_{abs} = effective capillary porosity
a_{Tabs} = tortuosity factor
r_h = $\varepsilon_{abs}/S_{abs}$ = hydraulic radius
S_{abs} = specific surface area
K = coefficient related to surface tension, contact angle, viscosity and temperature of the liquid

Due to the highly heterogeneous microstructure of the hydrated cement paste with pore radii varying over at least six orders of magnitude and an unknown turtuosity, A_w always has to be determined directly through experiments. Then, eq. (2) may be simplified to:

$$\frac{w}{w_1} = \left(\frac{t}{t_1}\right)^n \qquad\qquad (4)$$

where

w = water absorption at time t $[cm^3/m^2]$
w_1 = water absorption at time $t = t_1$, e.g. 1 hour $[cm^3/m^2]$

Transport of gasses or liquids in a gaseous state e.g. water vapor or of ions by d i f f u s i o n may in the most general way be described by Fick's second law of diffusion:

$$\frac{\partial H}{\partial t} = \frac{\partial}{\partial x}\left[D(H)\frac{\partial H}{\partial x}\right] \qquad\qquad (5)$$

where

H = internal relative vapor pressure at location x
$D(H)$ = diffusion coefficient $[m^2/sec]$ at a relative vapor pressure H

For some inert gases the diffusion coefficient is independent of H so that D = const.

The p e n e t r a t i o n o f g a s e s is described by eq. (6):

$$V = K_v \cdot \frac{A}{l} \cdot \frac{(p_1 - p_2)}{\eta} \cdot \frac{p^*}{p} \cdot t \qquad (6)$$

where

V = volume of gas $[m^3]$ flowing during time t [sec]

$p_1 - p_2$ = pressure gradient $[N/m^2]$

p^* = average pressure in capillary pores = $(p_1 + p_2)/2$

p = pressure at which V is observed $[N/m^2]$

A = penetrated area $[m^2]$

l = thickness of member [m]

η = viscosity of the gas $[Nsec/m^2]$

K_v = coefficient of gas permeability $[m^2]$

If in eq. (6) the viscosity η and the average pressure p^* are not taken into account a coefficient of gas permeability K_s in $[m^2/sec]$ has to be used.

From the preceeding sections it follows that one or several of the following materials characteristics govern the probable durability of a concrete skin:

- coefficient of permeability K_L [m/sec] of a liquid, particularly water, K_w, acc. to eq. 1;
- coefficient of permeability K_v $[m^2]$ or K_s $[m^2/sec]$ of a gas such as air or O_2 acc. to eq. 6;
- coefficient of water absorption A_w or w_1 $[cm^3/m^2]$ acc. to eqs. 2 or 4, together with the power, n, in eq. (4);
- diffusion coefficient D $[m^2/sec]$ for water vapor, gasses such as CO_2 or ions such as Cl^- acc. to eq. (5);

315

Depending on the particular exposure conditions several of these materials characteristics may be decisive for the durability of the structure. However, in the in-situ testing of concrete not all, in most instances only one of these coefficients can be determined. Therefore, interrelations between these characteristics have to be known. So far they have been determined only to a rather limited extent, e.g. there is an approximately linear relation between the logarithms of the diffusion coefficient and the permeability coefficient of O_2 in non-carbonated concrete [10]. In [18], [19] it is shown that a correlation exists between air permeability K_s and the coefficients w_1 and n, describing water absorption acc. to eq. 4. Nevertheless, there is a pronounced lack of information in this area, and it is an underlying, though reasonable, assumption in the subsequent sections that such interrelations actually exist.

3 Parameters influencing transport characteristics

Concrete permeability, diffusivity or absorption characteristics are governed by the microstructure of the concrete and in particular by the pore structure of the hydrated cement paste as well as by microcracks. A dense pore structure generally results in a slow rate of transport of most media in the liquid or in the gaseous state. A theoretical correlation, supported by experimental data between pore size distribution and permeability is given in [25]. It is this interrelation between microstructure of a particular concrete and transport coefficients which allows an estimate of the potential durability of a concrete from the measurement of transport characteristics.

Nevertheless, transport characteristics are also influenced by external parameters which in turn influence the moisture state and the temperature of the concrete. Whereas temperature is in many instances a minor parameter for the temperature range relevant in this context, the moisture content of concrete may influence some transport characteristics even more than the microstructure of the concrete. Fig. 1 shows schematically the influence of the internal relative humidity on transport parameters: For the transport of gasses by diffusion or permeation both the diffusion co-

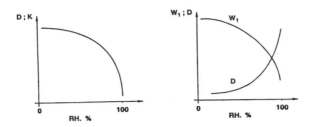

Fig. 1. Effect of moisture content on transport coefficients - qualitative

efficient and the permeability coefficient decrease by orders of magnitude with increasing relative humidity in the pore system of the concrete.
For the transport of water the diffusion coefficient increases with increasing rel. humidity whereas the rate of water absorption decreases.
Water permeability generally decreases with increasing rel. humidity of
the concrete though the mechanisms leading to this behavior are not as
yet clarified.

From this very brief review it follows that a control of the moisture
state of the concrete is decisive when measuring transport coefficients
on site or in the laboratory.

4   Test methods

4.1 Principles and general requirements

The test methods developed so far for direct or indirect in-situ measurements of the permeability of the skin concrete in a structure can be subdivided into three categories:

- water absorption
- water permeability
- air permeability

In the following only the general principles of these test methods will
be discussed. They will be demonstrated on the basis of three examples
which will be described in more detail.

In order to evaluate the suitability of a particular method certain general requirements have to be formulated which are generally valid for most in-situ test procedures of concrete. These are in particular:

1) The characteristic which is measured is a reliable indicator of the property to be controlled;

2) The method is selective i.e. it distinguishes clearly between "good" and "bad";

3) The obtained results are little influenced by external parameters, or means are provided to take such effects into account;

4) The results can be obtained in a sufficiently short period of time;

5) The test method is just simple enough to yield reliable results; the equipment is robust and can be operated by a person of medium skill;

6) The test method is non-destructive or at least semi-non-destructive;

4.2 Water Absorption

The rate at which a dry concrete surface may absorb water can be taken as a measure of the water permeability of the concret skin. Based upon this principle the so-called ISA-Test has been developed by M. L e v i t t [1]. Based upon this initial work a British Standard BS 1881 has been issued [2]. This general approach has been applied in a number of subsequent investigations such as in [11], [12], [16], [22], [23]. Fig. 2 which is taken from [23] describes the test set-up: A cap is sealed onto the concrete surface to be tested and then filled with water with a small pressure head. After closing the tap, the rate of water absorption can be derived from the rate at which the meniscus in the scaled glass capillary retracts. The test results may be evaluated on the basis of eq. (2). Differentiating eq. (2) with respect to time we obtain the following relation between rate of water absorption, coefficient of water absorption

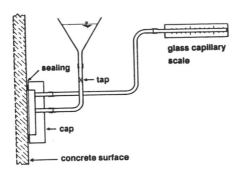

Fig. 2. Initial surface absorption test (from [23])

and the so-called ISA-Value.

$$v_W(t_s) = \frac{A_W}{2\rho_W} \cdot t_s^{n-1} = ISA(t_s) \qquad (7)$$

where

v_W = rate of water absorption at time t_s
t_s = time t at which $ISA(t_s)$-value is taken

The power n   0.5. For t_s frequently a value of 10 min. is chosen.

In Fig. 3 the results of an investigation reported in [23] are shown. There, the $ISA(t_s)$-value is given as a function of the water/cement-ratio and the duration of curing for the surfaces of concrete walls exposed unsheltered to the weather. At an age $120 < t < 360$ days the readings were taken approx. 1 1/2 days after the last rain fall. There is a pronounced increase of the ISA-value with increasing water/cement-ratio, however, the influence of curing is much less pronounced. Rather close correlations between ISA-values, water/cement-ratio and duration of curing have been observed for laboratory specimens by Dhir et al. [12].

A modification of the ISA-test is described in [12], and referred to as the CAT-test. It differs from the regular ISA-test in so far as a test

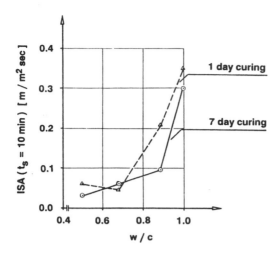

Fig. 3. Effect of w/c-ratio and curing on ISA(t_s)-value (from [23])

hole, diameter 30 mm, depth 50 mm, is drilled into a concrete surface, and water absorption is determined by filling water into the test hole.

In order to evaluate the various test methods it is essential that relations between the parameter measured, in this particular case the ISA-value, and concrete durability properties are available. In [16] predictions of depth of carbonation based on ISA-values are given and substantiated by a limited number of test results.

4.3 Water Permeability

In 1973 a method has been proposed by   F i g g   to measure the water permeability of a concrete skin [3]. The method is closely related to a test procedure for the measurement of the air permeability of a concrete skin which is described in more detail in the subsequent section. For the water permeability measurement a hole, diameter 5.5 mm, depth 30 mm, is drilled into a concrete surface. After thorough cleaning the hole is plugged up to a depth of 20 mm from the outside surface by polyether foam and sealed with a silicone rubber. After the rubber head has hardened a hypodermic needle is pushed through the silicone rubber plug. Then water

is forced into the assembly and a water head of 100 mm is applied. The
time for the meniscus in a capillary to travel 50 mm is recorded and is
taken as a measure of water permeability. The method has been improved
slightly e.g. in [4], [12] and a somewhat similar approach, however,
using considerably higher pressures is described in [5]. Considering the
low pressure head of 100 mm the results obtained by the Figg-test should
be closer related to water absorption than to actual water permeability.

4.4 Air Permeability

There are two different approaches to measure air permeability of a concrete skin. In the first group of the available test set-ups a hole is
drilled into the concrete surface. Either a pressure is applied to the
air in the hole or a vacuum is generated in the hole. In both instances
the variation of air pressure in the hole is taken as a measure of concrete air permeability. This approach originally has been proposed by
F i g g [3] and has been slightly improved by various investigators [4],
[6], [7], [8], [17], [20]. This method is described in the following taking the work of   H o n g   and   P a r r o t   as an example [20]
(Fig. 4). A hole, diameter 20 mm, depth 35 mm, is drilled into the concrete surface and sealed with a silicone rubber plug so that the uncovered circumference of the hole has a depth of 15 mm as shown in

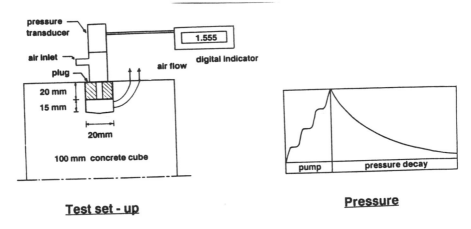

Fig. 4. Measurement of air permeability of cover concrete,
        Hong - Parott [20]

Fig. 4. By means of a pressure transducer and a pump, an initial pressure in the range of 1 to 3 bar is generated inside the hole and read on a digital indicator. The pressure decreases as the air permeates through the cover concrete into the atmosphere. From the loss of pressure during a given time interval a coefficient of gas permeability can be calculated acc. to eq. (6) if a value of A/1 is estimated from the geometry of the cavity and under the assumption that air escapes uniformly over a surface area with a diameter of 70 mm around the center of the hole. In [20] also a method is proposed to measure the rel. humidity inside the hole at the time of testing permeability.

Some of the results obtained with this method and reported in [20] are shown in Fig. 5 where air permeability of concretes cured between 1 and 28 days is given as a function of drying time of the concrete. As the duration of curing is increased from 1 to 28 days the coefficient of air permeability increases by 2 orders of magnitude. An increase of the duration of drying has a much smaller effect. In other test series Hong and

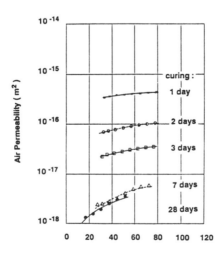

Fig. 5. Air Permeability of cover concrete - effect of drying,
        Hong - Parott [20]

Parott showed that there exists a unique relation between permeability after 80 days of drying and the weight loss after 10 days of drying for various types of concrete.

In the second group of test set-ups the permeability of a concrete skin is determined by placing an instrument directly on the concrete surface without drilling a hole. Such test procedures have been developed e.g. by S c h ö n l i n [13], [18], L y d o n [15] and H u d d et al. [24]. Whereas Schönlin and Hudd generate a vacuum as a pressure differential, Lydon causes air permeation by an external pressure. The overall approach is described on the basis of the test set-up by Schönlin (Fig. 6): a vacuum chamber, inner diameter 50 mm, is placed on the concrete surface. A rubber gasket is attached to the vacuum chamber. By means of a vacuum pump the pressure inside the vacuum chamber is reduced to a value p_o 20 mbar. Then the stop cock between vacuum pump and vacuum chamber is closed and the increase of pressure inside the vacuum chamber is observed. From the increase of the pressure inside the vacuum chamber during the time interval t_1 - t_o a coefficient of air permeability can be calculated on the basis of eq. (6). Since both the penetrated area, A, and the length through which air has to travel, 1, are unknown, Schönlin estimated a permeability index, M, acc. to eq. (8) which is proportional to the coefficient of air permeability under the assumption that the ratio, A/1, is constant. Since he did not take into

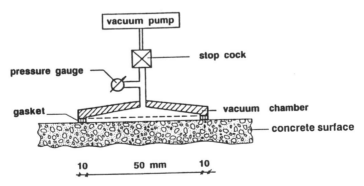

Fig. 6. Test set-up to measure surface air permeability,
Schönlin [13], [18]

account the viscosity of the air, the permeability index has the dimensions $[m^3/sec]$.

$$M = K_s \cdot \frac{A}{1}$$
(8)

where

M = permeability index $[m^3/sec]$
K_s = air permeability coefficient $[m^2/sec]$
A = penetrated area
1 = length over which air has to travel

An integral part of this particular test method is the preconditioning of the concrete surface. Since the method has been developed to determine the point in time at which curing can be commenced, it is assumed that initially the concrete surface is in a wet or moist state. Then the concrete surface has to be dried in a specified way for a period of 5 min. The depth of drying ranges between 2 and 4 mm from the concrete surface depending on the permeability of the concrete. Some of the results obtained with this method are shown in the following. Fig. 7 gives the cor-

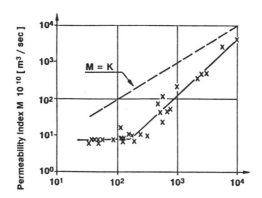

Fig. 7. Surface permeability index and air permeability coefficient, Schönlin, [18]

324

relation between permeability index and an air permeability coefficient
which had been determined on specimens for which the ratio A/1 was known.
Apparently, permeability index and permeability coefficient are not proportional, and a reduction of the permeability coefficient below a value
of approx. 2×10^{-8} m^2/sec. does not cause a further reduction of the
permeability index. Proportionality between both coefficients is obtained, however, if the duration of drying prior to determining M is substantially increased. Fig. 8 gives the effect of the duration of curing
and of curing temperature on the permeability index. Similar to the results reported by Hong and Parott an increase of the duration of curing
from 1 to 7 days results in a reduction of M by an order of magnitude.
The same effect can be observed if the temperature during curing is increased from 10 to 30° C. In Fig. 9 the square of the depth of carbonation of concretes made of different types of cements, water/cement-ratios
and cured for different durations of time after 1 year of storage in a
constant environment, 20° C, 65 % r.h. is given as a function of the permeability index. The depth of carbonation increases with increasing permeability index, however, for a given permeability index the depth of
carbonation is substantially higher if the concretes are made of blast
furnace slag cements (HOZ 35 L) compared to concretes made of regular
Portland cement (PZ 35 F) or a mixture of Portland cement and 20 % fly
ash (PZ 35 F + FA). Based on theoretical considerations the depth of
carbonation can be estimated from eq. (9):

$$d_c = d_{co} \sqrt{(\frac{M}{M_o})^{0.38} \cdot \frac{t}{t_o}} \qquad (9)$$

where

d_c = depth of carbonation
d_{co} = depth of carbonation at time t_o ($t_o = 1$)
t = duration of carbonation
M = permeability index
M_o = dimensional coefficient which depends on the type of cement

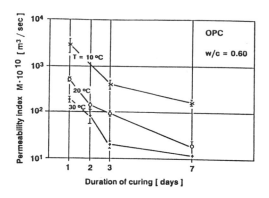

Fig. 8. Effect of duration of curing and curing temperature on surface
permeability index, Schönlin [18]

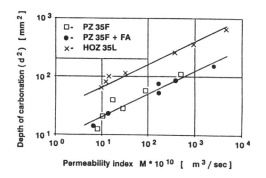

Fig. 9. Depth of carbonation after 1 year of exposure at 20° C, 65 % RH
and permeability index, Schönlin [18]

Close relationships have also been observed between permeability index
and characteristics of the microstructure of the cement paste.

5 Discussion

In evaluating the various test methods emphasis should be placed on the
following aspects:

- characterization of the surface layer or of the entire concrete cover
- sensitivity of the test methods with regard to technological parameters
- sensitivity of the test method with regard to environmental parameters, in particular moisture state
- ease of handling; reliability; non-destructiveness
- relation with durability properties of the concrete

The ISA-test as well as some of the air permeability tests [13], [15], [18], [24] characterize the outer 2 - 5 mm of the concrete cover whereas the air- and water permeability tests using the Figg-approach may characterize the properties of the concrete up to a depth which can be controlled by the depth of the hole and of the plug placed into the hole. Measurements of the surface properties are particularly vulnerable to changes in the moisture state due to drying. They may be hampered by a rough surface texture or by curing compounds so that in some instances a surface preparation may be necessary. Furthermore, the question arises whether the properties of the outermost layers of the concrete are a reliable indicator of the properties of the entire concrete cover. If the results given in Fig. 9 are generally valid this is indeed true. On the other, hand when measuring air permeability on the basis of a Figg-approach it is likely that the results are primarily controlled by the properties of the deeper layer which may be denser and normally has a higher moisture content than the surface layer.

All methods are reasonably sensitive with regard to technological parameters in particular curing conditions and water/cement-ratio. Whereas the ISA-Test clearly distinguishes between sufficient and insufficient curing of laboratory test specimens this may not be true for mature concrete in a structure [22].

All test methods are sensitive with regard to the moisture content at the time of testing though to different extents. The ISA-test as well as the surface-air permeability tests are affected most by the moisture content. Therefore, when employing these methods, a definition of the moisture state is mandatory. This can be done e.g. by starting out from a saturated condition and subsequent drying in a defined way as is proposed in

[13] and [18]. Also suitable methods may be developed which allow the rapid characterization of the moisture state of the surface layer in connection with calibration relations so that the actual moisture state can be taken into account. When drying the surface layer in a defined way, it should be kept in mind, however, that the depth of drying will depend on the diffusivity of the concrete, permeable concretes resulting in larger depths of drying than impermeable concretes so that the ratio A/l in eq. (8) is not a constant.

Air permeability measurements based on the Figg-approach are less sensitive to drying as has been shown e.g. in [20], however, as has been pointed out already, they may not take into account the properties of the drier surface layers.

All methods are reasonably simple in handling. The surface permeability or the ISA approaches are less distructive than the Figg-methods. With regard to reliability and repeatability of results more objective round robin tests are required before a conclusive answer can be given.

There is a pronounced lack of information regarding relations between various transport coefficients as well as between transport coefficients and durability of concrete. There are sufficient data available to state that depth of carbonation of laboratory specimens can be related to surface air permeability (Fig. 9). However, type of cement and additions have to be known for a reliable estimate. No experimental results are known in which such relations have been established under site conditions. Some results are available with respect to chloride penetration [18]. It is unlikely that permeability and frost resistance can be related without additional information such as air entrainment. However, it is likely that permeability may be used to estimate the rate at which a concrete reaches a critical degree of saturation.

The questions put forward in this brief discussion are dealt with in RILEM Committee TC 116-PCD "Permeability of Concrete as a Measure of its Durability". The committee also plans round robin tests and will try to

establish criteria with regard to permeability which have to be satisfied
by the concrete in a structure for given exposure conditions.

6 Summary

In this paper the various mechanisms for the transport of liquids, gasses
and ions in concrete are summarized, and generally accepted relations are
given which describe the transport of such substances by diffusion, permeation or capillary suction.

Test methods which are suitable for site applications to determine permeability and water absorption of concrete are described and discussed with
respect to their sensitivity to technological and environmental conditions. It is pointed out that the most crucial problem in applying such
methods and interpreting the results obtained is the moisture state of
the concrete at the time of testing.

There is a lack of information relating permeability measurements to concrete durability characteristics.

Most of the open questions dealt with in this paper will be treated in
RILEM Committee TC 116-PCD.

References

[1] Levitt, M.;
     "Non-destructive testing of concrete by the initial surface absorp-
     tion method",
     Symposium on non-destructive testing of concrete and timber, London,
     June 1969, Institution of Civil Engineers, 1970, paper 3 B, pp.
     23-26.

[2] British Standard Institution;
     "Methods of testing hardened concrete for other than strength",
     Test for water absorption BS 1881, Part 5, 1970.

[3] Figg, J.W.;
    "Methods of measuring the air and water permeability of concrete",
    Magazine of Concrete Research, Vol. 25, No. 85, Dec. 1973.

[4] Pihlajavaara, S.E., Parroll, H.;
    "On the correlation between permeability properties and strength of
    concrete",
    Cement and Concrete Research, Vol. 5, No. 4, July 1975, pp. 321-327.

[5] Steinert J.;
    "7erstörungsfreie Ermittlung der Wassereindringtiefe in Kiesbeton am
    Bauwerk",
    Forschungsbeiträge für die Baupraxis, Festschrift K. Kordina, Verlag
    W. Ernst u. Sohn, 1979.

[6] Kasai, Y., Matsui I., et al;
    "Air permeability and carbonation of blended cement mortars",
    Fly Ash, Silica Fume, Slag and other Mineral By-Products in Con-
    crete, American Concrete Institute, ACI-SP 79, August 1983, Vol. 1,
    pp. 435-451.

[7] Cather, R., Figg, J.W., Marsden A.F., O'Brien, T.P.;
    "Improvements to the Figg method for determining the air permeabili-
    ty of concrete",
    Magazine of Concrete Research, Vol. 36, No. 129, Dec. 1984.

[8] Kasai Y., MatsuiI., Nagano, M.;
    "On-site rapid air permeability test of concrete",
    In situ/nondestructive testing of concrete, American Concrete Insti-
    tute, ACI-SP82, 1984, pp. 520-541.

[9] Hansen, A.J., Ottosen, N.S., Petersen, C.G.;
    "Gas-permeability of concrete in situ: theory and practice",
    In situ/nondestructive testing of concrete, American Concrete Insti-
    tute, ACI-SP82, Detroit 1984, pp. 543-556.

[10] Lawrence, C.D.;
"Transport of oxygen through concrete",
The British Ceramic Society Meeting on "Chemistry and chemically re-
lated properties of cement", Imperial College, London, 1984.

[11] Dhir, R.K., Chan, Y.N., Hewlett, P.C.;
"Near-surface characteristics of concrete: an initial appraisal",
Magazine of Concrete Research, Vo. 38, No. 134, March 1986.

[12] Dhir, R.K., Hewlett, P.C., Chan, Y.N.;
"Near-surface characteristics of concrete: assessment and development of in situ test methods",
Magazine of Concrete Research, Vol. 39, No. 141, Dec. 1987, pp.
183-195.

[13] Schönlin, K.F., Hilsdorf, H.K.;
"Evalulation of the effectiveness of curing of concrete structures",
Concrete Durability, Katharine and Bryant Mather International Conference, American Concrete Institute, ACI SP100, Detroit, 1987, pp.
207.

[14] Tanahashi, I., Ohgishi, S., Ono, H., Mizutani, K.;
"Evaluation of durability for concrete in terms of waterlightness by
permeability coefficient test results",
Concrete Durability, Katharine and Bryant Mather International Conference, American Concrete Institute, ACI-SP100, 1987, pp. 189-205.

[15] Lydon, F., Odaallah, M.;
"On surface relative permeability test for concrete",
Construction and Building Materials, Vol. 2, No. 2, 1988,
pp. 102-108.

[16] Rostasy, F.S., Bunte, D.;
"Assessment of durability of concrete surfaces",
Durability of non-metallic inorganic building materials - DFG - Ab-
schlußkolloquium - Schriftenreihe des Instituts für Massivbau und
Baustofftechnologie, Universität Karlsruhe, Heft 6, Oktober 1988.

[17] Paulmann, K.;
"In-situ-Permeabilitätsmessungen an Betonoberflächen",
Deutscher Ausschuß für Stahlbeton, 22. Forschungskolloquium,
Braunschweig, 1989.

[18] Schönlin, K.F.;
"Permeabilität als Kennwert der Dauerhaftigkeit von Beton",
Schriftenreihe des Instituts für Massivbau und Baustofftechnologie,
Universität Karlsruhe, Heft 8, 1989.

[19] Hilsdorf, H.K.;
"Durability of concrete - a measurable quantity?",
Durability of Structures; IABSE Symposium 1989, Report pp. 111-123.

[20] Chen Zang Hong, Parrot, L.J.;
"Air permeability of cover concrete and the effect of curing",
Cement and Concrete Association Services, Report, October 1989.

[21] Reinhardt, H.W., Mijnsbergen;
"In-situ measurement of permeability of concrete cover by over-
pressure",
Seminar on Life of Structures, Paper No. 29, Brighton, U.K., 1989, .

[22] Rostasy, F.S., Bunte, D.;
"Evaluation of On-site conditions and durability of concrete panels
exposed to weather",
Durability of Structures, IABSE Symposium 1989, Report pp. 145-149.

[23] Bunte, D., Rostasy, F.S.;
"Test methods for on-site assessment of durability",
Durability of Structures, IABSE Symposium 1989, Report pp. 335-340.

[24] The Institute of Concrete Technology, Loughborough University of
Technology; "In-situ measurements of concrete permeability",
Workshop Dec. 12th, 1989.

[25] Reinhardt, H.W., Gaber, K.;
"From pore size distribution to an equivalent pore size of cement
mortar", Materials and Structures, Vol. 23, 3-15, 1990.

25 PERMEABILITY TESTING ON CONCRETE

J.G.A. van HULST
DHV Consultants, Amersfoort, The Netherlands

Abstract
The paper describes the tests executed on concrete to gather information on the quality of the reinforcement cover; this was done by determining the permeability for air with Figg-apparatus.

Information is also gathered with regard to the influence of the conditioning of the samples on the permeability.

1   PREAMBLE

The major threat regarding the durability of reinforced concrete structures is corrosion of the reinforcement. Reinforcement is liable to corrode if sufficient oxygen and water are present and/or if corrosive elements (e.g. chlorides) are able to penetrate the passivated layer of the bars. Since oxygen, water and in many cases also aggressive components penetrate from the outside of a concrete structure, the thickness and the quality of the concrete cover are very important for the durability of the structure as a whole.

In December 1986 the national standard "Regulations for concrete Technology" (VBT 1986) was introduced in the Netherlands. In addition to regulations regarding the strength of the concrete, this standard also concerns the durability of the concrete.

A problem was (and still is) to determine this durability: besides a method to be found, the parameters which influence the durability must also be defined.
In most cases cores drilled from the construction are tested in laboratories for their strength and porosity to get some information about this phenomenon .

For some time permeability has been considered to be a parameter which gives more accurate information about the durability of materials.
The difference between porosity and permeability is shown in figure 1 .

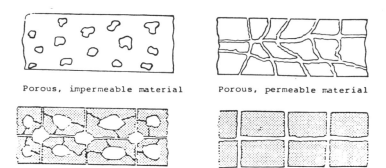

Porous, impermeable material   Porous, permeable material

High porosity, low permeability   Low porosity, high permeability

Figure 1   <u>Permeability and porosity (after Bakker)</u>

Under the auspices of CUR (Centre for Civil Engineering Research, Codes and Specifications) an extensive research programme was started in the Netherlands in 1986 with regard to the repair and protection of concrete. One of the objects of this programme was to establish the quality of the concrete cover in situ.
After a literature study by DHV Consultants with regard to different testing methods for establishing this quality several were selected.
Most attention was paid to the method developed by John Figg of Ove Arup & Partners, London. In comparison with other test methods (Karsten, ISAT) with this method the <u>permeability</u> (and not for instance the absorption) is measured. This is done for the <u>concrete cover</u>, whereas other methods are restricted to the quality of the <u>surface</u> or the <u>skin</u> of the concrete.

A big advantage of this method is also the possibility to execute the test on the concrete structure itself in situ, without drilling cores.

In the following chapters the Figg method is elaborated and the results of the tests executed in our laboratory are presented.

Part of the investigation was to find out whether the conditioning of the samples (mainly the moisture content) will influence the permeability values.

2   FIGG'S METHOD FOR ESTABLISHING THE PERMEABILITY

For the Figg method a hole has to be drilled. In this respect it is a destructive method, although in a very limited way. Figg himself wrote in "Building materials" Nov. 1989 "Concrete surface permeability": "The initial inspiration for the techniques came from observation of the aeration stone often used in an aquarium. This porous natural stone is attached to the air pump tube to produce a continuous series of air bubbles to replenish the oxygen content of the water. The interconnected pores of the stone diffuse the air flow in a way that appeared to be analogous to the movement of atmospheric air through the surface layers of concrete".

In the experiments by Figg the original diameter of the hole was 5.5 mm.
In later experiments hole sizes of 10 mm and 13 mm were used. The hole
depth was originally 30 mm and later 38 and 40 mm.
Until recently the hole was sealed by a foam plastic plug, covered by
silicone rubber cast in situ. From microbiological experience, it was
known that silicone rubber has the property of making an air-tight seal
when penetrated by a hypodermic needle.
It is rather difficult to apply the foam plug and the silicone rubber,
especially under field circumstances and on vertical surfaces.
Therefore, a specially moulded silicone rubber plug has been developed
recently.

Figg developed a gas permeability method and a water permeability method.
The gas permeability method was intended to give a correlation with the
progress of carbonation, wheras the ingress of water would correlate
with the penetration of solutions, in particular sulphate and chloride.
For the air permeability, initially vacuum was used. Because the test
method was developed for in situ test it was practical to use a hand
vacuum pump. This makes it impossible to reach a vacuum situation.
Nowadays almost all users apply an underpressure of 0.55 kPa.

The time (in seconds) for the underpressure to fall from -55 kPa to -50
kPa is measured. This time is the Figg value and is a measure of the air
permeability for concrete.

The testing apparatus has also been improved and the Figg value in
seconds will show automatically after applying the underpressure.

In Figure 2 the system is shown in detail.

In the test executed in the Netherlands a hole with a diameter of 20 mm
and a depth of 40 mm (after sealing) was used. The dimensions of the blind
hole differ from those used by (for instance) Figg: since the maximum
grain size of the coarse aggregate used in the Netherlands is higher
than the maximum grain size used in the U.K. and the grain size most
likely influences the permeability, the dimensions of the hole in the
Dutch tests have been adapted.

Figg introduced a classification system (see table 1) and found an
acceptable correlation between the compressive strength of the concrete
and its (air) permeability for different types of aggregates.

It should be noticed that originally Figg experimented under laboratory
conditions, with concrete which was also made in the laboratory.

Table 1   Classification of concrete quality as introduced by Figg

quality category	time (sec.)	quality of material	type of materials (examples)
0	< 30	poor	porous mortar
1	30 - 100	moderate	20 MPa concrete
2	100 - 300	fair	30 - 40 MPa concrete
3	300 - 1000	good	densified, well-cured concrete
4	> 1000	excellent	polymer-modified concrete

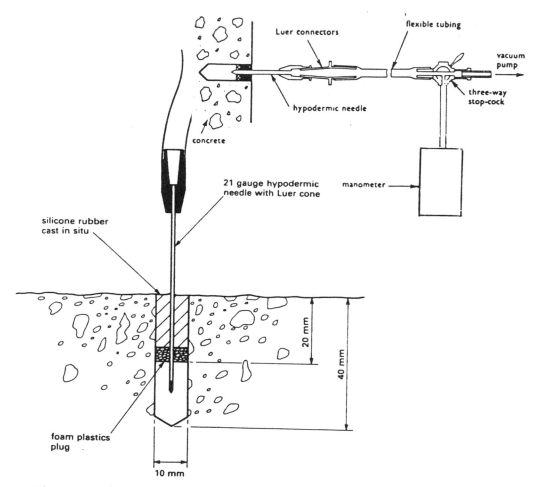

Figure 2   Figg apparatus for determining the air permeability

3   DUTCH EXPERIMENTS

3.1   General

As already stated, the permeability tests were executed on old concrete.
The original research programme also included permeability tests on
laboratory made (new) concrete, in order to determine factors like w/c
ratio, cement content, type of cement etc. For financial reasons this
part of the programme has not been executed up till now.

3.2   Results of the experiments

Two series of tests were executed.
In both in the first and second series cores were drilled or sawn from
concrete structures.

All the test specimens of the first series were over 10 years of age.
The values for both the permeability for air and the permeability for
water were established. In this paper only those for air are presented
since these values are more reliable.

The test specimen were conditioned as follows:
for two months they were stored at a temperature of 20°C and a RH of
respectively 50, 80 and 95%.

The following data can be presented of the types of concrete investigated.

Table 2 - Data of the investigated types of concrete

Type of Concrete	A	B	C	D
Type of Cement	Portland	Portland	Portland	Portland Blast furnace[1])
Cement Content (kg/m³)	370	355	450	340
Mass by volume (kg/m³)[2])	2247	2259	2359	2299
Porosity (% v/v)[3])	14.0	13.6	9.5	11.3
Compressive strength (Schmidt rebound hammer; N/mm²)	42.0	50.5	58.0	40.4

[1])   slag content >65%
[2])   mass of the samples dried at 105°C (contant weight)
[3])   total porosity

The results are presented in Table 3 .

Table 3 <u>Results of the permeability tests (in seconds) stored under different conditions</u>

Type of concrete	R.H.	50%	80%	95%
A	1	197	71	114
	2	103	95	58
	3	123	148	315
	4	177	23	45
	5	97	45	366
	6	178	144	--
Average		146	88	180
Standard deviation		43 (30%)	51 (60%)	150(83%)
B	1	568	1090	574
	2	476	1725	127
	3	790	254	296
	4	253	1050	444
	5			
	6			
Average		522	1030	360
Standard deviation		222 (43%)	603 (59%)	193(54%)
C	1	96	92	48
	2	--	226	833
	3	34	302	--
	4	146	40	--
	5			
	6			
Average		92	160	440
Standard deviation		56 (61%)	113 (71%)	
D	1			108
	2			171
	3			107
	4			133
	5			140
	6			226
Average				147
Standard deviation				45(31%)

In Figure 3 these results are presented graphically.

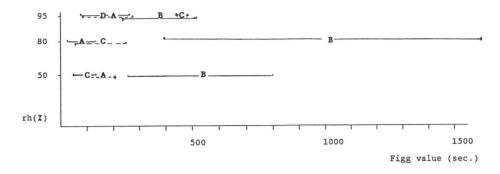

A: (OPC, 370 kg/m3)
B: (OPC, 355 kg/m3)
C: (OPC, 450 kg/m3)
D: (PBF, 340 kg/m3)

Figure   3   Relation between the permeability (in seconds) and storing
               condition (RH) for different types of concrete

In Figure 4 the Figg permeability values found for air are compared with
the compressive strength values (by means of the Schmidt rebound hammer)
of the different types of concrete (samples stored at 95% RH and 20°C).

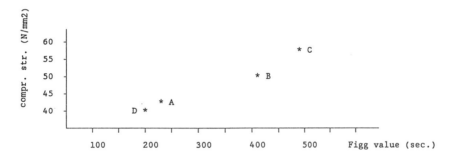

A, B, C, D: different types of concrete

Figure   4   Relation between permeability and the compressive strength

Although a <u>linear</u> relation between the air permeability values and the
compressive strength seems to be indicated, this type of relation is
highly unlikely: the deviation in the results of one test specimen is
considerable.
The influence of the humidity in the test specimen could not be established; most likely this influence is less than expected.

Since it was also the intention of these tests to set up a frame of
reference for the permeability and the compressive strength, it was
concluded from the results of this first series that the range in the
concrete quality was not large enough: in other words, all the concrete
tested was of a fair to good quality (see Table 1).

Therefore, in a second test series concrete samples were collected from structures which were expected to be of a (much) poorer quality.

For this series, the conditioning was stopped after an initial conditioning of the specimen. Ten days before establishing the permeability the samples were wetted and afterwards kept under laboratory conditions until tested.

The results are as follows with regard to the air permeability and compressive strength.

Table 4   Results of permeability and compressive strength tests

	RH 50%	A	80%	A	95%	A	compress. strength (N/mm²) A	carbonation depth (N/mm²) A		age (years) (mm)
	permeability (sec)									
1	7	3			3		22	45		60
2	42,46	–			32,87		39	0		6
3	12,14	25	51,51	51	–	60	33 36	6	3	6
4	14,23	–			–		36	2		6
5	29	–			43,53		39	16		6
6	18,17	19	–	14	–	48	36 36	2	9	6
7	12,19	14,14			–		33	10		6

A: Average value of the "group of results"

Core 1 was drilled from a 60 years old concrete structure exposed outdoors (protected against rain).
Core 2 t/m 4 were drilled from concrete elements which were exposed outdoors and core 5 t/m 7 from concrete elements which were stored in a cellar during 6 years.

4   CONCLUSIONS AND RECOMMENDATIONS

The following conclusions can be drawn from the experiments executed so far:

1.   The influence of the moisture content and of the conditions under which the test specimen were exposed on the air permeability of concrete is limited. This is important since it is not possible in situ to condition the structure to be checked.

2.   The relation observed between the permeability and the compressive strength is not significant. Concrete of a (very) poor and (very) good quality (defined on the basis of compressive strength) can however be distinguished by establishing the air permeability.

340

3. The air permeability test gives reliable information about the
   durability of a concrete structure.

4. It is necessary to exclude the influence of cracks and inhomogenities
   during the air permeability test. The sealing of the drilled hole
   has to be checked for leakages.

It is recommended to investigate both concrete structures that are
considered to be of bad quality (on the basis of visual inspection or
because maintenance is frequently required) and structures which have
proved to be of good quality. The Figg air permeability of these structures of different qualities should be determined in order to set up a
frame of reference. This frame of reference can be used to establish the
durability of a concrete structure by measuring its air permeability.

Together with measuring the thickness of the cover, using the Figg
apparatus for establishing the quality of the cover some time (any time)
after the completion of the works, the quality of the structure as a
whole can be checked and if necessary measures can be taken in order to
reduce repair and maintenance costs.

LITERATURE

1. The Concrete Society
   Permeability of concrete and its control; papers for a one-day
   conference, London, 12 December 1985.

2. Cather, Figg, Marsden and O'Brien
   Improvements to the Figg method for determining the air permeability
   of concrete.
   Magazine of concrete research (December 1984, vol. 36 no. 129).

3. Figg
   Concrete surface permeability: measurement and meaning
   Chemistry and Industry, 6 November 1989

4. In situ meetmethoden naar de kwaliteit van de betondekking met
   betrekking tot de duurzaamheid
   Croes, H
   T.U. Delft, oktober 1987

5. De kwaliteit van de betondekking in situ (not published)
   Nuiten, P
   DHV Consultants, mei 1988

6. Smolczyk H., Romberg H.
   Der Einfluss der Nachbehandlung und der Lagerung auf die Nacherhar-
   tung und Porenverteilung von Beton (Teil 1, 2)
   Tonindustrie Zeitung, Okt. 76

7. Hurling H.
    Oxygen Permeability of Concrete
    RILEM seminar, Hannover 84

8. Graf H., Grube H.
    The influence of curing on the gas permeability of concrete
    RILEM seminar, Hannover 84

9. Levitt M.
    In situ permeability of concrete
    Symposium Concrete Society, London 84

10. Figg J.W.
    Methods of measuring the air and water permeability of concrete
    Magazin of Concrete Research, Dec. 73

11. Kasai Y., Matsui I., Nagano M.
    On site rapid air permeability test for concrete
    A.C.I., S.P. 82

12. Hansen A.J., e.a.
    Gas permeability of concrete in situ; theory and practice
    A.C.I., S.P. 82

13. Montgomery F.R., Adams A.
    Early experience with a new concrete permeability apparatus
    Int. Con. Structural faults, London 85

14. Grube H., Lawrence C.D.
    Permeability of concrete to oxygen
    RILEM Seminar, Hannover 84

15. Schönlin K., Hilsdorf H.
    Evaluation of the effectiveness of Curing of Concrete Structures
    ACI SP-100, volume 1, pp 207-226

16. Chen Zhang Hong, Leslie J. Parrott
    Air permeability of cover concrete and the effect of curing

26 EFFECT OF CURING CONDITIONS AT EARLY AGE ON DURABILITY OF GYPSUM-SLAG CONCRETE

Y.V. ZAITSEV and K.L. KOVLER
Polytechnic Institute (VZPI), Moscow, USSR
V.N. GUSAKOV
Research Institute of Building Materials, Kraskovo, USSR
A.P. ZAKHAROV
Ural Polytechnic Institute (UPI), Sverdlovsk, USSR

Abstract
Effect of curing conditions on short-term and long-term strength and deformation of gypsum-slag concrete, which is demoulded at early age and does not need any heat-wet treatment, is discussed. It is found that the strength may be greater in interior layers of products in spite of great moisture content. It is shown that the protection of small-size elements may be recommended against quick drying but not against moistening. The material long-term strength and durability prediction based on fracture mechanics seems to be valid.

1 Introduction

The well-known economical and technological advantages of gypsum-based products (high productivity, low production cost, unlimited raw materials) stipulate for wider gypsum-containing material application in construction. However, low water resistance and "catastrophical" gypsum creep on moistening, especially, limit the possibility of its usage in loaded structures.

Gypsum-lime-slag binder (GLSB) is one of the varieties of cementless composite gypsum-based binders. Unlike pure gypsum binder, GLSB has the property of hydraulic hardening which promotes the structure development and gypsum-products quality. At the present GLSB-concrete blocks are manufactured according casting-like technology, these blocks being used in interior and exterior walls of rural buildings up to 4-5 storeys high. Blast furnace granulated slag is mostly used in the concrete as an aggregate. The average values of compressive strength in accordance with concrete composition and GLSB activity are equal to 5 - 15 MPa. The concrete mean density is 1600 - 1750 kg/m^3.

In spite of the long-standing (more than 40 years) experience of gypsum-slag concrete wall structures usage, the physical-mechanical properties of the material are insufficiently studied, especially while taking into account moisture effect. The study of the long-term concrete strength and durability in wet state is an important problem because of the relatively high exploitation concrete moisture in exterior wall

343

structures (as well as in high moisture conditions in rooms). Thus according to rough estimation, the specimens being loaded for 3 - 5 years, the relative limit for long-term strength of non-water-resistant concrete based on pure gypsum binder in moistening reaches zero.

The aim of this paper is to investigate the strength and deformation of gypsum-slag concrete while taking into account the curing condition effect as well as moistening and age of the materials influence and to estimate on this basis the admissible area of its usage in loaded structures. At the same time the clearly manifested dependence of GLSB concrete mechanical properties on its moisture is of interest for the material stress-strain state definition from fracture mechanics points of view, and in particular the long-term strength limit. The problem of curing conditions effect of the properties investigated arose in connection with gypsum-slag concrete demoulding in early age (in 20 - 30 minutes after moulding - because of quick gypsum component strength rise) and due to the absence of necessity in their heat-wet treatment.

2 Raw materials. Methods of testing

Prefabricated GLSB and blast furnace granulated slag (aggregate) with the fraction of no more than 20 mm were used as raw materials to manufacture concrete. Normalized GLSB composition (mass percentage) is as follows: gypsum binder of α-modification - 68-72; milled granulated blast furnace slag (acid) - 23 - 31; milled quicklime - 1 - 3. Characteristics of GLSB under investigation are in short: normal consistence of paste - 44 %; setting time: the beginning - 7 min, the end - 15 min; compression strength: at the age of 2 hours - 8.0 MPa, in dried state - 20.0 MPa; density - 900 kg/m^3. Density of the slag used as an aggregate is 1020 kg/m^3.

Binder/aggregate mass ratio is 1 : 1.3 proceeding from the expected compressive strength not less than 5 MPa. Water/binder proportion is 0.52 - 0.55; mixture mobility on cone slump (used to denote the workability of mortars) is 9.5 - 10.8 cm.

Concrete mix was produced in laboratory forced mortar mixer. Specimens (100 mm cubes and 100 x 100 x 400 mm prisms) were made without special compaction and were demoulded in 1.5 - 2 hours after their moulding.

Right after demoulding all the specimens were weighed, part of the cubes was tested for compression at the age of 2 hours and in dried to constant mass state (7 day drying at 55 ± 3°C). All the other samples were cured in different conditions till the beginning of testing (see below).

The samples were made simultaneously in series for short-term and long-term testing at different age (τ), in different curing conditions and at different concrete moisture values (W). As a rule a series comprised:

a) 3 cubes and 3 prisms to determine the cube strength (R) and prism strength (R$_b$), the initial elasticity modulus (E) and moisture content;

344

b) 2 - 8 prisms for testing in long-term axial compression, including relative long-te
strength limit evaluation;

c) 1 - 2 prisms for mass control (loss of moisture), including sealed prisms to estimate
waterproof reliability;

d) 1 - 2 prisms to measure shrinkage and swelling;

e) 3 - 6 cubes and 3 - 6 prisms for short-term testing at intermediate age (as regards
long-term loaded specimens).

Besides, an additional series of cubes was made (100 mm and 150 mm), aimed at
determining strength dependence on concrete moisture in its full range of variation
(from 0 to complete water saturation with intervals 0.5 - 3.0 %), and at preliminary
estimation of element section moisture gradient influence on the strength of certain
concrete layers.

Samples of different series were divided into three groups according to pre-testing
curing conditions:

1. air-dried curing at relative moisture of the environment $\varphi_a = 45 - 60 \%$ with
   quick drying (in two weeks concrete moisture did not exceed 1.5 - 2.0 %);

2. "normal" curing at $\varphi_a \geq 85 - 90 \%$ (right after demoulding) or at sealing with
   no less than 5 % residual moisture;

3. intermediate group: quick drying in air-dried conditions of up to $W = 5 - 7 \%$
   and later on - decelerated drying at $\varphi_a = 75 - 85 \%$; right after demoulding
   curing at $\varphi_a = 80 \pm 5 \%$, drying being of average intensity.

In all cases the air temperature was 20 ± 2 °C.

Short-term testing for axial compression was performed on a hydraulic press of 200
kN capacity. The prisms were loaded by steps amounting to 0.1 of predicted ultimate
load, the speed of the feeding being constant, with load exposure on steps for 5 min.
The loading was accomplished in axial compression spring devices. Samples were
sealed in the process.

3 Testing results

The main results of the short-term testing are given in Fig. 1 and 2. Concrete
moisture changing from 4.5 % to water saturation state (12 - 14 %), the values of R,
R_b and E differ insignificantly. That is why the data on series with different moisture
in the given interval are united ("wet" concrete). The test results of both air-dried

345

(W ≤ 0.3 %) and dried to constant mass (W = 0) samples are regarded as "dry" concrete. This accounts for the 1st group "dry" concrete strength reduction in 7 days. We failed to reveal strict dependence in concrete properties while changing its moisture from 0.3 to 4.5 %.

Fig. 3 illustrates the influence of curing conditions, their difference being due to moisture gradient in cube-sample section. The 150 mm cube-sample was cured 3 months in free water-change conditions, with $\varphi_a = 80 \pm 5$ % (the 3rd group), and was sealed 2 weeks before testing in order to even the moisture over the section. Local strength, moisture and density values were obtained on 25 micro-cubes sawn out of

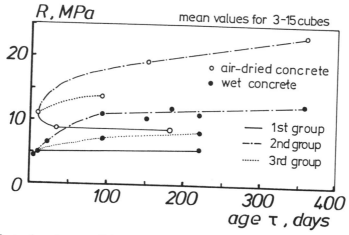

Fig. 1. Effect of curing conditions and specimen age on cube compressive strength

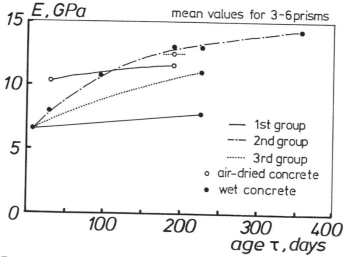

Fig. 2. Effect of curing conditions and specimen age on elasticity modulus

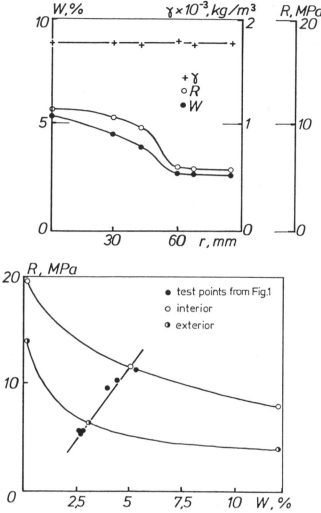

Fig. 3. The distribution of strength (R), moisture (W) and density (γ) in 150 mm GLSB concrete cube section: a) the dependences of R, W and γ on radius of microvolume removal from the sample core; b) R-W dependences for interior (I) and exterior (E) layers of the specimen

the 150 mm specimen medium part.

Main results of long-term compressive testing are shown in Table 1 and Fig. 4. Old-age concrete samples ($\tau > 220$ day) cured in different conditions were moistured to a certain value of W directly before testing. The planned load levels (σ/R_b) are given as relations to sample-twins average prism strength for corresponding moisture and age. Real σ/R_b-values may have deviations due to strength readings discrepancy.

Table 1 The main results of sealed gypsum-slag concrete long-term compressive testing, including to failure (∗)

Sample No.	Age τ, days	Moisture W, %	Stress level σ/R_b	Stress σ MPa	Test time t-τ, days	Total strain $\epsilon(t\text{-}\tau)\cdot10^5$	Creep characteristics $\varphi(t\text{-}\tau)$	$\varphi(\infty)$
				1st group				
3.1.3	28	0.25	0.30	2.81	696	325	14.8	21.3
3.1.4	28	0.25	0.30	2.81	696	360	16.4	21.5
2.2.1	220	4.5	0.30	1.50	24 ∗	1768	96.1	—
2.2.2	220	4.5	0.30	1.50	24	919	54.0	—
2.2.6	220	4.5	0.40	2.00	2 ∗	1317	37.2	—
2.2.9	220	4.5	0.40	2.00	2	295	10.9	—
2.2.3	220	4.5	0.50	2.50	1	458	13.7	—
2.2.4	220	4.5	0.50	2.50	1 ∗	865	22.9	—
2.2.5	220	4.5	0.70	3.50	0.1∗	556	12.8	—
2.2.8	220	4.5	0.70	3.50	0.1	280	4.6	—
				2nd group				
1.0.1	7	5.5	0.30	1.11	327	640	44.8	44.9
1.0.6	7	5.5	0.30	1.11	327	456	32.7	32.9
1.0.2	7	5.5	0.20	0.74	327	273	28.7	28.8
1.0.5	7	5.5	0.20	0.74	327	280	32.2	32.5
1.1.3	28	9.5	0.30	1.59	315	60	4.1	4.4
1.1.5	28	9.5	0.30	1.59	315	58	4.7	5.1
2.1.2	28	4.5	0.30	2.25	315	160	6.8	7.3
2.1.5	28	4.5	0.30	2.25	315	157	6.6	6.8
3.2.2	90	7.5	0.30	2.86	353	141	4.2	4.3
3.2.4	90	7.5	0.30	2.86	353	159	4.9	5.1
4.2.1	220	9.5	0.25	2.00	426	33	2.7	3.2
4.2.2	220	9.5	0.25	2.00	426	45	2.8	3.3
4.2.3	220	9.5	0.42	3.34	570	351	11.6	—
4.2.4	220	9.5	0.42	3.34	570	148	5.6	5.9
4.2.5	220	9.5	0.57	4.66	4 ∗	952	19.4	—
4.2.7	220	9.5	0.57	4.66	4	389	7.8	—
4.2.6	220	9.5	0.66	5.35	1.5	342	5.7	—
4.2.8	220	9.5	0.66	5.35	1.5 ∗	495	7.7	—
6.2.1	360	12.0	0.30	3.16	321	126	4.0	4.2
6.2.2	360	12.0	0.30	3.16	321	103	3.3	3.4
G-19	360	0.5	0.97	12.61	555	368	2.5	2.7
G-5	360	0.5	0.95	11.70	548	323	2.4	2.5
				3rd group				
7.2.1	220	4.5	0.30	1.85	393	132	6.8	7.1
7.2.3	220	4.5	0.30	1.85	393	63	3.5	4.0
7.2.2	220	4.5	0.50	3.12	521	220	8.7	8.8
7.2.5	220	4.5	0.50	3.12	521	770	26.1	26.7
7.2.6	220	4.5	0.70	4.31	521	522	15.0	15.3
7.2.7	220	4.5	0.70	4.33	521	828	22.5	23.0
1.2.1	220	7.5	0.30	2.22	411	96	4.9	5.7
1.2.3	220	7.5	0.30	2.22	411	43	2.5	3.0
1.2.2	220	7.5	0.50	3.69	2 ∗	587	11.2	—
1.2.6	220	7.5	0.50	3.69	2	318	9.2	—
5.2.1	220	10.5	0.25	1.85	438	69	5.8	6.3
5.2.3	220	10.5	0.25	1.85	438	98	6.8	7.0

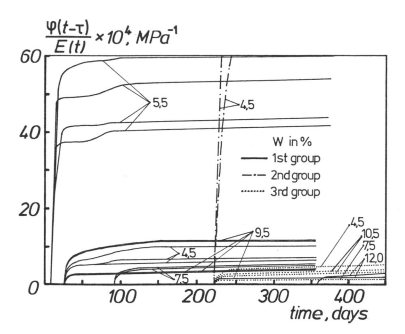

Fig. 4. Creep characteristic/elasticity modulus ratio vs. loading time at $\sigma/\mathrm{R}_b = 0.30$

Fig. 5 shows the results of the estimation of the long-term strength limit for the 1st and 2nd groups wet concrete. Experimental-theoretical curves are drawn according to the eq. obtained in [1,2] on the basis of fracture mechanics approach:

$$\eta(t,\tau) = \frac{m(t,\tau)R(t)}{R(\tau)} \sqrt{\frac{E(\tau)}{E(t)[1 + \varphi(t,\tau)]}} \tag{1}$$

where $\eta(t,\tau)$ - relative stress level corresponding to specimen failure time; $m(t,\tau)$ - the function of preceding loading influence on concrete strength; $\varphi(t,\tau)$ - creep coefficient. At the same time $\varphi(t,\tau)$ essentially depends on load level due to non-linear creep. Taking into account the data obtained in calculations according to eq. (1), creep characteristics were assumed at $\sigma/R_b = 0.3$ for both 7 and 28 day concrete and for the 1st group also. For old-age concrete of the 2nd group the values of $\varphi(t,\tau)$ are assumed at $\sigma/R_b = 0.4$ proceeding from the best accordance with fracture points in Fig. 5. In all cases the non-linear influence is reflected in numerical values $\varphi(t,\tau)$.

Kinetics of the 2nd group wet concrete strength and initial elasticity modulus changes were approximated by equations:

$$R_b(t) = R_b(7) + A\lg(Bt) \tag{2}$$

$$E(t) = E(7) + [E_o - E(7)](1 - e^{-Ct}) \tag{3}$$

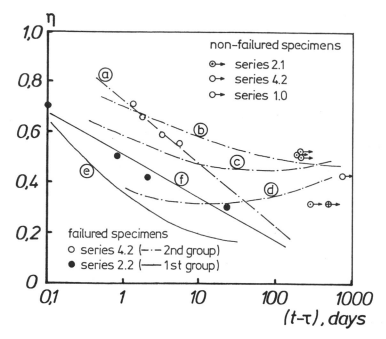

Fig. 5. Effect of curing conditions and specimen age (for 2nd group) on wet gypsum-slag concrete relative long-term strength η

(a) 4.2; $\tau = 220$ days; $\eta = 0.71 - 0.2481 \log(t - \tau)$ (b) 4.2; $\tau = 220$ days; Eq. (1)
(c) 2.1; $\tau = 28$ days; Eq. (1)   (d) 1.0; $\tau = 7$ days; Eq. (1)
(e) 2.2; $\tau = 220$ days; $\eta = 0.50 - 0.0821 \log(t - \tau)$ (f) 2.2; $\tau = 220$ days; Eq. (1)

where $R_b(7)$ and $E(7)$ - prism strength and initial elasticity modulus of concrete at the age of 7 days, MPa (increase in strength for the age from 2 hours to 7 days is not determined); A, B and C - constant coefficients, experimentally selected; E_o - initial elasticity modulus of old age concrete, MPa. Of course, some other functions can be used, for example functions described by N.J. Carino and R.C. Tank in this volume.

Having substituted experimentally obtained values into (2) and (3), we have

$$R_b(t) = 3.5 + 4.023 lg(0.2t) \tag{4}$$

$$E(t) = 6500 + 8000(1 - e^{-0.0107t}) \tag{5}$$

For the 1st group concrete we assume: $E(t) = E(\tau)$ and $R_b(t) = R_b(\tau)$. Cube strengths are equaled too: $R(t) = R(\tau)$.

While calculating $\eta(t,\tau)$ values for loading at the age of 7 and 28 days - $\tau > 1$ day, $m(t,\tau)$ - values are assumed to be constant and equal to 1.15, according to the results of parallel short-term testing of the 2nd group loaded and non-loaded sample-twins. Due to the absence of the experimental data we assume: $m(t,\tau) = 1$ at $\tau = 220$ days and for the 1st group concrete as well.

4 Discussion

Let us scrutinize the model experiment on the investigation of strength distribution in sample section due to curing conditions differences in its micro-volumes (Fig. 3). In spite of the authors' intention to even moisture over sample section, they failed to avoid the gradient of W. If we re-construct the obtained local strength dependence on radius of micro-volume removal from the sample core (Fig. 3a), we shall at first sight get an abnormal result: strength increases with an increase of moisture. Considering that compacting coefficient of concrete in the specimen core and the one near its border are practically the same (γ-values don't change in section), this paradox is explainable only by various curing conditions in micro-volumes of the sample due to moisture gradient in section.

It would be interesting to note that short-term wetting of micro-samples adjusted to the centre (r < 60 mm) and of those at a distance (r ≥ 60 mm), and their drying to constant mass allow us to draw "strength-moisture" dependences of traditional form, but considerably removed from each other. The above shows that we have to do with concretes of practically different properties in specimen sections.

Thus, the effect of moisture may be conventionally divided into two kinds. The "long-term" moisture effect takes place right after the concrete product formation, with hydraulic properties of the material being realized during the period (depending on curing conditions). The "short-term" moisture effect takes place during the exploitation period and is of cyclic character (e.g. exterior wall surfaces moistening by oblique rains and their subsequent drying).

In real working conditions, when "long-term" moisture effect is over, the material hydraulic properties have been realized. There is no hydraulic sealing of exterior wall surface constructions thus causing much more dryness of exterior concrete layers than in the above model experiment. Therefore we may as a rule expect a less strength gradient in section if the direct "short-term" moisture effect on the wall surface is evaded.

The adduced data show that sample curing conditions effect quantitatively the gypsum-slag concrete properties. If the concrete dries quickly up its hydraulic properties neither have time "to work" nor to manifest themselves in future after moistening of the samples which were for a long time cured in air-dried conditions. This is most obviously seen in Fig. 4. Unlike early age samples, cured under load, there was no noticeable hydration of the 1st group concrete in 24 days of load duration. This led to sample fracture. On the whole, the 1st group concrete properties still resemble the corresponding values of the concrete based on pure gypsum binder.

The concrete under testing is characterized by retarded strength and rigidity rise in up to 28 days age as compared to cement one. However, later on the process actively goes on for a long period (see Fig. 1 and 2).

Correlation between the 1st and the 2nd group concrete strength is confirmed by the distribution of strength over the concrete sample thickness as well. If we take into account the moisture gradients in working construction element sections (exterior layers being the drier ones), we may, as a rule, expect an approximately equal strength in the whole section depth due to great strength of the material in a dry state and far higher degree of hydration in interior wet layers. Thus, we see in Fig. 1 that in 40 days the 2nd group wet concrete strength exceeds that of the 1st group air-dry concrete. However, the above doesn't concern deformation under long-term load and, as a result, the long-term strength limit.

The character of the age influence upon the wet GLSB concrete creep is identical to the corresponding influence upon cement concretes, but the total deformation is considerably higher, especially at early loading age. The stress levels being up to 0.3, we couldn't reveal moisture effect (the former changing from 4.5 % to complete water saturation) upon creep characteristics. In some cases, the wetter concrete is even lower than that of the less water-saturated one. And this can be explained by the faults in the pre-determined loading levels. Their visible influence is possible in this case due to the essential non-linear creep, as the accepted long-term strength relative limit, i.e. far higher than its working level. This explains the deviations in regular creep lowering with concrete age (Fig. 4). Considerable discrepancy in the 3rd group concrete results is connected with a wide range of that group specimens curing conditions.

It was experimentally determined that the concrete long-term strength limit depends on its moisture content and its curing conditions. This is also confirmed by eq. (1), according to which the relative material long-term strength limit is to be lowered along with the material creep rise. The air-dried GLSB concrete creep is rather insignificant, just like the dry "pure" gypsum sample creep, with φ_a not exceeding 60 %. The data in Table 1 on a considerable creep of the 1st group concrete with W = 0.25 % are due to testing pecularities: air-dried sample sealing. On the one hand, this sealing apparently caused a uniform distribution of even small quantities of moisture within samples; on the other hand this sealing is insufficient to avoid the water vapour penetration when the moisture in testing chamber increases.

We may state the higher long-term strength limit for air-dried concrete (in medium of φ_a up to 70 %). While testing load levels of 0.90 - 0.97 none of the air-dried specimens were fractured for more than 550 days, creep deformation attenuating (Table 1). According to pre-evaluation the long-term strength relative limit doesn't exceed 0.4 for even the 2nd group wet concrete.

Fig. 5 shows that the long-term strength determination by the traditional method - that of linear regression extrapolation of experimental points coordinates η-lg(t-τ), which correspond to sample fracture, for the guaranteed period of the structure service life, e.g. t = 100 years, - considerably reduces its value. Thus, for the 1st group wet concrete the zero long-term strength is extrapolated with loading period of less than 3 years, which is physically pointless.

Drawn in Fig. 5 according to eq. (1), the 2nd group (τ = 220 days) wet concrete dependence is close to experimental points. For early-age loaded concrete there are no experimental points of fracture at all. However, with $\tau = 7$ days it is no difficulty to determine that the accepted level $\sigma/R_b = 0.3$ was close to the long-term strength limit at this age, because on the 2nd day non-hardened samples (the 1st group) are fractured at the level of 0.4. Thus, it won't be enough time for early age concrete and rigidity rise at the load level of more than 0.30 - 0.35.

The experimental-theoretical curves (not claiming the qualitative estimation for want of data) correctly reflect the tendancy for attenuation of dependences η-lg(t-τ) for the 1st group samples also. In the latter case considerable specimen fracture time shortening is due to the fact that creep characteristic values should have been applied at stress levels of no more than long-term strength limit (for the series in question it is probably no more than 0.10 - 0.15).

5 Conclusions

1. Curing conditions considerably effect the properties and durability of gypsum-slag concrete. While normalizing its characteristics one should take into account the massiveness of bearing structures. In case of need (small-size elements) the product protection may be recommended against quick drying but not against moistening, if a long-term moisture influence on the products is eliminated in the working period.

2. The long-term GLSB concrete strength relative limit, as well as the creep characteristic, depends on material moisture and amounts, according to pre-estimation to 0.4, the concrete moisture being no less than half of complete saturation moisture content, and to 0.85 as a minimum in air-dried state.

3. The strength and E-modulus distribution in structure sections is irregular and their values may be greater in interior layers in spite of greater moisture content. Taking into account this factor, we are able to rationally design the loaded wall structures.

4. The material long-term strength evaluation approach from the fracture mechanics point of view seems to be rather perspective, and especially when considering various factors: concrete age, moisture level, composition etc. One of the investigation tasks in future is the working out of concrete non-linear properties reliable consideration methods.

References

[1] Zaitsev, J.W. Simulation of concrete strength and deformations by fracture mechanics methods. Moscow, Stroyizdat Publishers, 1982 (in Russian)

[2] Zaitsev, J.W., Wittmann, F.H. Festigkeit und Verformung poröser Baustoffe unter Kurzzeitbelastung und Dauerlast. Deutscher Ausschuß für Stahlbeton, Heft 232. Berlin, 1974

27   IN-SITU AND IN-LAB TESTING OF SKIN CONCRETE

T.P. TASSIOS, S. KOLIAS and K. ALIGIZAKI
National Technical University of Athens, Greece

Abstract

The importance of skin concrete quality, both for flexural strength and (above all) for durability, cannot be overemphasized. The paper is dealing with several attempts to develop in-situ or in-lab methods of assessing the quality of skin concrete at early ages. The following methods have been used on concrete slabs compacted with three different methods: a) Pulse velocity, b) Water permeability under pressure, c) Porosity and d) Thin core-discs splitting. Curing effects are similarly followed up.

1   INTRODUCTION

The "near-surface" properties of concrete (skin concrete or cover concrete) are of great importance in characterising its behaviour versus environmental influences as well as under ultimate flexural loading. Recently, several research works (1-5) were carried out in order to advance the knowledge on the subject and to develop reliable test methods by which these properties may be measured.

In this paper an attempt is made to compare in-situ and in-lab test methods of assessing the quality of skin concrete, particularly at early ages. The following methods were used: a) Ultrasonic pulse velocity, b) Water permeability under pressure c) Mercury porosimetry, d) Total porosity measured by water absorption under vacuum, e) Splitting strength of thin core discs and f) Nail extraction-force measurement.

In an attempt to examine the influence of the compacting methods on the skin properties of concrete and the consequences of the compaction on the quality of young concrete, three different compaction methods were used i) vibration ii) rodding and iii) slight tapping.

Curing effects were similarly followed up for 28-day specimens compacted by vibration.

2 EXPERIMENTAL PROCEDURE AND DETAILS

Initially, 10 batches of concrete were prepared designated A,B,C,D,E,F,G,H, I,K, according to the method of compaction and curing, as shown in Table 1. From each batch, two small slabs 250mmX250mmX150mm were cast.

TABLE 1. Concrete batches and methods of compaction and curing

BATCH-SPECIMEN DESIGNATION		COMPACTION METHOD	SLUMP (mm)	CURING METHOD
A	A1,A2	Vibrating table	10	Wet room
B	B1,B2			
C	C1,C2	Tapping	10	Wet room
D	D1,D2			
E	E1,E2	Rodding	10	Wet room
F	F1,F2			
L*	L1 L2 L3	Vibrating table Tapping Rodding	10	Wet room
G	G1,G2	Vibrating table	10	Indoor atmosphere
H	H1,H2			
I	I1,I2	Vibrating table	10	Wet room
K	K1,K2			

(*) Supplementary batch, see para 3.2

The concrete mix proportions were kept constant through-out the research and are given in Table 2, together with the grading of the crushed lime-stone agreggate used. The workability of the mix, (see Table 1), was chosen very low (1-1.5 cm) in an attempt to increase the influence of the compaction method on the measured properties.

TABLE 2. Concrete mix proportions

AGREGGATE GRADING		
ASTM sieve	Opening mm	% Passing
3/4	19.05	100
1/2	12.70	97
3/8	9.53	86
No4	4.76	57
8	2.38	16
16	1.19	23
30	0.59	15
50	0.297	12
100	0.149	7
Water/cement 0.58 Cement content 300 Kg/m^3		

2.1 Testing of young concrete

All specimens were wet cured (RH>95%) until all measurements were completed. Table 3 summarizes the procedure followed from the time of demoulding until final testing.

TABLE 3: Testing of specimens

AGE	TESTING
1 day	Ultrasonic pulse velocity testing Water permeability under pressure
2 days	Core cutting
3 days	Sawing of 15mm thick slices from cores Testing the slices in diametral compression (splitting tension test)
7 days	Mercury porosimetry Porosity measurement by water immersion under vacuum (2 hrs under vacuum + 24 hrs water immersion)
28 days	Reference tests on cores in compression

2.2 Testing of mature concrete – Curing effects

The lab-cured specimens (temp. 22 to 24 °C) were put on metal supports with minimum contact to the specimen base. Table 4 summarizes the testing sequence followed.

TABLE 4. Testing sequence for examining curing effects

AGE	TESTS
1 day	Permeability under pressure
28 days	Ultrasonic pulse velocity testing Permeability under pressure Compression tests on cubes
38 days	Cutting of 100mm dia. cores Sawing 15mm thick slices from cores Testing the slices in diametral compression (splitting tension test) Compression tests on cores

2.3 Experimental methods

The following non-standard methods need some detailed description.

Water permeability under pressure
This method had been successfully used for in-situ measurements on old reinforced concrete dock buildings which had shown damage due to corrosion of reinforcement (6). The simple testing set-up consists of a metal plate (100mmX150mmX1mm approximately) which is glued onto the concrete surface using a fast hardening polyester glue. The metal plate is connected at the

center with a small metal pipe, 12mm diameter, through which water pressure
of 0,5 MPa is applied onto the concrete surface using a hand operated
hydraulic pamp. The drop of pressure in-time is measured by means of a
simple pressure gauge.

Total porosity measurement by water immersion under vacuum
The method is almost identical to the RILEM method CPC 11.3. However, since
the concrete specimens were fragments taken from the disc splitting test,
with dimensions not exceeding 30mmX20mmX15mm, the 750mm Hg underpressure was
kept only for 2 hours, while the water immersion lasted 24 hours.

Nail extraction
Standard nails (d=4mm, l=45mm) are driven by means of commercially available
gun used to fix nails in hardened concrete structures.
The extraction force is measured after 10 min and is correlated to the
compressive strength of concrete.

However, in this particular research on young concrete, splitting occures
during nail driving. The method needs to be appropriately adapted for the
purpose (smaller nail diameter and less powerful percussion cups).

3   TEST RESULTS ON YOUNG CONCRETE
3.1 Concrete density
The apparent density of the concrete slabs, compacted by different methods,
was measured. Despite the scattering and the small number of measurements,
it can be said that the average density is reduced in the following order:
vibrating table, rodding and tapping.

3.2 Ultrasonic pulse velocity test
Direct transmission measurements were carried out horizontally along the top
layer and the bottom layer, and also vertically as shown in Fig. 1. It can be
seen that, regardless of the method of compaction, higher velocities are
observed near the bottom of the specimen, a fact which is in accordance with
earlier results (see i.a.(8)); the concrete at the bottom of a lift is
expected to be subject to higher compaction and less bleeding, thus it will
be denser and of higher strength.

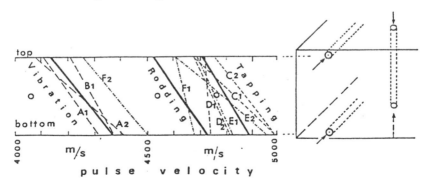

Figure 1. Pulse velocity values across the slab. Dotted and full lines
represent horizontal measurements (individual and mean values respectively).
Isolated full circles correspond to vertical pulse velocity measurements.

357

MOUNT PLEASANT LIBRARY
TEL. 051 207 3581 Ext. 3701.

It can also be seen that the pulse velocity of the specimens compacted by
vibration had consistently lower values in all directions than the velocity
of the specimens compacted by either tapping or rodding. In order to check
this result and exclude the possibility of an error in batch preparation,
an additional batch was prepared (batch L, Table 1), with the same mix
proportions, from which three specimens were cast and each one was compacted
by either vibration or rodding or tapping. The ultrasonic pulse velocity was
measured,after demoulding,by the direct transmission method (Table 5) and
also by the surface transmission method (Fig. 2).

TABLE 5. Ultrasonic pulse velocities from batch L.

Compaction Method	Pulse velocity m/sec	
	Direction	
	Horizontal	Vertical
Vibration	4348	4360
Tapping	4458	4460
Rodding	4346	4190

It can be seen from Table 5 that the pulse velocity of the specimen compacted
by tapping is higher than the pulse velocity of the specimens compacted by
vibration or rodding. When the pulse velocities between specimens compacted
by vibration and by rodding are compared, no difference in the horizontal
direction is observed while the pulse velocity of the rodded specimen on
the vertical direction is lower than that of the vibrated specimen, a result
which is given with some reservation.

It is believed that additional work is needed in order to clarify this point
before any explanation is attempted.

The results of the surface transmission measurements, plotted in Fig. 2 in
terms of pulse time against transducers dinstance, do not allow a distinction to be made between the velocities measured on different surfaces.
It is therefore concluded that this method is not sensitive enough to pick-
up differences in properties of young concrete.

3.3 Water permeability under pressure
The water permeability of the concrete was measured on the bottom and on the
lateral surface of the specimens and the results are plotted in Figure 3
in terms of pressure decrease versus time. Comparing the "water pressure
curves" between bottom and side surface of the same specimens,it can be said
that there is no consistent difference whether the permeability is measured
at the bottom or at the lateral surface of the specimen. It is noted,however,
that there is an indication of higher permeability of the side surface as
compared to that of the bottom surface in the case of specimens compacted by
tapping, which may be an evidence of horizontal compaction planes allowing
easier water movement.

For the comparison of the permeability between specimens compacted by
different methods,an "impermeability index" (P 10.) is introduced, defined
as the value of the residual water pressure 10 minutes after the application
of the initial pressure of 0.5 MPa. The average values of P10 and their
standard deviations are given in Table 6 for the different compaction
methods.

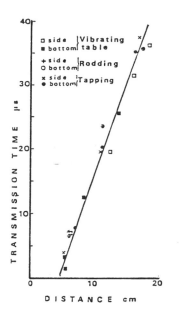

Figure 2. Surface transmission
measurements of ultrasonic
pulse

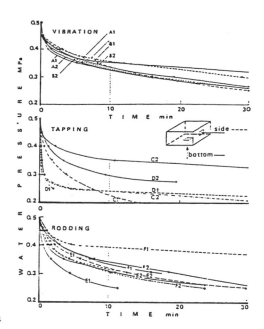

Figure 3. Permeability measurements

TABLE 6. Average values of impermeability index.

METHOD OF. COMPACTION	IMPERMEABILITY INDEX		
	MEAN MPa	STANDARD DEVIATION MPa 10^{-2}	COEF.VARIATION %
Vibrating	0,33	1,32	3,9
Tapping	0,27	5,14	19,0
Rodding	0,32	3,80	11,7

It is noted that the impermeability index increases with the order: tapping rodding, vibration, in agreement with the sequence of densities whereas, variabilities exhibit the inverse sequence: vibration, rodding, tapping.

It is believed that the high moisture content in the specimens after demoulding, regardless of the method of compaction, does not allow greater differences to be observed with an initial pressure of 0.5 MPa. Higher initial pressures will probably increase the number of bond failures between metal plate and concrete; longer observation times will make the test time consuming and therefore less practicable. However, the results seem to be promising.

3.4 Estimation of tensile strength

The vertical cores were cut in approximately 15mm thick slices and each slice was tested in diametral compression in an attempt to estimate the splitting tensile strength of the concrete. The results are presented in Fig. 4 and are reported with some reservation for the following reasons: a) local damages at the interface between aggregate and mortar may have been caused by sawing the young concrete b) the width of the discs was not always constant in each specimen c) the maximum aggregate size exceeded the thickness of the specimen. Notwithstanding these reservations, and despite the high variability of the results, the following observations may be made:

a. Specimens compacted by vibration: It seems that the strength increases towards the bottom.
b. Specimens compacted by tapping and specimens compacted by rodding: In both cases there is a clear indication that the strength increases from the bottom towards the interior of the specimen.

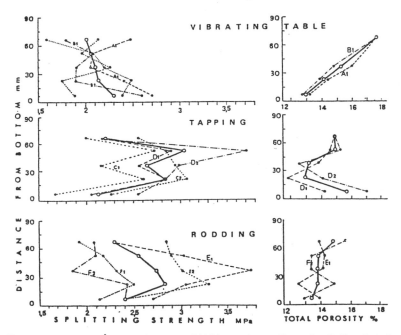

Figure 4. Variation of the splitting strength and of the total porosity in relation to the distance from the bottom of the specimen.

3.5 Total porosity by water immersion under vacuum

The results are presented in Fig. 4 in relation to the distance from the surface of the specimen. The concrete pieces used for porosity measurements had relatively small dimensions compared to the aggregate size; besides, the total porosity values included also the porosity of the aggregates. Thus, a high variability of results is expected. Consequently, the porosity variations within the specimen as shown in Fig. 4 have only an indicative significance. A larger number of tests is needed in order to definetely conclude whether the variations shown in Fig. 4 are really representative of the compaction method. Nevertheless the porosity variations are in most cases in accordance with the splitting strength variations, a fact which shows the potentiality of the porosity measurement as an indication of concrete quality.

3.6 Mercury porosimetry

The scope of the tests was to obtain an indication of the pore size distribution of the concrete and to correlate the results to the total porosity measurements. The measurements were carried out at the same age with the measurements of the porosity by water immersion under vacuum. Only fragments from the second slice of only one core per batch were used for mercury porosimetry. The results are shown in Fig. 5. It can be seen that the pore-size distribution differs according to the compaction method. In concrete compacted by vibration, most of the pores have sizes between 25 Å and 1000 Å while their percentages do not differ very much. On the contrary, the pore size distribution for concretes compacted by tapping or rodding exhibits a clear peak between 200 Å and 400 Å, while the rodded concrete exhibits also a comparatively increased number of pores between 1000 Å and 10000 Å.

The correlation between the mercury porosimetry values and the total porosity by water immersion is very good, as it can be seen from Fig. 6. It is also noted that the porosity measured by water immersion is higher than the porosity measured by mercury porosimetry.

3.7 Compressive strength

The core compressive strengths at 38 days given in Table 7, show that specimens compacted by tapping had lower strengths than specimens compacted by either rodding or with vibrating table. Rodding produced specimens with the higher values of standard deviation, as in the case of permeability tests.

TABLE 7. Compressive strength of cores

Compaction Method	Core compressive strength, MPa	
	Mean	St. Deviation
Vibrating table	29,8	2,0
tapping	27,8	1,7
rodding	30,3	4,7

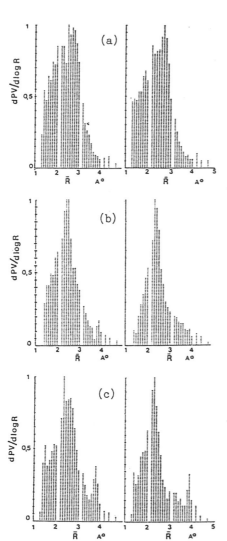

Figure 5. Results of mercury porosimetry

a) vibrating table
b) tapping
c) rodding

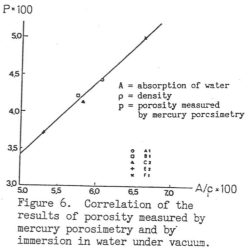

Figure 6. Correlation of the results of porosity measured by mercury porosimetry and by immersion in water under vacuum.

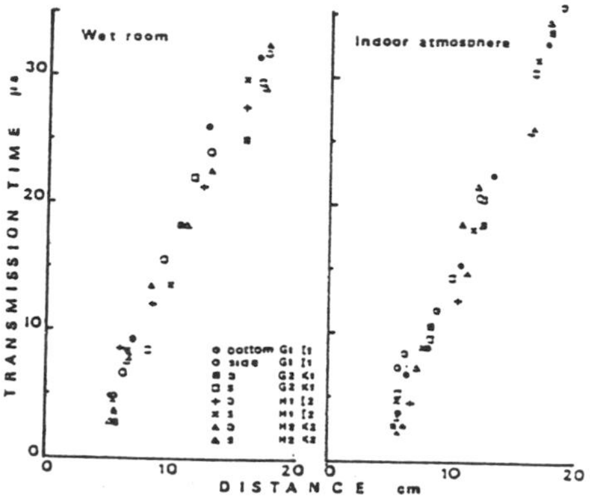

A = absorption of water
ρ = density
p = porosity measured by mercury porosimetry

Figure 7. Results of surface transmission measurements of ultrasonic pulse on differently cured concrete.

4 TEST RESULTS OF MATURE CONCRETE

4.1 Density

The core apparent-densities, estimated as described in para 3.1,are given in Table 8 and show slightly higher values for wet cured specimens.

4.2 Ultrasonic pulse velocity

Pulse velocity measurements were carried out at 28 days, and included direct transmission measurements, as well as, surface transmission measurements on the top, bottom and side surface of the specimen. The results are given in Table 8. As anticipated, the pulse velocity of the wet cured specimens is higher than that of the laboratory cured specimens, with the exception of direct transmission near the bottom surface of the specimen, where the picture is not so clear. The results of surface transmission measurements are plotted in Fig. 7 and it can be seen that this method of measurement is not sensitive enough to pick-up differences in mechanical properties along the various surfaces of the specimen; and it is quite so even between differently cured specimens as old as 28 days. This conclusion is in accordance to that for young concrete drawn in para 3.2.

TABLE 8. Effect of curing method in pulse velocity measurements

| CURING METHOD | PULSE VELOCITY m/s direction | | | | APPARENT DENSITY kg/m^3 |
	4 (bottom)	5 (center)	6 (top)	7 (vert)	
Indoor atmosphere	5301	5144	5026	5414	2362
wet room	5298	5233	-	5589	2372

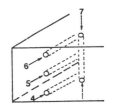

4.3 Water permeability under pressure

Water permeability measurements were carried out at 1 day and at 28 days onto the bottom and the lateral surface of the specimens. The 28-day measurements included, in addition, measurements onto the top surface of the specimen. The results are summarized in Table 9 in terms of impermeability index. Table 10 gives the values of impermeability index measured on the same specimen at 1 and 28 days, for different curing conditions.

TABLE 9. Influence of curing conditions on the impermeability index.

AGE days	CURING CONDITIONS	IMPERMEABILITY INDEX MPa	STANDARD DEVIATION MPa	COEFFICIENT OF VARIATION %
1		0.29	0.0431	14,7
28	wet room	0,314	0,0210	6,7
28	laboratory	0,285	0,0326	11,4

The limited number of test results and the relatively high variability do not allow definite conclusions to be drawn. It seems however that the laboratory cured specimens have higher and more scattered permeability values than the wet-cured specimens.

No significant differences in the permeability at 1 day and 28 days àre observed although there is some indication of a small increase of permeability at 28 days (see Table 10).

TABLE 10. Impermeability index at 1 and 28 days measured on the same specimen.

SPECIMEN No	IMPERMEABILITY INDEX		CURING CONDITIONS
	1 day (MPa)	28 days (MPa)	
G1	0.31	0.29	Laboratory
H1	0.24	0.28	Laboratory
I1		0.24	Wet room
K1	0.32	0.31	Wet room

4.4 Estimation of tensile strength

The procedure was the same as in the case of young concrete (para. 2.1). The results are presented in graphical form in Fig. 8 and in terms of strength versus distance from the bottom of the specimen.

It is noted that the wet-cured specimens have significantly higher splitting strengths than the laboratory cured specimens; this is in accordance with the core compressive strengths (Table 11). It can also be seen that in most cases the strength increases from the outside towards the center of the specimen.

TABLE 11. Effect of curing conditions on core compressive strength.

CURING CONDITIONS	CORE STRENGTH MPa	
	Mean Value	Stand Deviation
Laboratory atmosphere	29,9	1,5
wet-room	32,7	5,3

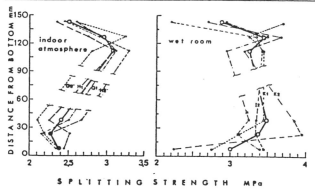

Figure 8. Variation of the splitting strength in relation to the distance from the bottom of the specimen for two different curing regimes.

5 OVER-ALL DISCUSSION AND CONCLUSIONS

The research work reported in this paper should be regarded as a preliminary investigation aiming at the assessment of the potential use of several testing methods in studying the "near surface" properties of young and mature concrete. The main conclusions are the following:

a. Ultrasonic pulse measurements: The direct transmission method is capable in picking-up differences in properties of skin areas in young and mature concrete. However, the diameter of the transducers impose a limitation on the thickness of the layer investigated. On the contrary the surface transmission method was not found successful.

b. Nail extraction: Some modification concerning the nail diameter and the power of the percussion cups is needed in order to avoid splitting of young concrete.

c. Water permeability under pressure: The method is very simple, it can equally be used in the laboratory and in-situ;it is capable in picking-up differences in properties of young concrete which are due to different compaction methods and of mature concrete which are due to different curing regimes. The variability of the results is high but this is a characteristic of all permeability testing methods of concrete and is due to the nature of the material. However, there are possibilities of improvement, such as increasing the area of concrete onto which water pressure is applied.

d. Splitting test: Despite the splitting test shortcomings, which are aggravated by the very small thickness of the discs (slices), the results are promising as a way of studying in the laboratory, the variations of concrete strength in relation to the distance from the surface of the concrete. Some practical limitations are imposed by the thickness of the slices and this subject needs further research. Also, due to the inevitable variability of the test, an increased number results is needed in order to draw definite conclusions.

e. Porosimetry: Total porosity measurement by absorption under vacuum as well as mercury porosimetry are applicable for in situ and in-lab investigations of young concrete in order to evaluate its performance related to durability or strength requirements.

ACKNOWLEDGMENTS

The authors wish to express their thanks to the Hellenic Cement Research Center (EKET) for carrying out the tests of mercury porosimetry, the TITAN cement company for lending additional equipment, and the Association of the Greek Cement Industry for their financial support.

REFERENCES

1. Kreiger, P.C. "The skin of concrete. Composition and properties".Materials and Structures. 17 (1984) No 100, pp 275-283.

2. Dhir. R.K. Hewlett, P.C. and Chan. Y.N. "Near surface characteristics of concrete:assessment and development of in-situ test methods", Mag.Concrete Res. 39 (1987) No 141, pp 183-143

3. Dhir. R.K. Hewlett. P.C. and Chan. Y.N. "Near surface characteristics and durability of concrete:an initial appraisal" Mag. Concrete Res. 38 (1986) No 134, pp 54-56

4. Figg. J.W. "Methods of measuring the air and water permeability of concrete". Mag. Concrete Res.,25 (1973) No 85, pp 213-219

5. Hong. C.Z. and Parrott. L.J. "Air permeability of cover concrete and the effect of curing". British Cement Association Publication c/5 (1989) pp 24.

6. Tassios. T.P., Aligizaki.K. "Durability assessment of old dock buildings in Pireus harbour" Research Report of the Laboratory of Reinforced Concrete of NTU Athens, 1989 (In greek), pp 188.

7. Tassios, T.P. and Demiris, C.A "A new non-destructive method for concrete's strength determination Publications from the National Technical University. Athens (1968) No 21 pp 113.

8. Tassios, T.P."Quelques applications de l'ausculation dynamique des materiaux et des constructions"(In greek) Technika Chronika 3 (1960) pp 3-13.

PART SIX

TESTS RELATED TO REINFORCEMENT AND PRESTRESSING

28 DETECTION OF VOIDS IN GROUTED DUCTS USING THE IMPACT–ECHO METHOD

N.J. CARINO
National Institute of Standards and Technology,
Gaithersburg, MD, USA
M. SANSALONE
Cornell University, Ithaca, NY, USA

Abstract
The impact-echo method was used to detect simulated voids in grouted
post-tensioning tendon ducts cast in a 1-m thick concrete wall specimen.
The study was part of a program to evaluate nondestructive test methods
based on stress wave propagation. The locations of the voids in the
ducts were not known by the authors until after their results had been
reported to the principal investigator of the project. The impact-echo
method was successful in locating the voids. However, the study showed
that additional research is needed to gain a complete understanding of
the interaction of stress waves with voids in cylindrical ducts.

1 Introduction

The authors participated in a study to evaluate the reliability of
nondestructive test methods based on stress wave propagation which are
used for locating defects within concrete structures. The study was
sponsored by the Canada Centre for Mineral and Energy Technology (CANMET)
and the objective was to determine whether existing methods could be
used to detect voids in grouted post-tensioning tendon ducts. The impact-echo method developed by the authors [1-13] was one of the methods
selected for evaluation by the principal investigator for the project.

The test specimen was designed by the principal investigator (B.H.
Levelton & Associates Ltd.) and was intended to simulate a portion of a
typical post-tensioned concrete ice-wall for an arctic offshore structure. During the grouting of the ducts, various artificially created
voids were incorporated into the ducts. The locations of the voids
within the ducts were not known by the authors prior to testing. After
the wall was tested and the results presented to the principal investigator, the locations of the voids were disclosed to the authors.

This paper describes the principle of the impact-echo method, explains
the instrumentation, signal processing technique, and test procedure
used in the investigation, and presents the results obtained by the
authors. The locations of voids identified using the impact-echo method
are compared with the known locations.

2 Background

2.1 Principle of impact-echo method

The principle of the impact-echo technique is illustrated in Fig. 1. A transient stress pulse is introduced into a test object by mechanical impact on the surface. The stress pulse propagates into the object along spherical wavefronts as P- and S-waves. In addition, an R-wave travels along the surface away from the impact point. The P- and S-waves are reflected by internal discontinuities or external boundaries. The arrival of these reflected waves at the surface where the impact was generated produces displacements which are measured by a receiving transducer. If the receiver is placed close to the impact point, the displacement waveform is dominated by the displacements caused by P-wave arrivals [1].

The displacement waveform can be used to determine the round-trip travel time, Δt, from the start of the pulse to the arrival of the first P-wave reflection. If the P-wave speed, C_p, in the test object is known, the distance, T, to the reflecting interface can be determined, as follows:

$$T = C_p \ \frac{\Delta t}{2} \tag{1}$$

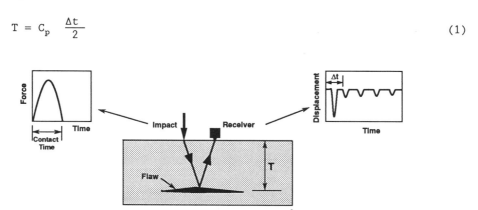

Fig. 1   Principle of the impact-echo method.

2.2 Impact-echo test system

An impact-echo test system includes three components: an impact source; a displacement transducer; and a waveform analyzer. The force-time history of the impact may be approximated as a half-cycle sine curve and the duration of the impact is the "contact time" (Fig. 1). The contact time is an important variable because it determines the frequency content of the input pulse. As the contact time decreases, smaller defects can be discerned or shallower depths can be measured. However, reducing the contact time decreases the penetrating ability of the stress waves. Thus the selection of the impact source is an important aspect of a successful impact-echo test system.

In the developmental work [1], impact was generated by dropping steel ball bearings, which resulted in contact times in the range of 30 to 60 μs. For this investigation, a commercially available impactor was adapted for use as the impact source. The components of this impactor are shown schematically in Fig. 2(a). The spherically-tipped mass is propelled toward the test surface by a spring, which permits testing on vertical surfaces. A 9-mm hardened steel ball was attached to the end of the device so that the mass did not impact directly on the concrete. The addition of the ball improved the repeatability of the impact and shortened the contact time. The impactor produced impacts with contact times ranging from 30 to 50 μs, depending on the surface characteristics of the concrete at the impact point. Thus frequency components as high as 20 to 30 kHz were contained in the propagating waves [4].

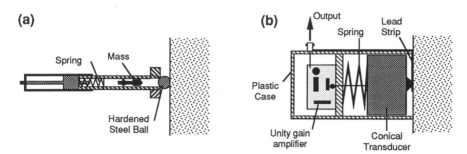

Fig. 2 Impact-echo test system: (a) schematic of spring-loaded impactor and (b) holder for NBS conical transducer.

The receiving transducer must be capable of accurately measuring surface displacement. A conically-tipped transducer, developed at the National Bureau of Standards (now the National Institute of Standards and Technology) as a secondary reference standard for calibrating acoustic emission transducers, is used in current work. The transducer has high sensitivity and responds to a broad range of frequencies [14]. For this investigation a housing was built to for the transducer so that it could be used on vertical surfaces (see Fig. 2(b)).

A waveform analyzer is used to capture the transient output of the displacement transducer, store the digitized waveforms, and perform signal analysis. Based on experience, the authors recommend a waveform analyzer having a sampling frequency of at least 500 kHz. The instrument should be capable of spectral analysis of the recorded waveforms.

2.3 Signal analysis

During the early stages of development of the impact-echo technique [1-3], interpretation of the recorded waveforms was performed in the time domain. This required determining the time between the start of the impact and the arrival of the first P-wave reflection or echo. While this was feasible, it was very time-consuming. Thus an alternative approach was adopted which is used almost exclusively in current work. This approach involves frequency analysis of the displacement waveforms.

The principle of frequency analysis is illustrated in Fig. 3, which shows a solid plate of thickness T subjected to point impact. The P-wave generated by the impact propagates back and forth between the top and bottom surfaces of the plate. Each time the wave arrives at the top surface it produces a characteristic downward displacement [4, 9-12]. The waveform is periodic, and the period equals the travel path divided by the P-wave speed, C_p. If the receiver is close to the impact point, the round-trip travel path is approximately equal to 2T. The frequency, f, of the characteristic downward displacement is the inverse of the period; therefore:

$$f = \frac{C_p}{2\ T} \tag{2}$$

If the frequency of the P-wave arrival can be measured, the thickness of the plate (or distance to a reflecting interface) can be calculated:

$$T = \frac{C_p}{2\ f} \tag{3}$$

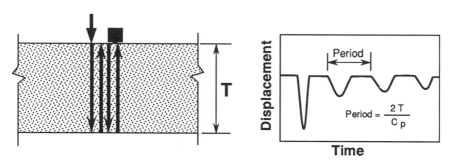

Fig. 3   Basis of frequency analysis to interpret impact-echo results.

The frequency content of a recorded periodic waveform is obtained using the fast Fourier transform (FFT) technique. The transformation from the time to the frequency domain is based on the idea that any waveform can be represented as a sum of continuous sine curves, each with a particular amplitude, frequency, and phase shift. A plot of the amplitude versus the frequency of each of these component sine curves is called an amplitude spectrum, which indicates the predominant frequencies in a waveform.

Figure 4 illustrates the use of frequency analysis. Figure 4(a) shows the amplitude spectrum obtained from an impact-echo test over a solid portion of a 0.5-m thick concrete slab; there is a peak at a frequency of 3.42 kHz. Using Eq. (3) and the known P-wave speed of 3530 m/s [1,4], this frequency corresponds to a depth of (3530 m/s)/(2·3420) = 0.52 m which is approximately the thickness of the slab. Figure 4(b) shows the amplitude spectrum obtained from a test over a portion of the same slab containing a disk-shaped void. The peak at 7.32 kHz corresponds to a depth of (3530 m/s)/(2·7320) = 0.24 m, which compares favorably with the known depth of the void of 0.25 m.

The resolution in the amplitude spectrum (the frequency difference between adjacent points) equals the sampling frequency divided by the number of points in the waveform. For example, for a sampling frequency of 500 kHz and 1024 points in the waveform, the frequency resolution is 0.488 kHz. This imposes a limit on the precision with which the depth can be calculated according to Eq. (3). Because depth and frequency are inversely related, the precision of the calculated depth improves as the frequency increases, i.e., as depth decreases. More information on digital frequency analysis and additional examples of its use in interpreting impact-echo waveforms are available [1,4-8,12,13].

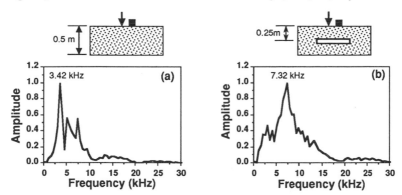

Fig. 4 Examples of amplitude spectra: (a) test over solid portion of slab and (b) test over a disk-shaped void in same slab.

3 Feasibility study

The feasibility of using the impact-echo method to detect voids in grouted post-tensioning tendon ducts was demonstrated by the authors in a controlled-flaw study [6]. A 100-mm diameter, sheet metal duct was cast into a 0.5-m thick concrete slab (P-wave speed 3910 m/s). The concrete cover over the duct was 150 mm. A portion of the duct was filled with grout and the remainder was hollow. Figures 5(a) and 5(b) show the amplitude spectra obtained for tests over the filled and hollow portions of the duct. In the spectrum obtained over the filled portion, Fig. 5(a), there is a predominant peak at 3.91 kHz. This frequency corresponds to multiple P-wave reflection between the top and bottom surfaces of the slab. In this case the P-wave travels through the filled duct [6]. (The peak at 1.46 kHz is due to resonance of the transducer and is disregarded in signal interpretation.)

In the spectrum obtained over the hollow portion of the duct, Fig. 5(b), there is a peak at 3.42 kHz and another peak at 14.6 kHz. The peak at 3.42 kHz is associated with the portion of the P-wave which propagates around the empty duct as it is multiply reflected between the top and bottom surfaces of the slab. The peak at 14.6 kHz is associated with multiple reflections between the top of the duct and the top surface of the slab. The depth of the duct is calculated as (3910 m/s)/(2·14600) = 0.13 m, which agrees closely with the known depth of 0.15 m. Thus this experiment showed that the impact-echo method is capable of distinguishing between a grouted and and an empty duct embedded in concrete.

4 Effect of reinforcing bars

The test specimen for the CANMET study was heavily reinforced on one face. Because of the large bar diameters and the short contact times that were used, it was expected that reflections from the bars would produce peaks in the amplitude spectra [1]. As explained elsewhere [1,13], the nature of reflections from a concrete/steel interface is different from that which occurs at a concrete/air interface. Due to this difference, the apparent frequency of the characteristic surface displacement produced by P-wave reflections from a reinforcing bar is one-half the value given by Eq. (2). This means that the apparent depth of a steel bar calculated using Eq. (3) is twice the actual depth. Thus reflections from a reinforcing bar can be incorrectly interpreted as reflections from a void at twice the depth. The displacement waveform can be used to distinguish between the two cases [13]. In applications where the locations of reinforcing bars are known, as in this investigation, the expected frequencies associated with reflections from the bars can be calculated. If measured frequency peaks were to occur at the calculated frequencies, it would be reasonable to assume that the peaks were due to the bars and not to voids at twice the depth.

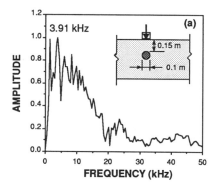 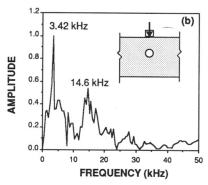

Fig. 5 Amplitude spectra from feasibility study: (a) test over grouted portion of duct and (b) test over hollow portion of duct.

5 Wall study

5.1 Procedure
Figure 6 shows an elevation and a plan of the wall specimen containing seven tendon ducts. The letters A through G were used to identify the ducts: ducts A through E were placed vertically and ducts F and G were placed horizontally. The locations of the ducts were known prior to testing. Impact-echo tests were performed at 100-mm intervals along the centerline of each duct. Test points were identified by sequential numbers beginning with 0; these numbers are shown along the perimeter of the wall in Fig. 6. The receiver was located directly over the test point and the impactor was located next to the receiver as shown in the enlarged sketch in Fig. 6.

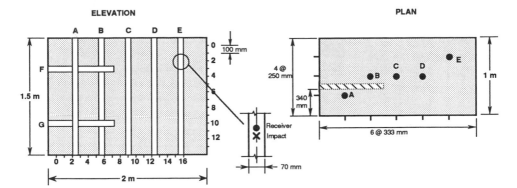

Fig. 6   Elevation and plan of wall specimen; numbers on perimeter are
the test locations along the centerlines of the ducts.

At each test point, tests were repeated until a representative amplitude spectrum was obtained. At points where the surface was in good
condition, three tests were sufficient. At points where the surface was
uneven, six to eight tests were required to establish a representative
spectrum. For each test point, a waveform and its corresponding amplitude spectrum were stored on a disk for subsequent analysis.

A sampling frequency of 500 kHz and a record length of 1024 points
were used to capture the waveforms. Thus the frequency resolution in the
computed amplitude spectra was 500/1024 = 0.488 kHz. This sampling
frequency was chosen because it had been used by the authors in most of
the previous studies. Because of the limited time available to test the
wall (two days), the authors were not able to experiment with other
sampling frequencies. Therefore, the sampling frequency used may not
have been the optimum.

5.2 P-wave speed
Two independent approaches were used to establish the P-wave speed in the
wall. First, an impact-echo test was performed on a standard (150 mm x
305 mm) test cylinder cast from the concrete used in the wall. A sampling frequency of 100 kHz was used to improve the resolution in the
spectrum. The amplitude spectrum had a dominant peak at a frequency of
6.93 kHz. Using Eq. (2), the calculated P-wave speed was 2·0.305·6930 =
4230 m/s. As a check, the ultrasonic pulse velocity across the 1-m
thickness of the wall was measured. The transit time was 213.6 μs,
which gave a pulse velocity of 4680 m/s. It has been shown [1,4,5] that
the P-wave speed obtained by the impact-echo method is about 90% of the
pulse velocity; that is, 4680 m/s·0.9 = 4210 m/s. Thus there was good
agreement between the two independent methods, and a P-wave speed of
4230 m/s was used for the investigation.

5.3 Data interpretation
To aid in interpreting the amplitude spectra, a table was developed
which listed the expected frequency values at different test points
assuming there were voids in the ducts directly below the test points.
Recall that stress waves generated by impact propagate along spherical

375

wavefronts. Therefore, voids in adjacent ducts, as well as a void in
the duct directly below the test point, can produce wave reflections.
Thus, for test scans along each duct, the frequencies corresponding to
P-wave reflections from voids in adjacent ducts were also computed.
Figure 7 shows the various reflection paths that were considered. Along
the three diagonal paths (#3, #5, and #7) the displacements in the S-
wave are large, and there is the possibility of multiple S-wave reflections from voids in these adjacent ducts. The frequencies corresponding
to these S-wave reflections were also computed. It was assumed that the
S-wave speed was 60 percent of the P-wave speed, which is typical of an
elastic material having a Poisson's ratio equal to 0.2 [1].

REFLECTION PATHS   **CALCULATED FREQUENCIES**

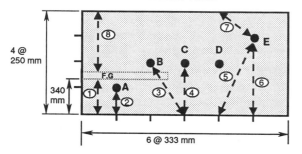

Path	Distance (m)	Frequency (kHz) P-Wave	S-Wave
1	0.340	6.22	
2	0.215	9.84	
3	0.572	3.70	2.23
4	0.465	4.55	
5	0.787	2.69	1.61
6	0.715	2.96	
7	0.396	5.34	3.20
8	0.590	3.58	

$C_p = 4230$ m/s $C_s = 2540$ m/s

Fig. 7 Possible reflection paths due to voids in ducts for different
scan lines; table gives the associated frequencies according to Eq. (2).

The computed frequencies corresponding to reflections from voids in
the ducts are shown in Fig. 7. The presence of these frequencies in the
amplitude spectra was interpreted to mean that there was a void in the
duct being scanned or in an adjacent duct. Because only certain frequency values are possible in the digital amplitude spectra (multiples
of 0.488 kHz in this case), the values given in Fig. 7 typically fell
between adjacent frequency values in the spectra.

5.4 Location of reinforcing bars
Figure 8 shows the locations of the reinforcing bars relative to the
ducts. The bar pattern was verified using a covermeter and a pulsed-
radar test system. Impact-echo tests were performed along the ducts from
both the unreinforced and reinforced faces of the wall to gain an understanding of the effects of reinforcement on the reliability of the results. Figure 8 also shows the locations of the test points with respect
to the centerline of the bars. For the scans of the vertical ducts, the
receiver was located directly above the horizontal bars. The test points
generally were not located directly above vertical bars.
     The reinforcement pattern resulted in six different cover distances
which were considered in calculating the expected frequencies associated
with reflections from the bars. The depths of the various bars were
calculated by assuming that there was 50-mm cover (based on information
from the principal investigator) over the outermost layer and that the
bars in the different layers were in contact with each other. The calcu-

lated frequencies ranged from 21.1 kHz for the outer layer to 8.8 kHz for the inner layer. Recall that the frequencies are one-half the values that would be associated with reflections from a void at the same depth. The frequencies associated with reflections from the inner layer of vertical bars (8.8, 9.6, and 10.6 kHz) may not be significant because of the shadowing effect of the outer layer. The authors had no previous experience testing specimens with three layers of steel bars. Therefore, it was not known whether the inner layer of vertical bars could be discerned by impact-echo testing. A separate study of a specimen containing only reinforcing steel needs to be carried out to study the effects produced by this type of reinforcing pattern.

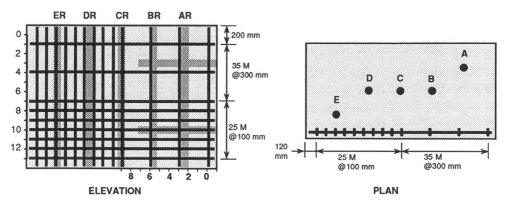

Fig. 8 Location of reinforcing bars relative to the scan lines along the centerlines of the ducts.

6 Results

The amplitude spectra for each scan line were plotted sequentially to produce "waterfall plots." For illustration, Fig. 9 shows waterfall plots for scan line ER (along duct E on the reinforced face) and scan line D (along duct D on the unreinforced face).

For scan line ER, the reflection path for a void in duct E is a type-2 path as shown in Fig. 7. Referring to Fig. 8, it is seen that scan line ER was located almost directly over the vertical bars and test points 1, 4, and 7 through 13 were also located over horizontal bars. Based on the design depths of the reinforcing steel and tendon duct, the frequency values expected in the amplitude spectra for reflections from various interfaces would be as follows:

Void in duct E:	9.77 or 10.25 kHz
Outer vertical bar:	20.98 kHz
Horizontal bar:	14.16 kHz
Inner vertical bar (at top):	9.77 kHz
Inner vertical bar (at bottom):	10.74 kHz

These are the digital values closest to the theoretical values for the 0.488-kHz resolution in the amplitude spectra. Note that the frequency

associated with reflections from the inner vertical bar is the same as
the possible frequency for reflections from a void in the duct. However,
as previously mentioned, the inner vertical bar is shadowed by the outer
bar and is not likely to produce strong reflections. In addition, the
inner bar is deeper and reflections from the bar, if it were not shadowed, would be weak. To confirm these assumptions, the waterfall plot
for scan line CR was examined. As can be seen in Fig. 8, the reinforcing
steel geometry relative to scan line CR was similar to that of scan line
ER. However, because duct C was deeper than duct E, frequencies associated with reflections from the steel did not overlap with frequencies
associated with reflections from a void in duct C. If reflections from
the inner vertical bar were to produce a frequency peak at 9.77 kHz,
this peak would have been present in the spectra of every test point
along CR. This was not found to be true. Therefore, it was concluded
that the inner vertical bar did not result in a large peak at 9.77 kHz.

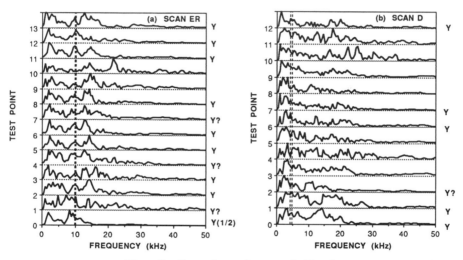

Fig. 9   Examples of waterfall plots

In Fig. 9(a), vertical lines are drawn at frequencies of 9.77 and
10.25 kHz. Test points whose spectra had peaks at 9.77 or 10.25 kHz were
identified as possible locations for voids in duct E. The points where
there was high confidence that a void was present are marked with a "Y"
(for "yes") on Fig. 9(a). For example, test points ER6 and ER11 have
large frequency peaks at 9.77 kHz and these points are classified as
"Y". At some points, there was a frequency peak close to the expected
value or there was a series of peaks in the vicinity of the expected
frequency. These points are marked as "Y?" which signifies that there
is an indication of a void, but the results are not as clear as points
marked with a "Y". For example, points ER1 and ER4 have multiple peaks
in the vicinity of 9.77 kHz. Based on the authors' developmental work,
it is believed that many of the points marked as "Y?" are near the boundaries of the voids.

Test point ER0 has a peak at 8.30 kHz which is close to the calculated
frequency of 8.46 kHz for a half-filled duct. Thus, at point ER0, the
duct is believed to be half empty and that is the reason for "Y(1/2)" in
Fig. 9(a).

Figure 9(b) shows the waterfall plot for scan line D. Reflections
from a void in duct D would be associated with the type-4 path shown in
Fig. 7. The permitted frequency values close to the theoretical frequency of 4.55 kHz are 4.39 and 4.88 kHz, which are plotted as vertical
lines in the figure. Test points D0 and D1 have peaks at 4.88 kHz, and
points D6, D7 and D12 have peaks at 4.39 kHz. Thus it appears that duct
D has voids at its ends and center.

The above procedure was used to interpret the waterfall plots for all
of the scan lines. The results are summarized in Fig. 10, where test
points identified as being over a void are shown in black. The test
points where there was a question about the presence of a void are shown
shaded. In drawing the extent of voids, it was assumed that, at each
point where a void was detected, the void extended 50 mm to either side
of the test point, i.e., the transducer location.

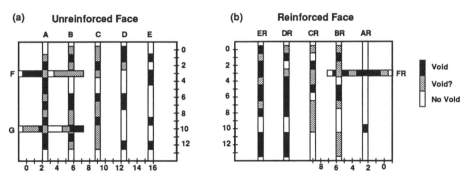

Fig. 10 Summary of results obtained from tests on (a) the unreinforced
face and (b) the reinforced face.

Figures 10(a) and 10(b) summarize the results obtained from tests on
the unreinforced face and the reinforced face. The results are not
identical; however, in most cases there was sufficient agreement to
positively identify those portions of the ducts which contain voids.
The following is a summary of the condition of each of the ducts as
obtained from the tests on the unreinforced and reinforced faces:

Duct A: <u>unreinforced face</u>: large voids at the two ends and a smaller
void at the center
<u>reinforced face:</u> only a portion of the void at the bottom of
the duct could be identified.
Duct B: <u>unreinforced face</u>: voids at the ends and center of the duct.
<u>reinforced face</u>: voids at top and center confirmed, but the
void at the bottom was difficult to identify positively.
Duct C: The results were similar to duct B.
Duct D: Voids were identified at the ends and middle of the duct.

The results from the unreinforced and reinforced faces were in good agreement.

Duct E: <u>unreinforced face</u>: voids at ends of the duct.
<u>reinforced face</u>: positive indications of these voids were also obtained despite the large depth of the duct from this face.

Duct F: <u>reinforced face</u>: voids seem to exist at points along the entire length of the duct. Similar results were obtained from the unreinforced face. It was known prior to testing that this duct was hollow.

Duct G: <u>unreinforced face</u>: there appear to be two voids.
<u>reinforced face</u>: time constraints did not permit testing from the reinforced face.

7 Comparison of results with actual locations

After submitting the test result to the principal investigator, the authors were given a sketch showing the actual locations of the flaws. In addition, the principal investigator drilled two cores from the specimen, and provided the photographs and descriptions of the cores. The following discussion compares the impact-echo results with the planned locations of the voids.

Figure 11 shows the location of the voids according to the sketch provided by the principal investigator. A comparison with Fig. 10 shows that all the voids were detected. However, there are two differences between the impact-echo results and the actual void configuration: (1) the size of the voids tended to appear greater in the impact-echo results; and (2) more voids were detected by impact-echo than those shown in Fig. 11. Explanations for these differences follow.

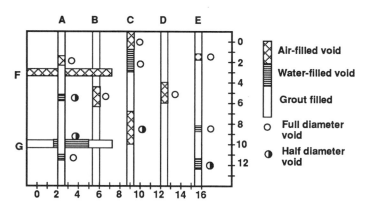

Fig. 11 Locations of voids as indicated in drawing provided by the principal investigator.

In impact-echo testing, the stress pulse penetrates into the test object along spherical wavefronts. As a result, points within the object which are not directly below the test point will interact with the propagating waves. The boundaries of an underlying defect will diffract the

380

incident wave as indicated in Fig. 12. Therefore, the edge of a defect will send out waves which travel along circular wavefronts. As a result, diffracted waves can arrive at the receiver even though the receiver is not located directly above a defect, as was shown in Fig. 1. This is an inherent characteristic of impact-echo testing, and it is both an advantage and disadvantage. It is an advantage because a defect can be detected when the test point is not located directly over the defect. Thus it is possible to detect defects which are smaller than the spacing between the test points. The disadvantage is that the boundaries of a defect cannot be determined precisely.

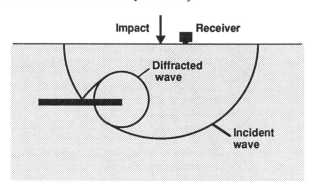

Fig. 12 Diffraction at edge of a flaw causes diffracted wave that travels along a circular wavefront.

Some of the test results indicated voids at locations which did not contain preformed voids. For example, in duct D there appeared to be voids at the top and bottom ends; whereas, according to the plans, the ends were fully grouted. A possible explanation for this discrepancy was provided by the description of a core drilled into duct A, 390 mm from the top of the specimen. The core was 275 mm long, and it penetrated duct A where it was fully grouted. In describing the core, the principal investigator stated that "shrinkage of grout has caused a separation of 0.05 mm between the grout and inside wall of the duct around 20 percent of the circumference." Such a shrinkage crack introduces an air gap between the grout and the tendon duct, and the duct will reflect waves as though it were empty. It is possible that similar shrinkage occurred in other ducts. Another possible reason for the greater number of voids according to the impact-echo results is poor consolidation of the concrete adjacent to the ducts. Lack of consolidation was noted at the exposed end of duct F. The core drilled into duct A cut through the duct and broke at the far side of the duct. The photograph of the fractured end of the core showed significant voids adjacent to the duct. From this evidence, it is believed that the impact-echo results indicated more voids because of shrinkage cracks between the grout and ducts or because of poor consolidation of the concrete adjacent to the ducts.

8 Summary

This investigation represented the most challenging application of the impact-echo method to date. The testing involved difficult conditions: the voids were small compared with their depth; the close spacing of the ducts, the overlapping of ducts, and the large amount of reinforcing bars resulted in complex spectra; the quality of the concrete surface at the test points was nonuniform; and time constraints did not allow optimization of the test procedure. In addition, past analytical studies have been limited to axisymmetric problems [1,6,9-12]; therefore, at the time of testing, the authors did not have a full understanding of the interaction of transient waves with cylindrical-shaped voids. However, the authors feel that the impact-echo method performed well under the circumstances. It has been demonstrated that the method has the potential of being developed into a reliable method for quality control of the grouting in post-tensioned concrete structures.

This investigation was also beneficial because it provided a focus for future work by raising questions that need to be answered through further research. For example, the impact-echo response due to reflections from voids in adjacent ducts needs to be better understood. By performing a test along one tendon duct it may be possible to detect voids in adjacent ducts, thereby reducing the amount of testing. The impact-echo response due to reflections from different types of discontinuities and the effects of voids within overlapping ducts also need further study. A better understanding of the interaction of stress waves with flaws of different sizes and shapes and the relationship between the contact time of the impact and the flaw size and depth are also needed. Finally, improved techniques for data reduction are needed. Further experimental and numerical work on locating voids in tendon ducts is currently being pursued [13], and methods are under development for automating the interpretation of tests results.

9 References

1. Sansalone, M., and Carino, N.J., "Impact-echo: a method for flaw detection in concrete using transient stress waves," NBSIR 86-3452, National Bureau of Standards, Sept., 1986, 222 pp. (NTIS PB #87-104444/AS)
2. Carino, N.J., "Laboratory study of flaw detection in concrete by the pulse-echo method," In Situ/Nondestructive Testing of Concrete, V.M. Malhotra, Editor, American Concrete Institute (SP 82), 1984, pp. 557-579.
3. Carino, N.J., Sansalone, M., and Hsu, N.N., "A point source-point receiver technique for flaw detection in concrete," Journal of the American Concrete Institute, Vol. 83, No. 2, April, 1986, pp. 199-208.
4. Carino, N.J., Sansalone, M., and Hsu, N.N., "Flaw detection in concrete by frequency spectrum analysis of impact-echo waveforms," International Advances in Nondestructive Testing, 12th Edition, W.J. McGonnagle, Ed., Gordon & Breach Science Publishers, New York, 1986, pp. 117-146.

5. Sansalone, M., and Carino, N.J., "Laboratory and field study of the impact-echo method for flaw detection in concrete," in <u>Nondestructive Testing of Concrete</u>, ACI SP-112, American Concrete Institute, 1988, pp. 1-20.
6. Sansalone, M., and Carino, N. J., "Impact-echo method: Detecting honeycombing, the depth of surface-opening cracks, and ungrouted ducts," Concrete International, Vol. 10, No. 4, April 1988, pp. 38-46.
7. Sansalone, M. and Carino, N.J., "Detecting delaminations in concrete slabs with and without overlays using the impact-echo method," ACI Materials Journal, V. 86, No. 2, March-April 1989, pp. 175-184.
8. Carino, N.J. and Sansalone, M., "Impact-echo: a new method for inspecting construction materials," To be Published in Proceedings of Conference on NDT&E for Manufacturing and Construction, Aug. 1988, Urbana, IL.
9. Sansalone, M., Carino, N.J., and Hsu, N.N., "A finite element study of transient wave propagation in plates," Journal of Research of the National Bureau of Standards, Vol 92, No. 4, July-August, 1987, pp 267-278.
10. Sansalone, M., Carino, N.J.,and Hsu, N.N., "A finite element study of the interaction of transient stress waves with planar flaws," Journal of Research of the National Bureau of Standards, Vol 92, No. 4, July-August, 1987, pp 279-290.
11. Sansalone, M., and Carino, N.J., "Transient impact response of thick circular plates," Journal of Research of the National Bureau of Standards, Vol 92, No. 6, Nov.-Dec., 1987, pp 355-367
12. Sansalone, M., and Carino, N.J., "Transient impact response of plates containing flaws," Journal of Research of the National Bureau of Standards, Vol 92, No. 6, Nov.-Dec., 1987, pp 369-381.
13. Sansalone, M. and Carino, N.J., "Finite element studies of the impact-echo response of layered plates containing flaws," <u>International Advances in Nondestructive Testing Vol. 15</u>, W.J. McGonnagle, Editor, to be published 1990.
14. Proctor, T.M., Jr., "Some details on the NBS conical transducer," Journal of Acoustic Emission, V. 1, No. 3, July 1982, pp. 173-178.

29 MEASUREMENT OF PRESTRESSING CABLE FORCES

T. JÁVOR
VUIS, Bratislava, Czechoslavakia

Abstract
The results of the measuring of prestressed cable forces of cable-stayed bridges in Czechoslovakia have shown that the ring-dynamometers using for example magnetoelastic system or the dynamometers with vibro-wire gauges allowed to receive precise results $\pm$ 2 % but we need to place the equipment in the structures during the erection. The frequency-vibration method is cheaper, the measurements can be repeated with the same accelerometer in various cables but the results show differences of max. $\pm$ 7 %. This accuracy is much higher if we repeat and compare our measured forces during traffic long-term to receive more information about the serviceability of cable-stayed bridges, as well as information about the change of cable forces during construction or during traffic.

1 Introduction

It it quite natural that in course of erection of the new cable-stayed prestressed concrete bridges in Czechoslovakia and for the long term service, the need arose of assuring not only the complex quality control of materials, but also the observation of the state of deformation and stress of these structures. The special tests and measurements were carried out and the aim of the research measurements is to verify the assumptions of the designers and to determine the real values. The strains are measured by embedded vibro-acoustical gauges and the magnitude of forces of the free cables are measured either with special dynamometers or by calculating the frequency change of cable oscillation to evaluate the cable forces.

2 Methodology of measurements

First of all we need to remark that our methodology of measurements of prestressed cable forces is related to the free cables of the cable-stayed bridges. Of course, various methods and equipments are very useful also for the measuring of prestressing forces of the free cables like the free prestressed tendons for reconstruction, but some equipments are possible to use for checking of prestressing in embedded prestressed reinforcements or strands and cables respectively.

The force measurements in suspender-cables is usually carried out by commercially produced special ring dynamometers, for example Maihak MDS 65, Telemac CV8, Hysteric, etc. The Maihak ring dynamometer applied vibrating-wire gauges and is generally produced for 500 kN, 1000 kN or 2000 kN with the sensitivity 0.02 % using the MDS 910 or MDS 920 reading equipment. The internal diameter is 60 mm, the external diameter is 130 mm. The Telemac CV 8 dynamometer for 250 t has an external diameter of 180 mm and the internal hole is ⌀120 mm. In our bridges we used the prestressing jack of fa Paul type TENSA 4800 kN and the force in anchoring was controlled also by Proseq WIGA 5050 dynamometer.

In case of our cable-stayed bridges we developed special ring dynamometers to the extent of force measurement of 2200 kN and can be used for cables of 23 to as many 36 strands ⌀15.5 mm of various qualities. Each dynamometer consists of 4 measuring and 1 compensation vibro-wire strain gauge. The material of the dynamometer is from high quality mangan-chrom-vanadium steel. The external diameter is 350 mm, the internal diameter 230 mm, the weight of the dynamometer is 30 kg, the height is 169 mm.

The second system and type of dynamometers, which was developed in Czechoslovakia for the possibility of measuring forces in cables and the stress in reinforcement, also consists in ascertaining the change in magnetic permeability of the cable when is stressed. We are speaking about the change of the magnetoelasticity of the prestressed reinforcement, where the electromagnetical field produced the current which is going through the coil fixed around the reinforcement or cable.

Both types of ring dynamometers are very sensitive for the changing of the cable forces, but we need to put in the dynamometers before placing the cables and before prestressing. Some examples with half-ring dynamometers have not given very good results. For this reason we used the most simple another method applied the direct measurement of oscillation frequency of the prestressed cables, or the cables of the cable-stayed bridges. The oscillation of the free cables is made by wind or very simple by hand, as we can see in Fig. 1.

For direct measurement of oscillation frequency of cables, we used one inductive accelerometer, for example BK 8206, which is useful for 0.2 to 1000 Hz. The accelerometer is connected to a frequency analyser. From the first signal using the Fourier transformation we can receive the first modus of the cable vibration.

For the control of the forces of the cable-stayed bridges during construction as well as during the traffic we used the BK 8206 accelerometer and the ONO-SOKKI CF-920 analyser, permitting the oscillation analysis of the 1st through the 5th harmonic quantity with the possibility of registering the vibration record, digitalisation and processing of value with a computer. Sometimes we also used a special type-recorder, see Fig. 2.

Fig. 1. The simple oscillation of the free cable with one inductive accelerometer BK 8206

Fig. 2. The oscillation of the cable-stayed bridge is registered by the measuring type-recorder

3 Results of the observation

For the calculation of the prestressing cable forces we used the very simple equation

$$F = 4 \cdot U \cdot l^2 \cdot f^2$$

where

F is the force, U the weight of cable, f the frequency and l the length of the measured free cable.

The value of the forces we received with the $\pm$ 30 kN accuracy from the measured frequency, but the frequency depends on the length, which is dependent on the damper of the cable-oscillation. The cable forces were measured by the analyser also with various accelerometers just after prestressing, after grouting of protective tubes for cables and after rectifications of cable forces. Of course, the method was used also during the statical load tests of the bridges and long-term during the traffic to receive the change of cable forces during loading.

Table 1 shows an example comparing various experimental methods received during the construction of the cable-stayed bridge over the Elba river near city Poděbrady. The 32,3 m highway bridge has three spans of length 61.6 + 123.2 + 61.6 m, the total length of the bridge being 253.0 m.

Table 1.

	Prestressed cables before grouting		Prestressed cables after grouting	
	Force		Force	
	kN	%	kN	%
Dynamometer Proseqe-WIGA	1990	100	1898	100
Magnetoelastic dynamometer	1996	100.3	1918	101.1
Frequency method	2014	101.2	2031	107.0
Prestressing jack-manometer	1965 226 Bar	98.7	1915 224 Bar	100.9

This bridge is formed by a box girder precast elements cantilevers supported. The 28.06 m high towers are formed by two twin cell steel columns, which are fixed in the deck. Each stay consists of two cables of 18 or 15 strands of 15.5 mm of ultimate strength 1800 MPa. The strands are encased in steel tubes completely filled with cement grout, in which during construction compression is created. Prior to grouting, which is carried out after casting the outer cantilevers, the cables are detensioned to a high-load level. After the grout has been cured, the cables are detensioned to a design level and the grout becomes a structural member. For this reason the cables are controlled during the erection many times by the frequency method. The very quick results were useful for the constructor during the rectification of the prestressing in the cables. The weight of one cable of 15 strands with grouting was 41.06 kg/m and without grouting 26.22 kg/m , for 18 strands without grouting it was 28.72 kg/m . The length of the damper was 3 m, the length of cables was over 20.0 m. The change of forces during the load test of the bridge was between 115 and 221 kN in one cable.

References

[1] Jávor, T. (1988) Inspection and experimental analysis of cable-stayed bridges during construction. Publication of the ING/IABSE International Seminar on Cable Stayed Bridges, I-80-90, Bangalore, India

[2] De Mars, Ph. and Hardy, D. (1985) Mesure des Efforts dans les Structures a Cables. Annales des Travaux Publics de Belgique, No. 6, 515-531

[3] Jávor, T. (1990) Experimental techniques used on concrete cable-stayed bridges. Proceedings of the 9th International Conference on Experimental Mechanics, Lyngby, Denmark, 81-1-9.

30 IN-SITU MEASUREMENT OF PRESTRESSING CABLE FORCES

G. DOBMANN
Fraunhofer Institute for Nondestructive Testing, Saarbrücken,
Germany

Abstract

This paper describes results to the research obtained in a joint project in the BRITE-programme of the European Community [1]. One specific part of the project has the objective to investigate the potential of micromagnetic NDT-techniques, developed elsewhere for stress analysis [2], for the insitu-measurement of prestressing cable-forces.

1 Introduction

Micromagnetic ndt-techniques, summarized under the keyword 3MA (Micromagnetic-, Multiparameter-, Microstructure- and Stress-Analysis) can be applied to magnetizable materials. The techniques are sensitive for the characterization of microstructure- and stress-material states (residual- and load induced stresses). Because the sensitivity against both influence parameters is in the same range the separation of the influences in the measuring quantities is a basical inspection task. For the separation the method of multiparameter regression is applied, more than one of the measuring quantities is used for the approach. The regression approach asks for a well defined set of calibration specimen which covers the variation of the microstructure of the material which is under inspection. The influence of stress then is measured under defined loads and is introduced in the regression algorithm.

2 3MA-Techniques

The research to the micromagnetic techniques was arranged in two parts. The first part has mentioned laboratory investigations to the sensitivity of the techniques for tensile stresses in prestressing steels. The second part was the automatic calibration of the measuring quantities during standard tensile tests in the workshop of a German prestressing cable manufacturer.

The basics of the 3MA-techniques were evaluated elsewhere [2,3], a prototype equipment was available. In this equipment two micromagnetic techniques are implemented. For the multiregression algorithm additionally a mini-computer (PC) was used.

The first technique is the excitation and the reception of the magnetic Barkhausen-noise. The material under inspection is magnetized dynamically in a hysteresis loop. The magnetizing frequencies are in the range 20 Hz - 90 Hz variable to adjust. The magnetization is introduced into the material by an u-shaped handy joke magnet, the field-coil of the joke is excited by a sinusoidal electric current delivered by a bipolar power supply. Symmetrically between the pole shoes of the joke an induction coil is arranged, the coil axis is perpendicular to the surface under inspection. A Hall-element is integrated in the coil for the measurement of the tangential magnetic field strength. This field strength measurement guarantees the control of the peak amplitude of the field, independent of the material which is tested. The magnetization of the material is realized microscopically by Bloch-wall jumps and rotational processes [4], these magnetization events induce in the receiving coil pulses of an electric voltage, the so-called Barkhausen-noise.

The voltage is amplified (80 dB), rectified and low-pass-filtered, i.e. the envelope of the Barkhausen-noise is formed. As function of the magnetic field H two characteristic values of this envelope curve M(H) are measured, the maximum M_{Max} and the field strength value H_{CM} belonging to this maximum. H_{CM} correlates with the coercivity value H_C of the hysteresis curve. Fig. 1 chematically shows the measuring quantities.

The second 3MA technique is a result of the frequency analysis of the magnetic field strength during one time period [5]. Because of the symmetrical behaviour of the hysteresis only upper harmonics with an odd number occur (1., 3., 5., 7., ..., harmonic). The complex amplitudes A_i, i = 1, 3, 5, 7, ... of the harmonics are calculated by a Fourier analysis of the time signal on the PC. A distortion factor K can be derived, which is the power of the fundamental wave:

$$K = \sqrt{(|A_1|^2 + |A_3|^2 + |A_1|^2 + \ldots)/(|A_1|^2)} \quad in\%$$

M_{Max}, H_{Max}, K, H_{co} are sensitive quantities for microstructure and stress. When the material under inspection has a positive magnetostriction (the material elongates under magnetic field) then M_{Max} and K increase with tensile stress and H_{CM} and H_{co} decrease. The laboratory investigations in the standard tensile test show for the prestressing steel ST 1470/1670 according to DIN not this normal behaviour though the material has a positive magnetostriction. Here M_{Max} and K decrease with tensile stress and H_{CM} and H_{co} increase. A deformation texture or a residual stress gradient should be the reason after manufacturing in the deep drawing process.

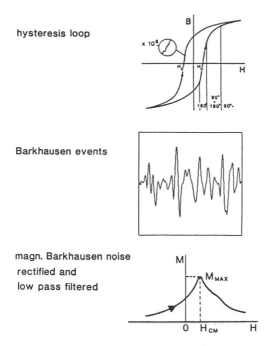

hysteresis loop

Barkhausen events

magn. Barkhausen noise
rectified and
low pass filtered

Fig. 1 Magnetic Barkhausen-noise and derived quantities M_{Max} and H_{CM} From the time signal of the upper harmonics also a coercivity H_{co} can be derived. H_{co} is the magnetic field value at the time point, where the time signal of the harmonics has a zero after H=0. Fig. 2 shows the analysis in principle.

3 Multiparameter evaluation

In the factory of a German prestressing cable manufacturer different charges of the prestressing steel quality ST 1470/1670 were investigated with the 3MA techniques named above. The variation in the charges are given by the variation in the chemical composition which is allowed by the DIN (C 0.7 - 0.9 %, Si 0.1-0.35 %, Mn 0.5-0.9 %). This results in tolerances of the tensile strength Rm (min. 1670 - 5 % = 1590 MPa, max. 1670 + 12 % = 1870 MPa). The variation in the charges were also observed directly in the 3MA quantities. Furthermore the deformation texture or the residual stress gradient is not constant along the length of a prestressing cable. Along a length of 20 m the following variations in the 3MA quantities were observed:

M_{Max} ± 7 %
K   ± 0.8 %
H_{CM} ± 8 %
H_{co} ± 1.4 %

391

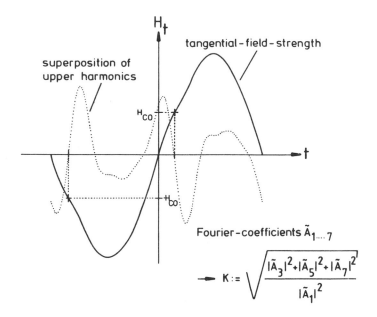

Fig. 2. Time domain signal of the magnetic tangential field strength and the time signal of the upper harmonics, derived quantities distorsion factor K and coercivity H_{co}

The results show that the Barkhausen-noise quantities are not reliable for the measurement of prestressing cable-forces. Therefore only for the quantities K and H_{co} a multiparameter approach was performed.

Fig. 3 shows the distorsion factor K and Fig. 4 the coercivity H_{co} as function of the tensile stress σ_1. Due to the charge variations the K^- and H_{co}-values cover the range between the two tensile strength classes Rm_1 and Rm_2 (1496 MPa s Rm_1 s 1725 MPa, 1849 MPa s 1870 MPa). The result is that neither K nor H_{co} gives a direct access for a unic prestressing-force evaluation. Fig. 5 shows the (H_{co}, K)-pairs in a histogram and by a linear regression we find the functional equations for the different tensile strength classes:

Rm_1:
H_{co1} (K) $= 1.107 + 78.999 \cdot (1/K)$; $r^2 = 99.9$ %

Rm_2:
H_{co2} (K) $= 17.376 + 1.0406 \cdot 10^5 \cdot (1/K^7)$; $r^2 = 99.9$ %

(r^2 is the square of the correlation coefficient r).

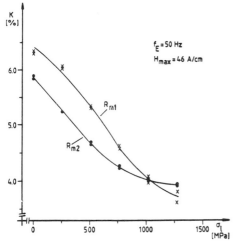

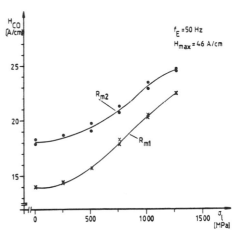

Fig. 3. Distorsion factor K as function of the tensile stress, charge variations cause a variation of the values between two tensile strength classes Rm_1 and Rm_2

Fig. 4. Coercivity H_{co} as function of the tensile stress

As function of the deformation texture or the residual stress gradient H_{co} varies only with $\pm$ 1 A/cm and therefore the tensile strength class $Rm_{1,2}$ can be determined uniquely by a H_{co}-measurement. Furthermore we find for prestressing-forces r 255 MPa calibration curves as function of the tensile stress σ_1:

Rm_1:
$$\sigma_1(H_{co}) = \text{-}3575.5 + 1019.2 \cdot \sqrt{H_{co}}; \; r^2 = 99.8 \%$$

Rm_2:
$$\sigma_2(H_{co}) = \text{-}5538.4 + 1365.1 \cdot \sqrt{H_{co}}; \; r^2 = 99.5 \%$$

Fig. 6 and Fig. 7 show these calibration curves compared with the real measured values. The maximum deviations are in Rm_1 with -47 MPa and in Rm_2 with +109 MPa.

4 Conclusions

The 3MA quantities K and H_{co} can be applied in order to measure the prestressing cable-forces at the prestressing steel ST 1970/1670. For this measurement in a first inspection task the influence of the charge variations has to be introduced in

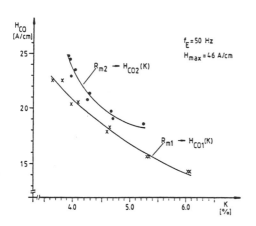

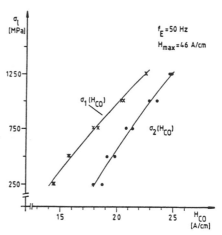

Fig. 5. Histogram H_{co} versus K for the estimation of the tensile strength classes

Fig. 6. After linear regression obtained calibration curves for the tensile stress as function of the quantity H_{co} and the different tensile strength classes

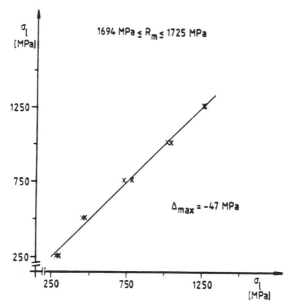

Fig. 7. Measured tensile stress values (ordinate) compared with the regression model value (abscissa), tensile strength class Rm_1

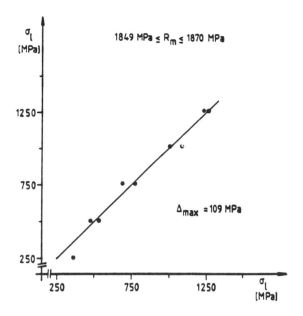

Fig. 8. Measured tensile stress values (ordinate) compared with the regression model values (abscissa), tensile strength class Rm_2

the multiparameter approach, i.e. the tensile strength class has to be estimated. For each class of tensile strength by the multiregression method calibration functions for the tensile stress (prestressing cable-forces) can be derived as function of the 3MA quantity H_{co}. The result should also be validated for other prestressing steel qualities. The inspection asks for an accessability to the prestressing cable at the surface of the concrete structure. A window with $10 \cdot 10$ cm² must be opened into the concrete; the duct and the mortar should be removed.

References

[1] Brite - Research No. RI 1B-0126-D(B)

[2] European community for coal and steel, Research No. P 640

[3] Federal Ministry For Research And Development of the FRG, Reactor Safety Research, Project No. 1500334

[4] Kneller, E. (1962) Ferromagnetismus, Springer Verlag, Berlin

[5] Dobmann, G., Pitsch, H. Patent DE 3813 739 AL

31 COVERMETER DEVELOPMENT IN JAPAN AND STUDY ON IMPROVING ELECTROMAGNETIC INDUCTION METHOD

S. KOBAYASHI and H. KAWANO
Public Works Research Institute, Ministry of Construction, Japan

Abstract
This paper describes the Japanese state of development of covermeters for measuring the location and cover thickness of a reinforcing-bar (re-bar) embedded in concrete and reviews the application of the covermeters in testing during concrete construction. It also shows the studies on improving the measuring method using the electromagnetic induction method.

1 Introduction

In Japan from a decade ago, the early deterioration of concrete structures became a big problem. The early deterioration was brought by mainly three causes, i.e., chloride attack, alkali silica reaction and poor execution. The Ministry of Construction carried out a comprehensive research project in order to deal with the problem, and through the investigation of the poor execution we found it very important to develop the technique to measure precious location, cover thickness and diameter of re-bars in concrete in-situ non-destructively.

There are two circumstances where the covermeter is necessary. One is for inspection of execution after the placing of concrete. The results of the investigation in the project show that a large number of cases of deterioration due to poor execution were brought by the lack of cover over re-bars. Fig. 1 and 2 show the examples where very small cover caused the corrosion of re-bars in a short period after completion of structures. To prevent this kind of damage of concrete structures, it is most effective to set the spacers appropriately. Fig. 3 [1] shows the relation between the number of spacers per unit area and the lack of cover thickness from planned one obtained from the investigation of the real structures. It shows that they don't use many spacers in many cases and that, if the spacers are used more than 2 pieces per square meter, the lack of cover is avoidable. Checking the cover thickness bevor and after the placing of concrete is important as well as appropriate setting of spacers. Proper treatment to the lack of cover would prevent the early deterioration of concrete structures.

Another case is for repairing. Before repairing of the deteriorated structures, it is very important to obtain information about re-bars. The blueprints of old structures are sometimes not kept, therefore in such cases the covermeter is essential.

Fig. 1. Corroded re-bars due to lack of cover in a bridge beam

Fig. 2. Corroded re-bars due to lack of cover in a bridge railing

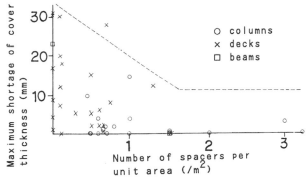

Fig. 3. Relation between number of spacers per unit area and lack of cover thickness [1]

2 Outline of Developed covermeters

Theories of already developed covermeters can be classified into the following categories:

- X-ray or radioactive ray

- Radar

- Ultrasonic

- Electromagnetic Induction

- Others

The characteristics of those covermeters are as follows:

2.1 X-ray method

In this method, an X-ray is irradiated from the one side of the structure to the film on the other side and the picture on the film shows the location, diameter and cover thickness of re-bars. Gamma-rays can be used instead of X-rays. As those three factors, location, diameter and cover thickness, cannot be obtained from one shot, the method with two shots from different locations of the X-ray tube is sometimes used to obtain those three factors. Fig. 4 [2] illustrates this method. Even in this method, scattering of the ray makes analysis difficult and some devices are used to avoid this influence. One is an X-ray grid. As shown in Fig. 5 [3], this grid cuts scattered rays to the film. Another is the computer picture treatment system which makes the contrast of the picture remarkable even to find cavities in the sheath of PC tendons, because it can show the hollow part in concrete [4]. Fig. 6 [5] shows the original picture and the result after of the picture analysis.

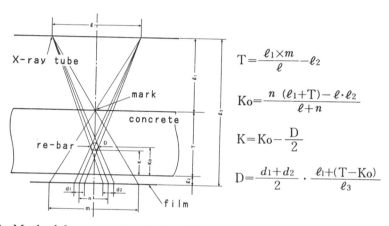

$$T = \frac{\ell_1 \times m}{\ell} - \ell_2$$

$$Ko = \frac{n\ (\ell_1 + T) - \ell \cdot \ell_2}{\ell + n}$$

$$K = Ko - \frac{D}{2}$$

$$D = \frac{d_1 + d_2}{2} \cdot \frac{\ell_1 + (T - Ko)}{\ell_3}$$

Fig. 4. Method for measuring the location, diameter and cover thickness with X-ray [2]

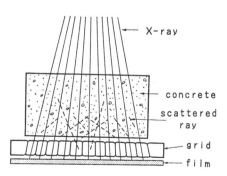

Fig. 5. X-ray grid [3]

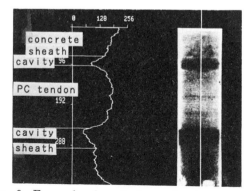

Fig. 6. Example of analysis result of X-ray photo [4]

398

The X-ray method is precise even with complex arrangements of re-bars regardless of the condition of concrete surface. However, this method can be used on structures with both sides opened, not thicker than approximately 50 cm. Furthermore, a license is required to use X-ray due to their dangerous affects. As the area of the film is not large, the cost per area to be measured is still quite expensive.

2.2 Radar
In this method, pulses of electromagnetic waves from the transmitting antenna reflect on the boundary of substances which have different electric characteristics. In this case the boundary is the surface of re-bars. The location and cover thickness of re-bars embedded in concrete are determined by mean of analysis of the reflected wave. The wave is harmless. This method cannot show the diameter of re-bars and it is difficult to distinguish sheathes from re-bars. It is also difficult to measure the thin cover thickness precisely because of mechanical reasons, and we need to measure thin cover precisely. This method is not applicable to complex re-bar arrangements at this stage.

2.3 Ultrasonic Method
In this method, the location of different substances and/or cavities in concrete is measured through analysis of the velocity and/or reflection of ultrasonic waves in concrete. This theory has been used in many other fields, for example geological and medical field. They try to develop the technique to measure the depth of cracks in concrete or the location of re-bars. When the ultrasonic wave spreads in concrete, much noise occurs and it makes the analysis difficult. This method is not applicable to complex re-bar arrangements at this stage.

2.4 Electromagnetic Induction Method
Some devices with this method are already on sale. But they measure the location and cover thickness of re-bars, the diameter of which is known, and cannot measure the cover thickness and diameter simultaneously. The theory of this method is that, when alternating current runs through the probe coil, electromagnetic field appears around the coil, and if there is a metal body in the field, it brings about a change in the voltage of the coil. The voltage change occurs according to the diameter and the cover thickness of re-bars. The measuring procedure is simple. The following introduces two current approaches in Japan using this method to develop the techniques to measure diameter and cover thickness of a re-bar simultaneously.

3 Method using the lag of phase of alternating current

Covermeters using the electromagnetic induction method on sale so far measure only the change of voltage of the probe coil. When alternating current runs through the probe coil, induction current, which is brought by the influence of re-bars, has different phase angle from original current. There is a lag between these two currents.

In this method, it is noticed that the lag angle is constant according to the diameter of a re-bar. Diameters of re-bars regulated in JIS (Japanese Industrial Standard) are discrete and if the angles for each diameter are measured in advance, as Fig. 7 [3], the diameter can be measured and then the cover thickness is also determined by the conventional way.

In real measurement, when the probe moves over a re-bar, the point which shows the intensity of inducted current moves on CRT as shown in Fig. 8a of which angle shows the diameter. When the probe is just on the re-bar, the point shows the maximum output. This point indicates the location and cover thickness. When two re-bars of same size are embedded, the output point goes and returns on the same line, and two re-bars are of different size, the output is like Fig. 8b. On the recording chart, X and Y components are recorded. As the maximum depth to measure and precision depends on the probe specification, changing probes according to cover thickness or pitch of re-bars brings more precise results.

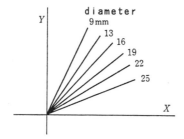

Fig. 7. Relation between diameter of re-bar and difference angle of AC phase on CRT [3]

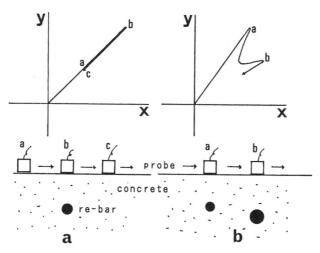

Fig. 8. Examples of output on CRT.

With this method, the influence of the kind of concrete, if ordinary aggregate is used, or influence of wet/dry condition of concrete is very small. But there is a thought that the quality of re-bar materials has some influence on the induction current and that the calibration line must be made to the very re-bar to be measured.

4 Method by analysis of output curve

This method was developed in PWRI using a conventional covermeter [6]. When the probe moves over the re-bar to be measured, the output curve like Fig. 9 can be obtained. The shape of this curve corresponds to the diameter of the re-bar. Thus we tried to apply this fact to measure the diameter and set a new value, Df as shown in Fig. 10, that means the distance of two points where the output level of induction voltage is half of the maximum. Fig. 11 shows the results of tests where Df ws measured with various sizes of re-bars and with various cover thicknesses.

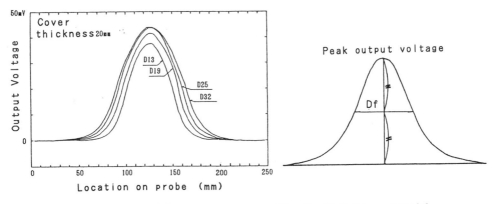

Fig. 9. Output curve [6]   Fig. 10. Definition of Df [6]

Fig. 11 makes it possible to measure the diameter and cover thickness of re-bars simultaneously. Practically, we input the curves of Fig. 11 into a computer and use an auto-scanning system.

Fig. 12 shows the result of a test in which the specimen had 8 re-bars of D19 (d = 19 mm, deformed bar) with pitch of 130 mm and planned cover thickness of 50 mm embedded in concrete. The average cover thickness measured through the holes was 52.2 mm. The probe was driven with the speed of 5 mm per second in the direction at right angles to re-bars and the output of induction voltage was recorded in every 1 mm. The output was the actually average of 10 outputs in 0.1 mm section before and after the point. The influence of temperature change and voltage shift of the device were automatically canceled by calculation. Measured diameters by this method

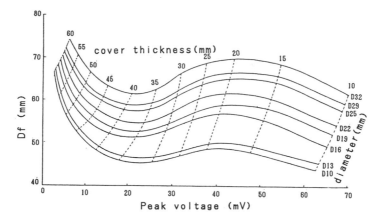

Fig. 11. Calibration curve [6]

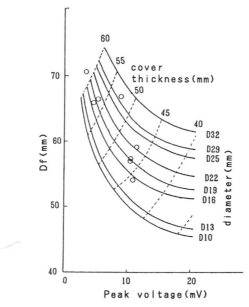

Fig. 12. Example of test result [6]

almost all ranged within 19 + 3 mm, and the measured cover thicknesses were within +20 %.

We have to carry out further tests to increase precision and to improve the analysis method for re-bars with small pitch and/or crossed re-bars and the analysis method when the probe cannot move at a right angle to re-bars.

5 Acknowledgement

Most of the tests shown in chapter 4 were carried out by H. Tanno, H. Tamura and A. Negishi.

References

[1] Fujii, M. and Miyagawa T. (1989) Nondestructive Testing Method for Grouting Condition in Prestressed Concrete Structures (in Japanese), Proceedings of JSCE No. 402, Vol. 10

[2] Maeda H. et al. (1988) **Nondestructive Test Method on Concrete Structures with X-ray** (in Japanese), Symposium on Evaluation Durability of Concrete Structures

[3] Ministry of Construction (1988) The Final Report of Research Project on "Development of Techniques for Improving Durability of Concrete Structures" (in Japanese)

[4] Mizunuma, M. (1986) Measuring Method of Location and Diameter of re-bar with X-ray multi-shot (in Japanese), 3rd Symposium on Application and Precision of Nondestructive Tests on Concrete Structures

[5] Tamura, H. (1988) Study on Improving of Precision of Covermeter with Electromagnetic Induction Method (in Janpanese), Technical Memorandum of Public Works Research Institute, No. 2676

[6] Wami H., Oda Y. and Hayashi N. (1989) Non-destructive Test Method for Concrete Structures - Location, Diameter and Cover Thickness of Reinforcing Bars (in Japanese), **Concrete Journal**, 27, No. 3

32 ACCURACY OF COVER MEASUREMENT

T. FEHLHABER and O. KROGGEL
Darmstadt University of Technology, Darmstadt, Germany

Abstract
This study investigates the application of nondestructive test methods to the determination of the diameter and position of reinforcement in concrete structures. An overview on the testing methods for this topic will be presented. Accuracy of cover measurement is discussed. A field test for the evaluation of the durability of a building structure is described, where a statistical approach was used. A new method called Electro- Magnetic-Reflection method (EMR) will be presented.

1 Introduction

Diameter and position of reinforcement in concrete structures are important parameters for the evaluation of the durability and the stability of building structures. A lot of samples are needed to assess the remaining life time and to determine the optimum date for the rehabilitation of a building structure. A reliable durability analysis can be yield only by nondestructive testing or semi-destructive testing methods. Even when detailed structural-drawings are available they often must be regarded as a useful tool by means of an additional information about the structure under investigation.

Nondestructive control of the cover and size of reinforcement in addition to other testing parameters like, for example, the permeability of the concrete, immediately after construction may help to improve the durability of the building structure in an early stage.

2 Overview of nondestructive test methods

The choice of a special test method to examine cover, size and location of reinforcement in concrete depends on the different conditions on site, like the test problem (Table 1), the accessibility and the costs. Mostly the amount of time or equipment increases with the difficulty of the test problem (A to E). In the past, a working group formed by the DAfStb named 'Nondestructive Testing Methods for Reinforced and Prestressed Concrete' provided an overview of useful test methods related to different types of test problems occuring on site (Table 2). The remarks in Table 2 give an estimate of the performance and the limits of the test methods. A rough estimate of the amount of money needed for the tabulated test methods are shown in Table 3.

Table 1: Different types of test problems.

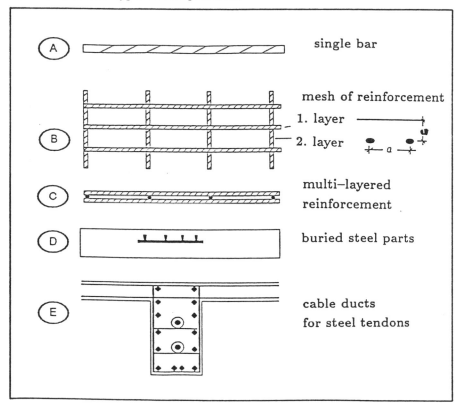

3 Accuracy of cover measurement

It must be indicated clearly that the overall accuracy of concrete cover measurements depends in an equal way on the combination of:

- The characteristics of the devices, in terms of: stability, insensibility to temperature changes, case of calibration and handling, display of data and mainly on the reliability of the measuring method itself.

Table 2: Test problem and related test methods.

Problem Test method	single bar A	2–layers B	3–layers C	buried steel parts D	cable ducts E
Covermeter Bore–hole probe	× up to 20 cm (×)	× c ≈ a ^{2}) (×)	×^{1}) ×	like 3–layers	–
X–Ray	×	×	×	×	×
EMR/Radar	×	× e ≈ a	?	×	× ^{3})
Induction- Thermography	× 30 cm	× 30 cm	–	–	×
Conductive- Thermography	(×)	(×)	–	–	×
Residual- Magnetism	(×)	(×)	?	?	?
Eddy Current	× up to 20 cm	× e ≈ a	^{1})	–	

^{1}) in special cases; great spacing of the 1st and 2nd layers is needed
^{2}) c = cover; a = spacing of rebars
^{3}) size of rebar great compared to that of the 1st and 2nd layer

× = yes   (×) = further development is needed

Table 3: Rough estimate of costs and personnel qualification

	Availability	costs of devices DM / piece	costs of test on site			personal- confication
			≤ 3000	3000 - 10000	≥ 10000	
Covermeter / Bore–hole probe	×	1000-5000	×			middle
X–Ray	×	high		×	×	high
EMR/Radar	×	≥ 150000 GSSR		×	×	high
Induction- Thermography	×	IRK 100000 80000		×		middle
Conductive- Thermography	(×)	high		×		middle
Residual- Magnetism	×	5000-10000		×		high
Eddy Current	×	20000 without Scan				high

× = yes   (×) = further development is needed

406

- The statement of the problem itself with respect to rebar configuration, preinformation and accessibility etc.

- The qualification of the personnel in terms of experience in covermeasurement and training on the equipment of the civil engineering know-how for the interpretation of data.

3.1 Classification of the problem of accuracy

In the first case, there is only the demand to check the concrete cover and combined with that the position of the rebars and the diameter of the rebars at a small number of discrete loci. The dowel setting for the mounting of assembly of the 'recalculation' of existing buildings are practical examples.

The second case is more focussed on a statistical statement of the amount of concrete cover of a large surface. This problem is strongly correlated with the problem of maintenance of existing buildings and quality assurance in the state of the building itself.

In the first case, it is essential to meet the requirements of the special problem by choosing a 'covermeter' or 'rebar locator' - as these devices are usually called providing an optimal match to the problem, in terms of depth information with respect to plane-resolution. These two quantities can only be optimized in the sense: 'depth resol. · plane resol. should be an optimum' for the problem. That is beside others the effect of the physical properties of the electro-magnetic field. Another effect of these properties is the 'shielding', produced by the first layer of reinforcement: it is simply impossible to get sufficient information about the third, fourth ... position and diameter of rebars because of this effect.

Until now only radiographic methods offer complete information in this field. In the second case, well-adopted statistical methods have to be applied and consequently the accuracy of the individual measurement must also be judged following this strategy, which is discussed in detail in [1].

The main question is whether the concrete cover of a building or of different parts of the building is sufficient, i.e. in accordance with the standards. In the sense of statistics this means: after having taken a sufficiently large number of samples (single measurements) it must be supposed that the distribution of concrete cover follows a Gaussian distsribution characterized by the mean value c and the standard deviation s_n. The accuracy of $\bar{c}$ and s_n depends on the number of samples n; it is increasing with $\sqrt{n}$. Details are given in all textbooks of applied statistics [2] and [3]. Plotting a large number of measured data (i.e. cover measurement of a facade) in the sense of probability distribution it may be found that the ordinary Gaussian distribution does properly not describe reality. There are two main reasons for this:

- the special geometry of the problem, especially the fact concrete cover = 0 has a probabilaity of 0, is more correctly represented by a logarithmic normal distribution.

- Areas of concrete cover being significantly different are contained in total number of data. This fact might even produce two peaks.

- There is a systematic error like mismatch of the covermeter or extraneous reading.

Provided that the ensemble of data is approximately represented by a Gaussian distribution, fractile values can easily be calculated by subtracting a corresponding multiple of the standard deviation s_n from the mean value $\bar{c}$.

Example 5 % fractile value: $c_{5\%} = c - 1.64 \, s_n$

Following the rules of error propagation the standard deviation s_n, which is derived directly from the measured data can be split in two parts

$$S_n^2 = S_c^2 + \sum_i^k S_i^2. \tag{1}$$

S_c = Standard deviation of the concrete cover itself.
S_i = Standard deviation produced by the measuring process.
Examples for the 'i':

    i = 1 Instability of the zero-point
    i = 2 Influence of temperature
    i = 3 Erraneous reading
    i = 4 Improper positioning of the sensor

to be continued.

The total value of $\sum S_i^2$ can be significantly reduced by well trained personnel and by improved technology of covermeters as computerized data capturing and evaluation.

Practical tests of different devices were performed, for example, by Schaab, Flohrer, Hillemeier [1]; Fehlhaber [4], Danßmann [5]; Some investigations were even considering the "human factor" by comparing the results of independently measuring personnel. Results shall not be discussed in detail, but it was found that the value $\sum S_i^2$ is unsatisfyingly high.

4 Case study on durability using cover measurement

The study was performed at an 18-year old faculty building of Darmstadt University of Technology where the facade consists of different types of reinforced concrete slabs (Fig. 1). More than 800 measurements of concrete cover and 180 samples of drilled cores, where the depth of carbonation reaction was investigated, were used to examine the condition of the building. A statistical approach was used to analyze the damage process.

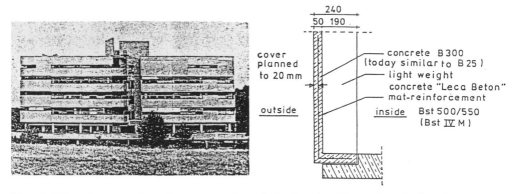

Fig. 1: View from south and cutaway view of the facade. Dimensions in [mm].

After a visual inspection, the conditions of the building were tested by measuring the cover of reinforcement and the carbonation depth at the sandwich slabs of the facade. The cover was measured with two different types of covermeters; the first was developed by a German company and the second by a British company.

The advantage of the first device was the ability to measure reinforcement of small spacing (a ≈ c) with a high accuracy in the range of 10 to 40 mm of concrete cover with maximum deviation of ± 2 mm, the easy handling and the low weight. The disadvantages were the frequently necessary calibration. The second covermeter is designed according to the British design standard for covermeters and has a comparatively large probe. The maximum deviation of the measurement was found to be less than ± 2 mm. Since the dimensions of the probe were large, the device was not suitable to examine the cover of reinforcement of small spacing without the application of an additional correction to the measured cover.

4.1 Analysis of Test Data

The depth of carbonation was measured at each side of the building facade. Figure 2 shows the influences due to orientation and height of the building. The influence in orientation corresponds very well to the climate conditions, where the west side is most exposed to rain. The examinations confirm that the rate of carbonation depends on the rate of carbon dioxyde diffusion, which slows down with increasing humidity in concrete.

Equation (2) describes the carbonation progress with y as the carbonation depth in mm, where D denotes the carbonation coefficient and t the time.

$$y = D \cdot \sqrt{t} \tag{2}$$

Figure 3 shows the progress of carbonation at the 3rd floor orientated to the south. The left hand part shows the depth of carbonation versus the exposure time starting after construction (1970) on a linear scale. The right hand part shows the cumulative distribution of concrete cover in probabilistic scale under the assumption that the

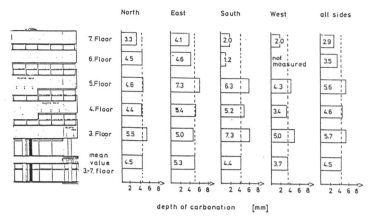

Fig. 2: Geometrical distribution of the depth of average carbonation at the facade slabs

distribution is Gaussian. Using the diagram one is able to predict the percentage of reinforcement which, at certain date, is no longer protected against corrosion by the alkalinity of concrete.

For example, in 1987 when the examination was made, the average carbonation depth was 7 mm at the south facade. The fractile of carbonation was calculated with a confidence level of 90 %. The progress of the damage potential can be read from the Figure 3, for example in the year 2000, the alkalinity surrounding of at least 63 % and at most 68 % of reinforcement will be destroyed and corrosion process can start. The confidence level for the damage progress, read from the figure, is $0.9 \cdot 0.95 = 0.86$.

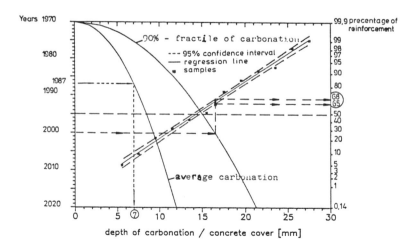

Fig. 3: Graphical solution of the damage progress of the facade slabs

5 EMR as a useful tool to examine concrete cover

Electro-magnetic-reflection (EMR) is a test method recently become accepted as a pavement- evaluation technique. It is also known as ground-penetrating-radar (GPR). This method can also be used for the examination of reinforcement in concrete structures. The radar system directs a brief pulse of electromagnetic energy, 0.8 nanoseconds long, into the concrete. When this energy encounters an interface between two materials of different electric properties, such as reinforcing steel, air, or moisture, a portion of the energy is reflected back to the radar antenna. The reflected energy is received by the transducer, amplified, and recorded by a personal computer. Simultaneously the signal is sent to the chart recorder, where a greyscale printout of the data is produced.

The electromagnetic pulse is repeated at a rate of 50 kHz and the resultant stream of time of flight measurements of radar pulses produces a contineous record of the concrete subsurface. During the collection of data, an oscilloscope displays the waveform reflected by the various interfaces within the concrete structure. This allows the operator to monitor the quality of the data and to note any anomalous signals. The data from site tests can be manipulated by using filter techniques and false colour representation. Fig. 4 shows such a representation for a concrete section with a buried single bar, investigated at a laboratory set-up.

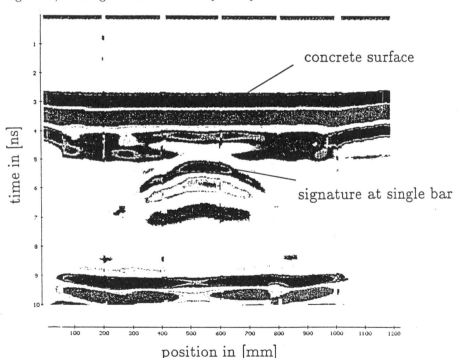

Fig. 4: Section of a single bar.

The accuracy of this test method depends on different conditions as the heterogeneity of the moisture content and the concrete itself of the subsurface. The pulse speed depends on the dielectricity of the concrete. Cause the concrete behaves heterogeneous to the radar pulses the accuracy of measuring the location of reinforcement in fact is a problem of determining the pulse speed in concrete from the diffraction hyperbola in Figure 4. Tests yielded to an accuracy of about 10 mm in vertical and lateral location of the reinforcing bars, examined in the laboratory tests.

In the future, investigations to improve EMR as reinforcement-detection-tool will be the design of an antenna, able to transmit radar pulses with a frequency content greater than 1 GHz, in order to get wavelengths in concrete as long as the size of reinforcing bars. The application of interpretation techniques, like migration and tomography, will be a useful tool to get a better accuracy than mentioned above.

6 Conclusion

Nondestructive testing of concrete cover is based on a manifold of physical phenomena. Convential methods using covermeters, where electromagnetic and magnetic properties are used, have physical limits, like penetration depth and resolution. Often the 3rd layer can not be examined, because of the shielding effect. Mostly a lack in lateral resolution is the reason, when covermeters are failing, when tests due to the 1st and 2nd layers are performed. New methods in testing concrete cover, as for example EMR, need for further development and in-situ tests, not at least to decrease costs on site.

7 Acknowledgement

The support by the "Geophysical Evaluation" Company Händel and Illich, D 7500 Karlsruhe, with respect to the performance of the EMR-laboratory tests is gratefully acknowledged.

References

[1] Schaab A., Flohrer C., Hillemeier B. Die zerstörungsfreie Prüfung der Betondeckung der Bewehrung. Beton und Stahlbetonbau, 84, 11 und 12, 1989.

[2] Papoulis A. Probability, Random Variables and Stochastic Processes. McGRAW-HILL, 1965.

[3] Owen D. B. Handbook of Statistical Tables. Addison-Wesley Publishing Company, 1962.

[4] Fehlhaber T. Electromagnetic Covermeasuring Devices. Darmstadt Concrete, 3, 19 pp., 1988. Editor: Inst. f. Massivbau TH Darmstadt.

[5] Danßmann J. Erfahrungen mit Überdeckungsmeßgeräten für Stahlbetonbauteile. Zerstörungsfreie Prüfung im Bauwesen, zfP Bausymposium 10, 1985, Berlin.

[6] Sohni M. Case Study on Durability. Darmstadt Concrete, 3, 199 pp., 1988. Editor: Inst. f. Massivbau TH Darmstadt.

PART SEVEN

EVALUATION OF MEASUREMENTS

33 A-POSTERIORI ESTIMATION OF CONVENTIONAL CONCRETE STRENGTH FROM IN-SITU MEASUREMENTS

T.P. TASSIOS
National Technical University of Athens, Greece

Abstract
The differences between concrete strength measured on standard specimens and strength measured directly in the structure are first enumerated, together with possibilities for mutual conversation. Subsequently, numerical data are presented regarding the additional variability of in-structure strengths, as well as the uncertainties related to the calibration of in-situ methods for strength evaluation. Finally, a couple of practical problems are discussed.

1   INTRODUCTION

There is an increasing practical need for in-structure concrete strength assessment by means of cores and or non-destructive tests. (In this context, the term "in-structure" is more restrictive than "in-situ"; separate cylinders or cubes cured near the structure are not covered by the term "in-structure"). In this paper an attempt is first made to describe briefly the main differences between the conventional strength and the in-structure strength of concrete, as well as to remind possible approximate relationships between them. Subsequently, the various experimental uncertainties of strength evaluation by means of cores or by means of non-destructive methods are enumerated. Finally, a formal numerical procedure is presented in order to evaluate the non-uniformity of concrete strength in-structure, as well as to estimate a potential characteristic strength.

2   THE DIFFERENCES BETWEEN CONVENTIONAL AND IN-STRUCTURE CONCRETE STRENGTH

It is well known that the conventional strength of concrete, measured on standard specimens taken at the mixer and cured in laboratory, is only a convenient (but inadequate) approximation of the real strength of concrete incorporated in the structure:

a) Standard specimens do n o t account for possible segregation of concrete during final transportation, nor for inadequate
compaction and real curing. The unfavourable consequences of
all these events and conditions are not checked by the conventional standard-specimens philosophy; these consequences are
meant to be limited by means of appropriate inspection only.

b) Standard specimens reflect only the non-uniformity of the
material per se. In-structure strength measurements show additional non-uniformities, due to the causes enumerated in §a, as
well as to the experimental techniques used for in-structure
assessment of concrete strength. Uncoupling of "material" scattering and "testing" scattering is not always possible.

c) Standard specimens are tested at 28 days; in-structure
strength measurements may be made practically at any age.
Transforming actual strengths to 28-day strengths is not an easy
task: Maturity factors are not easily assessed, whereas possible durability effects cannot be quantified a posteriori.
Consequently, conventional and in-structure strengths are of
considerably different nature and generally, they cannot be
related to each other. Of course, a radical change of design
strength specifications ("in-structure" specified strength,
instead of "standard" specimens strength philosophy), would eliminate the problem.
However, for the time being, practical needs do justify every
effort towards establishing possible relationships, approximative as they may be, between strengths measured with each of
these two methods. Such relationships (conversions) are necessary since design specified by current Codes is based on conventional concrete strength.
In doing so, one should "free" the in-structure strength values
from all influences which do not affect conventional strength:
(i) Systematic differences in mean values, due to:
    - Possible segregation after mixing
    - Inevitable lower degree of compaction (upper parts of
      slabs, areas of dense reinforcements)
    - Worse curing conditions (especially for thinner elements)
    - Differences in age.
(ii) Additional randomness influencing the variability of indi-
     vidual strength values: Compaction of standard specimens
     is made by one specialized technician under well specified
     conditions; site conditions are much less uniform. Addi-
     tional non-uniformity is normally induced by site-curing
     conditions.
(iii)Additional experimental variables are introduced in the
     techniques of in-structure strength measurements which may
     adversely influence b o t h the mean strength value and
     the variability of individual values: Sample disturbance
     through drilling of cores, influences due to curing and
     capping of cores (these influences are more serious than in
     the case of larger standard specimens), large uncertainties
     related to the calibration of non-destructive methods, etc.

All of the above mentioned influences should be quantified and
taken into account when converting in-structure strengths
$f_{c,situ}$ to conventional strengths $f_{c,stand}$.
The large number of these influences and their corresponding

uncertainties make the conversion extremely difficult.

3 APPROXIMATE CONVERSION FACTORS

It is not within the scope of this paper to elaborate on the theoretical feasibility of such conversions, nor on rational values of conversion factors reflecting each of the influences previously described. However, it is worthwhile to discuss the order of magnitude of some of these factors. Obviously, these indicative values cannot be used in specific cases without additional information.

3.1 Systematic Decrease of Mean Values

Within this context, the in-structure strength $f_{c,situ}$ (when obtained from cores) is assumed to be already corrected by the appropriate g e o m e t r i c a l factors to match the shape and size of the corresponding standard specimen (on which a strength $f_{c,stand} = f_{c,situ}/\mu$ would be expected).

a) Segregation during final transportation: Very unclear and highly dependant on local conditions. Normally $\mu_s \sim 1$.

b) Compaction: The top layers of building elements (which are not covered by any form) are i n e v i t a b l y affected by bleeding and incomplete compaction. Therefore, a top layer of 20 to 30 mm should never be considered when seeking compliance with standard strength. This top layer should be discarded from cores top surface and ND-measurements should not be carried-out on such surfaces (compare also Ref. 13). However, such a top layer of slabs may be the only compressive zone...
A systematic correction, due again to the i n e v i t a b l e non uniformity of compaction, is needed in case of columns.
Fig. 1 illustrates some experimental findings showing an average decrease of strength equal to 10% on top, and an average increase of 15% in the bottom of columns, as compared to a mean value. (Compare also the findings of Bellander (5) regarding "on top" and "under top" characteristic strengths). Therefore, $\mu_{co} \sim 0,85$ to 0,95 for top and $\mu_{co} \sim 1,05$ to 1,25 for bottom locations of columns, depending on the fluidity of fresh concrete.

c) Curing conditions: Depending on the general in-situ curing conditions, the following values have been roughly estimated, combining i.a. the findings of Bloem (6) and Johansen et al. (7):
$\mu_{c1} = 1,0$ (moist curing)
$\mu_{c1} = 0,9$ (normal good curing)
$\mu_{c1} = 0,8$ (dry curing conditions).
For given curing conditions, the thickness of the structural element may further modify its actual curing.
Therefore, for thicknesses smaller than say 250 mm, an additional factor might be needed [e.g. $\mu_{c2} = 0,95$ according to (8)].

d) Reference age: When relationships are available between strength of the specific concrete and its maturity index

$I_m = \int_0^t (T-T_o).dt$, (t in hours, T in °C), it is possible (see i.a. Malhotra (9)) to use a function like

$$f_c = f_{co}.log_{10} \frac{I_m}{k}$$

However, this conversion is a highly unpredictable operation: Real curing temperatures are not known, poor curing may hinter further hydration (Ref. 5), environmental agents may have adverse influences on strength, etc. However, a practical rule gives the following rough relationships for in-structure-cured concrete uncovered and in open air; figures are applicable when normal portland cement was used.

$f_{c, 3 \ months} = 1,10 \ f_{c, 28d}$
$f_{c, 6 \ months} = 1,20 \ f_{c, 28d}$
$f_{c, 1 \ year} = 1,25 \ f_{c, 28d}$

Indoor concrete or covered by renderings exhibit doubtful increases of strength with age.

Obviously, in case of an assessment of in-structure concrete strength for r e d e s i g n purposes, the problem of conversion for age is less important, although the use of existing values of safety factors foreseen by Codes is connected with standard strengths at 28 days (Ref. 10, p. 277).

e) Drilling and testing cores: Extensive experimental research [(1), (11), (12)] has proved that for good quality diamond bits, the following conversion factors may be used as a function of the maximum aggregate size D_{max} and the diameter d of the core (horizontal drilling on concrete of compressive strength higher than 20 MPa).

$\mu_d = 0,95$ for $d/D_{max} = 6$ to 7, (d = 150 mm)
$\mu_d = 0,90$ for $d/D_{max} = 3$ to 4, (d = 100 mm)
$\mu_d = 0,75$ for $d/D_{max} = 2$ (d = 50 mm)

These factors assume that the drilling torque is appropriate, and that the drilling equipment is in good mechanical condition. These conversion factors increase towards unity when vertical drilling is performed.

It has been reported (2, 13) that there is a strength decrease of the order of 5% when cores are loaded perpendicularly to the layers of concrete.

The effect of the presence of steel bars in cores is a highly controversial subject. Combining the results of Ref. 3 and Ref. 4, Fig. 2 has been compiled. Strength reductions (%) depend on the location of bars in the core.

Capping of cores and high moisture content at the moment of testing may lead to further core strength decreases of the order of 2% to 20%; however, ground end-faces and humid air curing of cores after drilling, seem to eliminate these experimental drawbacks.

All these conversion factors apply in converting core strengths to standard strengths. For non-destructive test methods, other factors related to the systematic decrease of mean strength values should also be taken into account when test results are converted to standard strengths.

3.2 Increase of Variability and Calibration Errors

Cores: The standard deviation of core strengths (d=100 mm) is greater than the standard deviation (s_{stan}) of conventional

specimens. Meininger (15) found a ratio equal to 4, Bloem (6) a
ratio equal to 2 to 3; and according to the Concrete Society
(13) this ratio may equal 1 to 3. The variability is also
affected by core size. Ramirez et al. (1) reported a two times
larger strength variance for 50 mm diameter cores compared to
00 mm diameter cores.

Core strength variations should be carefully considered since
they may include two components of variability: a) Additional
scattering (Δs_{cor}) due solely to the experimental techniques.
The decrease of dispersion of core strengths when their diameters increase (from say 50 mm to 150 mm) proves the existence of
such an induced variability which does not reflect true in-
structure variability, b) The real variability (s_{situ}) of
in-structure concrete strength.

Uncoupling of these two components would be needed when attemp-
ting to draw conclusions on structural reliability. The induced
additional dispersion (Δs_{cor}) may be estimated by drilling cores
very close to each other at one "location"; the standard deviation of their strengths will be very little affected by the in-
situ non-uniformity of concrete. On the basis of data included
in Ref. 16, values of Δs_{cor} equal to $(0,01$ to $0,05)f_c$ are used
in this paper.

In addition, there is another systematic source of potential
uncertainty: the expected inaccuracy of correction factors
accounting for shape and size of cores. This inaccuracy may
result in an unknown strength difference $\pm\Delta f_{c,geom}$. The
magnitude of this uncertainty may only be guessed by comparing
the various correction factors proposed by different Standards;
discrepancies of the order of 5% exist. Therefore, $\Delta f_{c,geom}$
may also be of the same order of magnitude.

The additional variability of local strengths induced by obtain-
ing and testing cores, and the uncertainty of converting core
strengths into strengths of standard specimens, contribute to an
"uncertainty range" for in-structure strengths determined by
means of cores (comp. Table III). Therefore it is difficult to
share the opinion that core strengths represent "true" in-
structure concrete strengths.

Non-destructive methods: The variability of strengths estimated
by means of non-destructive methods also contain an additional
large variability induced by the corresponding experimental
techniques. This "induced" variability, let us call it "Inherent error of the method" (IEM), expresses the lack of reproducibility of results when the s a m e concrete area is tested.
This error decreases with the skillfulness of the operator, with
increasing number of local individual repetitions of N.D-measu-
rements, as well as with improved contact conditions of the
tested concrete surface in case of rebound or pulse-velocity
tests.

Based of findings of Logothetis (16) and on own experience, we
have used the following values in this paper:

• Individual rebound hammer tests in one "location" in-situ,
  result in a standard deviation $s_{o,M}$ of corresponding compressive strengths equal to 2 to 7 MPa. Similarly, in ultrasonic
  tests in-situ, this standard deviation equals 1,0 to 2,5 MPa.
  Therefore, if n individual hammer tests or ultrasonic measurements are carried out in each "location", their corresponding
  mean strength values (the "local" strengths) are expected to
  have an additional standard deviation

MOUNT PLEASANT LIBRARY
TEL. 051 207 3581 Ext. 3701.

$\Delta s_{ND} = s_{indiv.} / \sqrt{n}$ equal to 0,5 to 2,0 MPa if $n_v = 4$ and $n_R = 15$, (where n_v and n_R denote the respective number of individual measurements).

- For a 95% probability level (and for a calibration curve supposed to be known deterministically), the corresponding inaccuracy $\Delta f_{c,IEM}$ of assessment of a local mean value of strength equals ±1,5 MPa to ±5,0 MPa (when $n_v = 4$ and $n_R = 15$).
- As in the case of cores, an additional uncertainty comes from possible systematic "calibration errors" (CE), reflecting the uncertainties of conversion of ND-test results into concrete compressive strengths; these errors decrease with better knowledge of concrete composition, age and curing conditions.

Table I contains indicative values of corrections to be made on rebound numbers "R" and on pulse velocity values "V", when additional information is available on composition, age and curing conditions of the concrete under investigation. These corrections are derived (as rough average values) from several research papers, i.e. Tomsett (17), Tanigawa et al. (18), (19), Bellander (5), Willetts (20), Grieb (21), Chung et al. (22), Mitchell et al. (23) and Logothetis (16). Both the list of corrections and the influencing parameters are clearly incomplete. Some other parameters are listed here below; however, the corresponding corrections proposed seem to be less accurate than those included in Table I:

Table I. CORRECTIONS ON "V" anr "R" VALUES measured on in-situ concrete under conditions different than "reference" concrete (on which calibration curves have been prepared).

Corrections	Aggregates			Transversal compression "σ"	Age (months)		Curing before testing	
	siliceous rounded	coarse			3	12	saturated	dry
		D_{max}	%					
ΔR	-7			If σ<5 MPa ΔR=+(5-σ) or +[5-(σ/2)]	-(1÷2)	-(2÷4) If carbonated concrete not removed, ΔR<-4	+ (2÷6)	-1
ΔV(m/s)	(-ΔV)	20mm +100 10mm +300		for lower percentages→+ΔV σ/f_c<0,60 ΔV = 0		(+ΔV)	- 100	+(100÷200)
Δf_c:f_c,ref strength estimated by combined R and V values		20mm +0,15 15mm +0,25	30%→+0,05 40%→-0,05				mutually counterbalanced effects	

The figures of this Table are valid for a "reference" concrete composed of crushed limestones, maximum grain size D_{max} = 30 mm, coarse aggregate abs. volume 35%, Portland cement 300 kg/m³, on which ND-measurements were made at 28 days, after humid air curing, and under transversal compressive stress>5 MPa)

- If full unloading of a compressed concrete area has taken place (from σ } $f_c/2$), a correction $\Delta V = +0,04V$ should be made after 2 cycles, or $\Delta V = +0,08V$ after more than 4 cycles; (Logothetis (16)).
- If slender slabs are tested by a Schmidt hammer type-N, the following corrections should be made to account for energy absorbed by the vibration of the slab as a whole:
$\Delta R = +4(0,05\ l_o/h - 1)$, for $l_o/h > 20$,
where l_o = the distance of concecutive contraflexure points, and h = the thickness of the slab; (Logothetis (16)).
- When a hammer test is performed on a top surface of segregated areas, $\Delta R = +(1\ \text{to}\ 4)$.
- Several correction factors for strengths evaluated by means of ND tests have been proposed, to account for types of cement and cement contents different than those used in "reference" concrete (from which the calibration curves were derived); see i.a. RILEM Recommendation (25), where indicative values of "cement-corrections" are included for strengths evaluated by means of combined hammer and ultrasonic tests:

Cement content (kg/m^3)	200	300	400
$\dfrac{f_c \text{ corrected for cement content}}{f_c \text{ from reference curve}}$ =	0,87	1,00	1,14

However, some contradictions do persist in the technical literature regarding not only the numerical values of "cement-correction" factors, but also the sense of the correction (especially for different types of cement).
- Finally, in case chemical or thermal damage has occurred, totally different calibration curves are needed, such as proposed by Chung et al. (22) and Logothetis et al. (28); no correction factors can be reliably conceived for this purpose.

Independently of the precision of all these correction factors, there is also a considerable amount of uncertainty accompanying the "mean" calibration curve of each ND-method, as well as (to a lesser extent) the calibrating nomograms of combined ND-methods. Whenever additional information is missing (regarding the influencing parameters described in Table I and further), the corresponding correction factors are transformed into expected calibration errors (CE).
Note that a pulse-velocity variation $\Delta V = \pm 100 m/s$ corresponds roughly to a strength variation $\Delta f_c \approx \pm 3$ MPa; similarly a $\Delta R = +1$ corresponds to $\Delta f_c \approx \pm 2$ MPa. Therefore, a cumulative calibration error $\Delta f_{c,CE}$ ranging between ± 3 to ± 10 MPa must be expected. As a consequence, if a square root rule is adopted for the superposition of inherent and calibration errors, a <u>range of uncertainty</u> equal to

$$\Delta f_{c,ND} = \sqrt{\Delta f_{c,IEM}^2 + \Delta f_{c,CE}^2} \approx \pm 4 \text{ to } \pm 11 \text{ MPa} \qquad (2)$$

is to be expected when estimating in-structure concrete strength by means of one ND-method. For concrete strengths ranging from 20 to 40 MPa, this $\Delta f_{c,ND}$ corresponds approximately to an

423

uncertainty percentage equal to ±15% to ±40%.
Extremely rough as they may be, these calculations seem to confirm the level of accuracy reported in literature for concrete
strength estimations made by means of ND tests:

Test	Author		
	Murphy (25)	Facaoaru (12)	Malhotra et al.(26)
R	±20*%	+40% ~ −17%	±25%
V	±15*%	+12% ~ −20%	±20%

(* under "favourable" conditions)

However, c o m b i n e d ND-methods may bring this inacurracy
down to ±10% to ±20% (Facaoaru (12), RILEM (24)), because of the
opposite effects of some factors (see Table I) on "R" and "V"
values. Numerical values regarding additional variability
induced by a combined method are more scarse than for single
N.D.-methods. Nevertheless, for the sake of completeness, we
will assume here that combined N.D.-methods lead to the same
additional standard deviation of local strengths as single
methods, but they are subject to half the calibrations uncertainty of the more precise of the two single methods combined.
The Tables II and III summarise the numerical data presented in
this chapter.

4   EVALUATION OF IN-STRUCTURE CONCRETE STRENGTH DATA

Under the light of the preceding chapters, the following conclusions can be drawn:
The in-structure concrete strength, measured on cores or estimated by means of N.D.-tests in-situ, c a n n o t be directly
compared with the "conventional" strength of standard specimens:

a) In the majority of cases, actual in-structure strengths are
("legitimately", so to say) influenced by conditions decreasing
concrete strength and increasing strength-variability.

b) Any experimental "operation" needed for in-structure concrete
strength assessment (cores drilling, testing and interpretation
of results, as well as non-destructive testing and interpretation), is bound to uncertainties concerning both the estimation
of strengths and their variability.

Factors ($\mu = f_{c,situ} : f_{c,stand}$.) converting (very roughly
though) in-structure strengths into conventional strengths, have
been summarized in Table II. Conversion factors accounting for
compaction and curing conditions (as well as for the age of concrete tested) are equally applicable for strengths found on
cores or assessed by means of N.D.-tests. Additional conversion
factors take into account the specific consequences of cores
drilling and testing (§3.1.e and Table II).
Regarding variability now, let us introduce the following nota-
tion for standard deviations (making the practical assumption

that in-structure strengths follow a normal distribution):
S_{situ} = real variability of local strength-values $f_{c,situ}$
throughout the structure (each of them being assessed by means
of an appropriate number of individual measurements carried-out
in the same "location"),

Table II. Local Strength Conversion Factors "μ_j"

Test	$\mu_j = f_{c,M}/f_{c,stand.}$					
a	1 compaction (Columns)		2 Curing (in-structure)			3 Thickness
CORES and/or	top	bottom	very moist	normal	dry	<250mm
N.D.TESTS	0,85 to 0,95	1,05 to 1,25	1,0	0,9	0,8	0,95

b	1 Drilling when d/D_{max}			2 Direction of loading		3 Steel bars	4 Inappropriate capping and moisture during testing
CORES only	6 to 7	3 to 4	2	$\perp$	$\parallel$		
	0,95	0,90	0,75	0,95	1,00	see Fig.2	(0,80) to 1,00

D_{max} = maximum grain size of aggregates
d = core diameter
$\perp, \parallel$ = perpendicularly or parallelly to the layers of concrete

The global conversion factor "μ" is equal to the product of all
relevant factors "μ_j":$\mu = \Pi\mu_j$ (j = 1, 2, 3,..., as in the Table)

Table III. Additional variability and uncertainties of in-
structure assessed strength values

No	Nature of uncertainty	C o r e s	Pulse velocity	Rebound Hammer
1	Additional standard deviation of "local" strength, due to experimental techniques	$\Delta s_{cor}=(1\%\div5\%)f_c$	$\Delta s_v=0,5\div1,5$ MPa	$\Delta s_R=1,0\div2,0$ MPa
2	Corresponding uncertainty (95%) of "local" strength assessment, $\Delta f_{c,TEM}$	(for n = 3) $\pm(2\%\div12\%)f_c$	(for n = 4) $\pm(1,5\div4,5)$ MPa	(for n = 15) $\pm(2,5\div5,0)$ MPa
3	Uncertainty due to calibration errors, $\Delta f_{c,CE}$	$\pm5\%f_c$	$\pm(3\div8)$ MPa	$\pm(5^*\div10)$ MPa
4	Overall uncertainty range $\sqrt{\Delta f_{c,TEM}^2 + \Delta f_{c,CE}^2}$ =	$\Delta f_{c,cor}\sim\pm(5\%\div13\%)f_c$	$\Delta f_{c,v}\sim\pm(4^*\div8)$ MPa	$\Delta f_{c,R}\sim\pm(6^*\pm11)$ MPa

n number of individual tests at each "location"

* first figures apply when adequate information about concrete composition, curing and age is available

It is assumed that "combined" V#R method is bound to the same $\Delta f_{c,TEM}$ level, but to a calibration uncertainty of the order of $2\div4$ MPa.

Δs_M = additional variability induced by the assessing experimental method "M" used,
$s_{M,situ}$ = overall variability of $f_{c,situ}$-values, as found experimentally by means of an in-situ method "M",
If we assume that

$$\sqrt{s_{situ}^2 + \Delta s_M^2} = s_{M,situ} \qquad (3)$$

a rough estimation of s_{situ} might be possible, making use of an appropriate expected value of Δs_M (e.g. from Table III). Hence, a characteristic value of in-structure local strengths might also be possible to be estimated; the level of this strength should be appropriately corrected by means of the factors included in Table II,b.
However, the usefulness of such probabilistic features is rather doubtful, (Murphy (25)). Not only because the distribution of real local strengths is not normal (several non-random parameters influence their values), but mainly because structural reliability requirements cannot be fulfilled in such a simple but uneconomical way: The expected minimum concrete strength on top of a thin slab is not necessarily critical for beams and much less for columns, since these building elements (in their really critical cross-sections) do avail of systematically higher concrete strengths than the top of the slab. Besides, even in the case of slabs, concrete strength might not be critical (under-reinforced cross-sections), especially when more sophisticated methods of elastoplastic analysis are putting in evidence the higher capacity of slabs to redistribute action-effects.
However, it is beyond the scope of this paper to elaborate on the subject of such a n e w reliability theory based on in-structure specified concrete strengths.
For the time being, this paper will conclude with a couple of more practical well known questions:

4.1 Reliability of Non-Uniformity Encountered

At a small "location" (meaning a region which may be considered as practically "uniform"), a local strength $f_{c,M}$ has been estimated by means of an in-situ method "M" (as a mean value of an appropriate number "n" of individual measurements presenting a standard deviation $s_{o,M}$).
May this local strength be considered as reliably different from the local strength $f_{c,M}$ found at another location (on the basis of "n" measurements having practically the same standard deviation $s_{o,M}$)?
Assuming that calibration errors are the same at these two locations, the following comparison should be made:

$$|f_{c,M} - f'_{c,M}| > 2,5 \cdot \frac{s_{o,M}}{\sqrt{n}} \qquad (4)$$

for a probability around P=90%. A better approximation might be achieved if, instead of the "average" values $s_{o,M}$ used in Equ. 4, the standard deviation of all 2n individual values were taken.
Thus, non-uniformity of in-structure concrete will be practically evaluated in a more rational, be it approximate, way.

4.2 Missing or disputed standard specimens

In case of missing or disputed results of conventional specimens, how could it be possible to estimate the strength of the
in-structure concrete, as if it were sampled after mixing and
tested conventionally, at 28 days? In other words, what is the
potential strength, according to the terminology of Ref. 13?
For this purpose it seems reasonable to take into account a
large representative number "v" of "local" strengths $(f_{c,M})_i$
from several locations, belonging to different building clements
(which have preferably been cast from one batch). All these
strength values should first be converted into potential conventional strengths by means of the conversion factors summarised
in Table II and discussed in §3.1. It is expected that the population of these converted values will fit better to a normal
distribution since, by means of factors μ_j appropriately chosen
for each location, the non-random influences on this population
have been significantly alleviated. Therefore we may consider
the mean value.

$$\bar{f}_{c,M,pot} = \frac{1}{v} . \sum_{i=1}^{v} [(f_{c,M})_i : \Pi\mu_j] \qquad (5)$$

where Π denotes the product of individual μ_j factors. The
standard deviation $S_{M,pot}$ of the populatiron of c o n v e r-
t e d in-structure strengths should also be calculated.
Now, in order to estimate a characteristic value (5% fractile)
of this population, it is reasonable to reduce the value $S_{M,pot}$
by the additional variability ΔS_M (see Table III) which was due
to the experimental method "M" used for the strength assessment
in-situ.
Thus, a potential characteristic strength of the concrete mass
under examination might be evaluated:

$$(f_{c,M,pot.})_k = \bar{f}_{c,M,pot.} - 1,7 \sqrt{S_{M,pot.}^2 - \Delta S_M^2} \qquad (6)$$

Note that $f_{c,M}$ values are also bound to the uncertainty $\pm\Delta f_{c,CE}$,
i.e. the calibration uncertainty (see Table III, line 3).
It is worth to remind here that this is not the characteristic
value of actual in-structure strengths (which in the introduction of §4 has been deemed of doubtful significance); the value
given by Equ. 6 is only a f i c t i t i o u s value to be
compared with the conventional specified characteristic
strength. Besides, due to the cumulative effects of several
assumptions and estimations made, such a comparison must be
submitted to extensive trial applications to verify that confidence can be put on the procedure proposed.

4.3 Redesign

For the specific purpose of r e d e s i g n of existing
structures, it is more advisable to limit this procedure to each
structural element separetely.

CONCLUSIONS

a) In-structure strengths are very different from the specified
standard (conventional) strength; they depend on real compaction

and consolidation conditions, as well as on actual curing history in-situ. Table IIa summarises some approximative relationships between estimated in-structure local strengths and the potential standard strength the same locations would show if that concrete were sampled, cured and tested conventionally.

b) These in-structure strengths cannot be precisely known; the assessing methods induce inevitable strength modifications: Core drilling and testing may produce strength decreases summarized in Table IIb. On the other hand, strengths evaluated by means of N.D.-methods are subject to possible calibration-errors (summarized in Table I), depending on our knowledge regarding composition, curing and age of the concrete under examination.

c) In addition to the uncertainties related to the previous paragraph, all assessing methods increase the apparent variability of the in-structure local strengths found by means of these methods. In Table III, line 1, some indicative values of additional standard deviations are presented which are experimentally induced to the local strengths when they are evaluated by means of cores, pulse velocity or rebound hammer measurements.

d) Reasonable estimators of "actual" local strengths may be found by correcting experimental local strengths to account for modifications induced by the assessing method.
However, these estimators are bound to some uncertainty
(labelled $\Delta f_{c,IEM}$ in Table III, line 2).

e) These actual local strengths may also be converted into "potential" standard strengths (as if concrete strength were measured conventionally). Here again, a calibration uncertainty still exists ($\Delta f_{c,CE}$, Table III, line 3).

NOTATION

d	= diameter of cores
D_{max}	= maximum grain size of aggregates
f_c	= concrete compressive strength in general
$f_{c,M}$	= in-structure local strength as estimated by means of a method "M"
$f_{c,M,pot}$	= potential concrete strength at a location in-structure, as if concrete were sampled, cured and tested conventionally
$\bar{f}_{c,M,pot}$	= mean value of several $f_{c,M,pot}$ corresponding to a number of locations throughout the structure
f_{co}	= constant strength, in maturity index calculations
$f_{c,ref}$	= strength estimated by means of a ND-method, out of a calibration curve prepared on a "reference" concrete
$f_{c,stand}$	= strength on standard specimens
$\Delta f_{c,CE}$	= uncertainty range of $f_{c,M}$, due to calibration errors
$\Delta f_{c,cor}$	= overall uncertainty range of local strengths found by means of cores
$\Delta f_{c,geom}$	= uncertainty range of local strengths found by means of cores, due to doubtfulness on shape and size correction factors
$\Delta f_{c,IEM}$	= uncertainty range of strength due to inherent errors of the assessing method

$\Delta f_{c,ND}$ = overall uncertainty range of local strengths estimated by of a N.D.-method
h = thickness of slab
I_m = maturity index
k = a constant in maturity index calculations, characteristic value
l_o = distance of consecutive points of contraflecture in a
continuous slab
$\mu = f_{c,M}$: $f_{c,stand}$, conversion factors against actual in-
structure conditions, or against strength decreases
induced by drilling and testing of cores.
n = number of individual measurements made by an assessing method "M" within a location (meant to be reasonably uniform), in order to determine a local
strength $f_{c,M}$
v = number of potential strengths $f_{c,M,pot}$ considered
throughout the structure
Π = symbol of product (multiplication of several factors)
R = rebound hammer indication (number)
s = standard deviation, in general
$s_{M,pot}$ = standard deviation of $f_{c,M,pot}$ values of different
locations
$s_{M,situ}$ = standard deviation of $f_{c,M}$ values of different
locations
$s_{o,M}$ = standard deviation of individual strength values
measured by a method "M" within a location
s_{situ} = standard deviation of actual in-structure local
strengths
Δs_{cor} = additional standard deviation induced to local
strengths
Δs_M = idem, by any assessing method "M" in general
Δs_{ND} = idem, by N.D.-methods
σ = transversal compressive strength when carrying-out
pulse velocity measurements
T = temperature
T_o = threshold temperature for maturity index calculations
t = time
V = pulse velocity

REFERENCES

1. RAMIREZ J.L., BARCENA J.L.: "Some data on concrete cores
strength evaluation", RILEM Symposium "Quality control of
concrete structures", Stockholm, 1979.
2. KASAI Y., MATUI I.: "Studies on concrete strength of structures in Japan", RILEM Symposium "Quality control of concrete structures", Stockholm, 1979.
3. KEMI T., HIRAGA T.: "On the distribution of the strength of
concrete structures", RILEM Symposium "Quality control of
concrete structures", Stockholm, 1979.
4. TAKAHASHI H., NAKANE S.: "Strength of concrete in structures
and factors contributing to strength differentials", RILEM
Symposium "Quality control of concrete structures", Stockholm, 1979.
5. BELLANDER U.: "Strength in in-situ structures (in Swedish),
Swedish Cement and Concrete Research Institute, Stockholm,
1976.

6. BLOEM D.L.: "Concrete strength in structures", ACI Journal, March 1968.
7. JOHANSEN R., DAHL-JORGENSEN E.: "Curing conditions and in-situ strength development of concrete measured by various testing methods", RILEM Symposium "Quality control of concrete structures", Stockholm, 1979.
8. ELOT 344: "Correlation between strength of in-situ and the conventional strength of concrete made of limestone aggregates", 20.5.79.
9. MALHOTRA V.M.: "Testing hardened concrete: Nondestructive methods", ACI Monograph 9, 1976.
10. CEB Gen. Task Group 12: "Assessment of concrete structures and design procedures for upgrading", CEB Bulletin 162, 1983.
11. CAMBELL R.H., TOBIN R.E.: "Core and cylinder strengths of natural and lightweight concrete", ACI Journal, April 1967.
12. FACAOARU I.: "The correlation betwen direct and indirect testing methods for in-situ concrete strength determination" RILEM Symposium "Quality control of concrete structures", Stockholm, 1979.
13. CONCRETE SOCIETY (GB): "Concrete core testing for strength", Techn. Report No 11, 1976.
14. PLOWMANN J.M., SMITH W.F., SHERIFF T.: "Cores, cubes and the specified strength of concrete", The Structural Engineer, Nov. 1974.
15. MEININGER R.C.: "Effect of core diameter on measured concrete strength", Journal of Materials, ASTM, June 1968.
16. LOGOTHETIS L.: "A contribution to the in-situ assessment of concrete strength by means of combined non-destructive methods" (in greek), Dr Eng. Thesis (Supervisor T. Tassios), Nat. Tech. University, Athens 1978.
17. TOMSETT H.N.: "The practical use of ultrasonic pulse velocity measurements in the assessment of concrete quality", Magazine of Concrete Research, March 1980.
18. TANIGAWA Y., KOSAKA Y., YAMADA K., BABA K.: "Estimation of concrete strengh by combined Schmidt hammer and ultrasonic pulse velocity methods", Trans. Japan Concrete Institute, 4/1982.
19. TANIGAWA Y., YAMADA K., KOSAKA Y.: "Combined non-destructive testing method of concrete", Trans. Japan Concrete Institute, 2/1979.
20. WILLETS C.H.: "Investigation of the Schmidt concrete test hammer", Misc. Paper 6-267, US Army Waterways Exp. Station, 1958.
21. GRIEB W.E.: "Use of the Suiss hammer for estimating the compressive strength of hardened concrete", "Public Roads", 30 No 2 Washington D.C., June 1958.
22. CHUNG H.W., LAW K.S.: "Diagnosing in situ concrete by ultrasonic pulse technique", Concrete International, Oct. 1983.
23. MITCHELL L., HOAGLAND G.: "Investigation of the impact type concrete test hammer", Highway Research Board, Bulletin 305/1961.
24. RILEM 41 CNDT: "Tentative recommendations for in-situ concrete strength determination by N.D. combined methods", Draft 1982.
25. MURPHY W.E.: "The assessement of concrete strength in structures", RILEM Symposium "Quality control of concrete

structures", Stockholm, 1979.
26. MALHOTRA V.M., CARETTE G.G.: "In-situ for concrete strength", Canada Centre for Mineral and Energy Technology, Rep. 30, 1979.
27. LOGOTHETIS L., TASSIOS T.P.: "In-situ assessment of concrete quality by means of three combined N.D.-methods", RILEM Symposium "Quality control of concrete structures", Stockholm, 1979.
28. LOGOTHETIS L., ECONOMOU C.: "The influence of high temperatures on calibration of N.D.-testing of concrete", Materials and Structures, No. 79, January-February, 1981.

34 STATISTICAL METHODS TO EVALUATE IN-PLACE TEST RESULTS

N.J. CARINO
National Institute of Standards and Technology,
Gaithersburg, MD, USA

Abstract
In-place testing is used to estimate the compressive strength of concrete
in a structure by measuring another property related to compressive
strength. Statistical methods are needed for reliable estimates of in-
place strength. Such methods should account for the uncertainties in
the measured property, the uncertainty of the correlation relationship,
and the variability of the in-place concrete. Standard statistical
procedures have not yet been adopted in North American practice. Recommendations are provided for developing the correlation relationship, and
a reliable, easy-to-use approach is presented to estimate in-place characteristic strength.

1 Introduction

On April 27, 1978, work was in progress on a hyperbolic reinforced concrete cooling tower located at Willow Island, West Virginia. At about
10:00 a.m., while concrete was being hoisted for the placement of the
next 1.5-m lift of the shell, there was a sudden failure of the concrete
placed on the previous day. The catastrophic accident resulted in 51
deaths, making it the worst construction accident in U.S. history. A
subsequent investigation concluded that the most probable cause of the
failure was inadequate concrete strength to support the applied construction loads [1].

This accident demonstrates the critical importance of considering construction loads and monitoring in-place strength development in concrete
construction. This incident also provided great impetus in North America
to modify construction practices and search for reliable methods for in-
place testing of concrete.

In-place tests are used to estimate the strength of concrete in a
structure by measuring a property which is related to compressive strength. The compressive strength is usually needed to assess structural
capacity using accepted design methods. In-place testing is more economical than testing cores drilled from the structure, and it provides
information that is more representative of the concrete in the structure

432

than is obtained from the traditional method of testing field-cured specimens.

In North American practice, in-place testing became a "recognized" alternative to testing field-cured cylinders by the addition of the following sentence to the section of the 1983 ACI Code dealing with form removal [2]:

> "Concrete strength data may be based on tests of field-cured cylinders or, when approved by the Building Official, on other procedures to evaluate concrete strength."

The Commentary to the Code listed acceptable alternative procedures, and further stated that these alternative methods "require sufficient data using job materials to demonstrate correlation of measurements on the structure with compressive strength of molded cylinders or drilled cores." Thus, to use any of the alternative methods, a correlation relationship must be developed to translate the in-place test results to equivalent compressive strength values. In addition, a procedure is needed to analyze in-place test results so that the compressive strength can be estimated with a high degree of confidence. Both steps require statistical analysis of test data.

Procedure for performing in-place tests have been standardized by ASTM [3] and other organizations, and a recent report by the American Concrete Institute [4] provides additional guidance on their use. However, there is no standard practice in North America for the statistical procedures to analyze test results. This paper deals with the key aspects of a suitable statistical procedure. The goal is to propose an initial framework for future efforts to develop a consensus standard for analysis of results. First, the objective of in-place testing during construction is discussed. This is followed by a review of the statistical principles to be considered in applying in-place testing.

2 Background

In designing a reinforced concrete structure, the designer uses the **specified compressive** strength of the concrete. However, the strength of concrete in a structure is variable and can be described by some type of probability density function. In North American practice, the specified strength is generally taken to represent the strength that is expected to be exceeded with about 90% probability, i.e., if 10 random samples were taken from the structure, 9 of them would be expected to exceed the specified strength. The specified strength is also called the **characteristic strength**, and this term will be used in subsequent discussion.

The objective of in-place testing during construction is to assure, with a high degree of confidence, that the concrete in the structure is sufficiently strong to resist construction loads. To have a safety margin during construction that is consistent with that under service conditions, one needs to know the in-place characteristic strength at critical stages of construction. How can we be assured of a reliable estimate of the in-place characteristic strength? The answer is by using statistical principles to account for the various sources of uncertainty.

The application of in-place testing for monitoring strength development during construction involves the following steps:

- Establish the correlation relationship between the in-place test result and compressive strength.
- Perform the in-place tests during critical stages of construction.
- Analyze the test results to obtain a reliable estimate of the in-place characteristic strength.

The statistical principles to be considered during these steps are discussed.

3 Correlation relationship

Some manufacturers of in-place testing equipment provide correlation relationships along with their devices. However, the ASTM standards [3] and the ACI report [4] recommend that a correlation relationship should be established for the specific test instrument using the concrete materials that will be used in construction. The following questions must be answered to develop a reliable correlation relationship:

- How many test points (i.e., strength levels) are needed?
- How many replicate tests should be performed at each strength level?
- How should the data be analyzed?

3.1 Number of test points
The usual procedure is to perform replicate in-place tests and replicate standard compression tests[a] at various strength levels. The average values of the results are used to establish the correlation relationship using regression analysis. Because the correlation relationship will be used subsequently to estimate compressive strength from in-place test results, compressive strength is treated as the dependent variable (Y-value) and the in-place result as the independent variable (X-value).

The first principle to consider in planning the correlation testing program is that the tests should have **coverage**. This means that the range of strengths should cover the range of strengths anticipated in the structure during the period when strength will be estimated. **The correlation relationship must never be used for extrapolation.** Therefore, if very low in-place strengths are to be estimated, such as might be required during slipforming, the testing program must include these low strength levels. The second principle to consider is that the test points should have **balance**, which means that they should be evenly spaced.

---

[a]Compressive strength is considered in this discussion, but correlation could also be established between in-place test results and other standard strength properties, such as flexural strength or indirect tensile strength.

The number of strength levels that should be used depends on economic considerations and on the acceptable level of precision. To have an understanding of the effect of the number of test points on the precision of the correlation relationship, it is useful to examine the residual standard deviation (standard error) of the best-fit correlation relationship. The residual standard deviation is the basic statistic used to quantify the uncertainty of the calculated relationship. For a linear correlation relationship, an unbiased estimate of the residual standard deviation is as follows:

$$S_e = \sqrt{\frac{\sum (d_{yx})^2}{n - 2}} \tag{1}$$

where,

S_e = residual standard deviation,
d_{yx} = deviation of each test point from the best-fit line, and
n = number of test points.

For the sake of discussion, assume that each test point deviates from the best-fit line by the same amount, δ, and that this deviation is independent of the number of points. The estimated residual standard deviation would equal:

$$S_e = \delta \sqrt{\frac{n}{n - 2}} \tag{2}$$

When the correlation relationship is used to estimate the mean value of Y at X, the width of the confidence interval for the estimate is related to the residual standard deviation by the following expression [5]:

$$W = 2 \, t_{n-2, \alpha/2} \, S_e \sqrt{\frac{1}{n} + \frac{(X - X_a)^2}{S_{xx}}} \tag{3}$$

where,

W = the $(1-\alpha)$ confidence interval for the estimated mean value of Y for the value X,
$t_{n-2, \alpha/2}$ = Student t-value for n-2 degrees of freedom and significance level = $\alpha/2$,
X_a = grand average of X-values used to develop correlation relationship,
S_{xx} = sum of squares of deviations of X-values from X_a.

The second term under the square root sign in Eq. (3) shows that the confidence interval increases as the value of X is farther from the grand average X_a.

To examine qualitatively how the width of the confidence interval is affected by the number of test points, consider the case where $X = X_a$, so that the second term under the square root sign in Eq. (3) equals zero. Substituting Eq. (2) into Eq. (3), we obtain the following expression for the confidence interval:

$$W = 2 \, t_{n-2, \alpha/2} \, \delta \sqrt{\frac{1}{n - 2}} \tag{4}$$

435

Figure 1(a) shows the variation W as a function of the number of points for a 95% confidence level. It is seen that, for a small number of test points, by including an additional test point there is a significant reduction in the confidence interval. However, for a large number of points, the reduction is small. Therefore, the appropriate number of strength levels is determined by considerations of economy and precision. Figure 1(b) shows the reduction in the confidence interval by using n test points compared with using (n-1) points. From this figure it is reasonable to conclude that the minimum number of test points is six, while more than nine tests would probably not be economic.

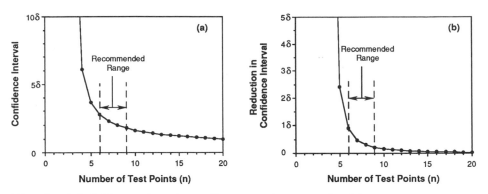

Fig. 1 (a) Effect of number of test points on the 95% confidence interval at $X = X_a$; (b) decrease in confidence interval for n test points compared with (n-1) test points.

In summary, in planning the correlation testing program, six to nine strength levels should be used. The strength levels should be evenly spaced, and the strength range should include the anticipated strengths to be estimated in the field.

3.2 Number of replications
The number of replicate tests at each strength level affects the uncertainty of the average values. The standard deviation of the average value varies with the inverse of the square root of the number of replicate tests used to obtain the average. The effect of the number of tests on the precision of the average is similar to that shown in Fig. 1(a). Thus the number of replicate tests is also governed by considerations of precision and economy.

The required number of replicate tests depends on the within-test variability of the method (also called repeatability), the allowable error between the sample average and the true average, and the confidence level that the allowable error is not exceeded. This number can be established from statistical principles (e.g., ASTM E 122, Practice for Choice of Sample Size to Estimate the Average Quality of a Lot or Process). Alternatively, the number can be based upon accepted practice. For acceptance testing, ACI 318 [2] considers a test result as the average strength of two molded cylinders. Therefore, for the correlation testing, two replicate standard compression tests are adequate for measuring the average strength at each level.

The number of companion in-place tests at each strength level is governed by the competing factors of precision and economy. The ACI report [4] suggests using a number of replicates such that the average values of the in-place test results have a comparable precision to the average compressive strength. If the coefficients of variation of the standard compression test and of the in-place test method are CV_y and CV_x, respectively, the ratio of the number of tests should be:

$$\frac{n_x}{n_y} = \left(\frac{CV_x}{CV_y} \right)^2 \qquad (5)$$

where,

n_x = number of replicate in-place tests, and
n_y = number of replicate compression tests.

Table 1 lists recommended values of within-test coefficients of variation for various tests [4]. These values may be used for planning a correlation testing program. For example, for correlation testing involving cylinders and the pullout test, the replicate number of pullout tests at each strength level would be $2 \cdot (8/4)^2 = 8$.

3.3 Test specimens

Ideally, it would be desirable to determine the compressive strength and the in-place result on the same specimen. Unfortunately, this is only possible with those methods that are truly nondestructive, such as pulse velocity and rebound number. For other methods, separate specimens are needed for measuring compressive strength and the in-place test result. The ACI report provides recommendations on the type of specimens for pullout and probe penetration tests [4].

Table 1   Within-test variability for various test methods [4]

Test Method	Within-Test Coefficient of Variation (percent)
Cylinder Compression (ASTM C 39)	4
Core Compression (ASTM C 42)	5
Pullout (ASTM C 900)	8
Probe penetration (ASTM C 803)	5
Rebound hammer (ASTM C 805)	12
Pulse velocity (ASTM C 597)	2

It is important that companion specimens are tested at the same maturity. This is especially critical for early-age tests, when strengths vary significantly with age and are dependent on the previous thermal history. The problem arises because of differences in early-age temperature rise in specimens of different geometries. A simple approach for moderating differences in specimen temperatures is to cure all specimens in water baths. The high heat capacity of the water will absorb the heat of hydration without a significant rise in temperature. Failure to perform compression and in-pace tests on companion specimens of equal maturity will result in an inaccurate correlation relationship, which will lead to systematic errors (or biases) when used to estimate the in-place strength in a structure.

3.4 Regression analysis
After the data are obtained, the correlation relationship is determined by regression analysis of the average test results at each strength level. Historically, most correlation relationships have been modeled as straight lines, and ordinary least squares (OLS) analysis has been used to estimate the slopes and intercepts of the lines. However, OLS-analysis is based on two assumptions:

- There is no error in the X-value.
- The error (standard deviation) in the Y-value is constant.

The first of these assumptions is violated because, as shown in Table 1, in-place tests (X-value) generally have greater within-test variability than compression tests (Y-value). In addition, it is generally accepted that the within-test variability of standard cylinder compression tests is described by a constant coefficient of variation [7]. Therefore, the standard deviation increases with increasing compressive strength, and the second assumption is also violated. The end result is that OLS-analysis leads to errors in the estimated parameters of the correlation relationship. There are approaches for dealing with these problems.

The problem of increasing variability of compressive strength as the average strength increases is discussed first. If test results having a constant coefficient of variation are transformed by taking their natural logarithms, the standard deviation of the logarithmic values will be constant [7]. Thus the second assumption of OLS can be satisfied by regression analysis using the average of the logarithms of the test results at each strength level. If a linear relationship is used, it would be as follows:

$$\ln C = a + \beta \ln i \tag{6}$$

where,
$\ln C$ = average of natural logarithms of compressive strengths,
a = intercept of line,
β = slope of line, and
$\ln I$ = average of natural logarithms of in-place test results.

By taking the anti-logarithm, Eq. (6) becomes a power function:

$$C = e^a I^\beta = A I^\beta \tag{7}$$

The exponent, β, determines the degree of non-linearity of the untransformed correlation relationship. If $\beta=1$, the correlation relationship is a straight line passing through the origin with a slope = A. If $\beta \neq 1$, the relationship is curved upward or downward, depending on whether β is greater than or less than one. Regression analysis using the natural logarithms of the test results provides two benefits: (1) it satisfies one of the underlying assumptions of OLS-analysis, and (2) it allows for a non-linear correlation relationship, if it is needed.

Regression analysis which accounts for X-error can be performed with little additional effort compared with OLS-analysis. One such procedure was proposed by Mandel [8] and was used by Stone and Reeve [9] to develop a rigorous procedure for estimating the in-place characteristic strength. Mandel's approach involves the use of a parameter, λ, which is defined as the variance (square of the standard deviation) of the Y-variable divided by the variance of the X-variable. For the correlation testing program, the value of λ would be obtained from the squares of the average (pooled)[b] standard deviations of the average compressive strengths and in-place results. If the number of replications have been chosen so that average values are measured with comparable precision, the value of λ should be close to one.

Mandel's procedure also allows for the possibility of correlation between the errors associated with the average X- and Y-values. For the case where the compression tests and in-place tests are performed on separate specimens, there should be no correlation between the errors. For the case where tests are performed on the same specimens (e.g., using the rebound method), there could be some correlation. In this study, it was assumed that there was no correlation between the errors.

Mandel's procedure is explained in Ref. [8] and will not be repeated here in detail; however, the principle is discussed. The parameter λ and the correlation test results (averages of the logarithms of the in-place results (X-values) and average of logarithms of compressive strengths (Y-values)) are used to calculate two constants k and b. These calculations use the usual sum of squares and cross-products used in OLS-analysis. The constants k and b are used to transform the X- and Y-values into U- and V- values as follows:

$$U = X + k Y \tag{8a}$$
$$V = Y - b X \tag{8b}$$

The key feature is that this transformation results in U- and V-values which satisfy the two assumptions for OLS-analysis, and a straight line is fitted to the transformed data using OLS-analysis. The results are transformed back to the (X,Y) scales, and one obtains estimates of a and β in Eq. (6) along with the standard deviations of the estimates. The estimate of β is equal to the constant b determined in the initial step of the analysis. The most important feature of Mandel's analysis is that the estimated standard deviation of the predicted value of Y for a new value of X accounts for error in the new X-value as well as the error of fit. The procedure can be implemented using a hand-held programmable calculator or a spreadsheet program as was done for this paper.

The difference between OLS-analysis and Mandel's procedure is illustrated in Fig. 2. In OLS-analysis, the best-fit straight line is the

---

[b] If the same number of replicate tests are used at each strength level, the average standard deviation is used. If the numbers differ because of rejection of outliers, the pooled standard deviation must be calculated accounting for the number of replicates at each strength level [8].

one which minimizes the sum of squares of the vertical deviations of the data points from the line, as shown in Fig. 2(a). On the other hand, Mandel's analysis minimizes the sum of squares of deviations along a direction inclined to the straight line, as shown in Fig. 2(b). The inclination of the direction of minimization depends on the value of λ. As λ decreases, there is more error in the X-values, and the angle θ in Fig. 2(b) increases.

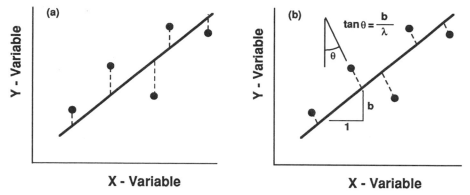

Fig. 2 Comparison of least-squares fit using (a) OLS-analysis and (b) Mandel's method [8].

Table 2 Average of natural logarithms of pullout-loads and cylinder compressive strengths; concrete made with 19-mm river gravel [10]

Pullout Load(P) (kN)	Compressive Strength (C) (MPa)	ln (P)	ln(C)
9.72	10.41	2.274	2.343
12.18	14.21	2.500	2.654
16.62	18.69	2.811	2.928
19.97	22.83	2.994	3.128
27.13	28.34	3.301	3.344
30.12	31.66	3.405	3.455
34.09	40.07	3.529	3.691
36.00	42.90	3.584	3.759
Pooled S.D.		0.112	0.037

Results from a pullout-test correlation study [10] performed at the U.S. National Bureau of Standards (NBS) will be used to illustrate the difference between the results of OLS-analysis and Mandel's procedure. Table 2 lists the averages of the natural logarithms of 11 pullout tests and five cylinder compression tests performed at each of eight strength levels. Also shown are the pooled standard deviations of the natural logarithms of the replicate test results. The value of λ is as follows:

$$\lambda = \frac{\dfrac{(0.037)^2}{5}}{\dfrac{(0.112)^2}{11}} = 0.24$$

In this case, λ is less than one because five replicate compression tests were performed at each strength level, which resulted in less error in the average compressive strength compared with the average pullout load.

The logarithm values in Table 2 were used to obtain the best-fit straight lines using OLS-analysis and Mandel's procedure with $\lambda = 0.24$. The results are shown in Table 3. The data and best-fit lines are plotted in Fig. 3(a). The lines obtained by the two regression methods are practically identical and cannot be distinguished in Fig. 3(a). So it might be argued that there is no need to resort to the more rigorous method of analysis since the same best-fit line is obtained. However, this argument does not consider the uncertainty associated with the estimate of Y for a new X. For example, consider the case where the average of the natural logarithm of 11 pullout tests is 3.0 and the standard deviation is 0.11. Table 3 shows that the estimated values of the logarithms of compressive strength are equal, but Mandel's procedure results in greater uncertainty in the estimate. The standard deviation of the estimate and the appropriate Student's t-value can be used to calculate the lower confidence limit of the average compressive strength. Table 3 shows the lower limits for a 5% significance level. Thus accounting for the X-error results in a lower value for the estimated Y at a particular confidence level.

Table 3 Results of regression analysis of data in Table 2

	OLS	Mandel
a	0.065	0.037
S_a	0.134	0.134
β	1.016	1.025
S_β	0.043	0.044
Y at X=3.0	3.112	3.112
SD_Y at X=3.0	0.020	0.054
5% lower limit	3.074 (21.6 MPa)	3.006 (20.2 MPa)

Figure 3(a) compares the lower confidence limits obtained with the two analysis procedures. In Fig. 3(b), the results have been converted back into real units. It is seen that the uncertainty of the estimated average compressive strength increases with increasing concrete strength. This is because the within-test coefficient of variation is assumed to be constant.

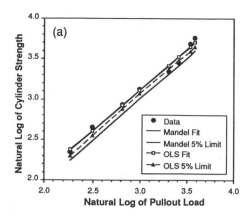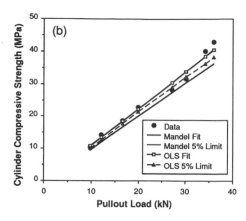

Fig. 3 Correlation relationship and lower confidence limits (0.05 significance level) using OLS-analysis and Mandel's analysis: (a) natural logarithm transformation and (b) real units.

In summary, it is recommended that regression analysis should be performed using the natural logarithms of the test results. This will accommodate the increase in within-test variability with increasing compressive strength. By so doing, the correlation data is fitted with a power function rather than a straight line. Use of the transformed data implies that concrete strength is distributed as a log-normal rather than a normal distribution. It has been argued that possible errors from this assumption are not significant [9]. Regression analysis should be performed using a procedure that accounts for X-error. Failure to account for X-error results in a low estimate of the uncertainty associated with estimates of average compressive strength.

4 Estimate of in-place characteristic strength

Having established the correlation relationship and its associated uncertainty, the next step is to use this relationship and the results of tests performed on the structure to estimate the in-place strength of the concrete. The objective is to estimate the characteristic strength which has a high probability of being exceeded. Thus it is necessary to account for the uncertainties of the correlation relationship and of the average value of the in-place test results.

The first question that has to be answered is how many in-place tests are required to estimate the characteristic strength in a given placement. As was the case in selecting the number of replicate tests for correlation testing, the number of in-place tests is influenced by the desired precision and by economy. However, there are now two sources of variability to consider: (1) the within-test variability of the test method and (2) the batch-to-batch variability of the concrete in the structure. The ideal case would be to perform enough tests at a single location so that the average in-place result is measured with sufficient precision at that location, and there should be sufficient test locations so that the average property of the concrete in the structure is measured with sufficient precision. For a test like the rebound number method, this approach could be used. However, for methods such as pullout, probe penetration, or break-off, such a scheme would result in a prohibitively expensive testing program.

To arrive at a practical answer to the question of how many tests, we can rely on the accepted practices for quality control based on standard-cured specimens. Codes of practice generally require a minimum number of standard tests per specified volume of concrete in the placement. Thus the number of in-place tests in a given placement can be determined by using Eq. (5) and Table 1. This concept has been discussed further in Ref. [10].

After the in-place tests are performed, the final step is to analyze the results and estimate the characteristic strength in the structure. This is probably the most critical phase of an in-place testing program, but it has not been studied as much as other aspects of in-place testing. One of the procedures that is being used is based on the concept of the tolerance limit [11]. The lower tolerance limit represents the value that is expected to be exceeded by a certain proportion of the population with a prescribed confidence level [5]. Thus the characteristic strength, i.e., the strength expected to be exceeded by 90% of concrete in the structure, would be:

$$C_{0.1} = C_a - K_{0.9,\gamma} \ S_{cf} \tag{9}$$

where,

$C_{0.1}$ = characteristic strength,
$\quad C_a$ = average compressive strength,
$K_{0.9,\gamma}$ = one-sided tolerance factor for proportion 0.9 and confidence level γ, and
$\quad S_{cf}$ = standard deviation of concrete strength in the structure.

In applying the tolerance limit approach to evaluate in-place test results [11], the average concrete strength is obtained by using the average of the in-place results and the correlation relationship. The value of S_{cf} is taken as the standard deviation of the individual strengths estimated for each in-place test result. The tolerance limit approach has been criticized [10] because: Eq. (9) does not account for the uncertainty in the correlation relationship, and the approach assumes that the variability of concrete strength equals the variability of the in-place test results. Thus it felt that Eq. (9) does not have a sound statistical basis for evaluating in-place test results.

To overcome the deficiencies of the tolerance limit approach, the NBS developed a rigorous statistical procedure to estimate the in-place characteristic strength [9]. Basically, the NBS procedure estimates the expected characteristic strength and its uncertainty. From these estimates, one determines the value of the characteristic strength that is expected to be exceeded with a desired level of confidence. The procedure is complex, and it is well-suited for implementation on a personal computer. The characteristic strengths computed by using the rigorous approach were compared with the values computed by using the tolerance limit method [10]. The comparison showed that the tolerance limit approach resulted in lower estimates of in-place characteristic strength, especially when the variability of the in-place results was high. While this may be acceptable for safety, it can lead to unnecessary delays in the construction schedule.

A simpler technique than the rigorous NBS method was developed which accounts for the various sources of uncertainty in estimating the in-place characteristic strength [12]. The simplified method was designed for use with spreadsheet software and was shown to be slightly conservative compared with the rigorous method. Although it was shown to be adequate, the simplified method was not rigorous because OLS-analysis was used, and an empirical approach was used to arrive at the procedure to compute the value of the characteristic strength for a 95% confidence level.

In this paper, a new method is proposed which retains the main features of the rigorous NBS method, but which can be readily implemented using spreadsheet software. The basic approach in the new method is illustrated in Fig. 4(a). Mandel's procedure is used to obtain the correlation relationship and its uncertainty. The results of the in-place tests are used to compute the lower 5% confidence limit for the estimated average in-place strength. The characteristic strength is determined assuming a log-normal distribution for the in-place concrete strength. Calculations are performed using natural-logarithm values, and the last step is to convert the estimated characteristic strength into real units.

The lower 5% confidence limit is obtained using Mandel's formula [8] for the standard deviation of an estimated value of Y for a new X. The formula has been modified according to Ref. [9] so that it incorporates the variability of the average in-place result from tests on the structure. The standard deviation is:

$$S_Y^2 = [\frac{1}{n} + (1 + k b)^2 \frac{(X - X_a)^2}{S_{uu}}] \ S_e^2 + b^2 \ S_X^2 \qquad (10)$$

where,

S_Y = standard deviation of estimated value of Y (average concrete strength),

n = number of points used to obtain the correlation relationship,

b = estimated slope of the correlation relationship,

k = b/λ, where λ is obtained from the within-test variability during correlation testing,

X = average of results of in-place tests performed on the structure,

X_a = average of X-values during correlation tests,

S_e = residual standard deviation of correlation relationship,

S_{uu} = sum of the squares of the deviations of the U-values (Eq. 8(a)) from the average, and

S_X = standard deviation of average of in-place tests performed on the structure.

It is seen that the S_Y is composed of two parts: the first part considers the uncertainty of the correlation relationship and the second part considers the variability of the in-place test results obtained from testing the structure. Because Eq. (10) is the sum of two variances which may have different degrees of freedom, a formula has been suggested for computing the effective degrees of freedom for S_Y [9]. For simplicity, this new method assumes that there are (m-1) degrees of freedom

for S_Y, where m is the number of in-place tests performed on the structure.

The lower confidence limit for the average concrete strength is as
follows:

$$Y_{low} = Y - t_{\nu,\alpha} S_Y \tag{11}$$

where,
$\quad Y_{low}$ = lower confidence limit at significance level α,
$\quad\quad Y$ = transformed average concrete strength for new value of X (the
$\quad\quad\quad$ average of the in-place results),
$\quad t_{\nu,\alpha}$ = Student's t-value for ν = m-1 degrees of freedom and signi-
$\quad\quad\quad$ ficance level α.

The next step is to estimate the transformed characteristic strength
based on the average strength obtained from Eq. (11). It is assumed that
the in-place compressive strength follows a log-normal distribution, so
that the characteristic strength is computed as follows:

$$Y_{0.9} = Y_{low} - 1.282 S_{cf} \tag{12}$$

where,
$\quad Y_{0.9}$ = logarithm of strength expected to be exceeded by 90% of the
$\quad\quad\quad$ population,
$\quad S_{cf}$ = standard deviation of logarithm of concrete strength in the
$\quad\quad\quad$ structure.

The value of S_{cf} is obtained from the assumption that the ratio of the
standard deviation of compression strength to the standard deviation of
in-place test results has the same value in the field as was obtained
during the laboratory correlation testing [8,9,12]. Thus the following
relationship is assumed:

$$\frac{S_{cf}}{S_{if}} = \frac{S_{cl}}{S_{il}} \tag{13}$$

where,
$\quad S_{cf}, S_{cl}$ = standard deviation of logarithms of compressive strength
$\quad\quad\quad$ in the field and laboratory, respectively, and
$\quad S_{if}, S_{il}$ = standard deviation of logarithms of the in-place results
$\quad\quad\quad$ in the field and in the laboratory, respectively.

The final step is to convert the result obtained from Eq. (12) into
real units by taking the anti-logarithm.

To evaluate the new method compared with the rigorous NBS procedure,
a simulation was performed using the pullout test. Ten replicate pullout
loads were randomly generated for three strength levels and for two
levels of variability [10]. These simulated test results were used to
obtain the characteristic strength (5% significance level) by the procedure described above (see also Fig. 4(a)). The calculations were performed for the four sets of correlation data reported in Ref. [10], and
the results are summarized in Table 4. Figure 4(b) compares the characteristic strengths estimated by the two procedures. It is seen that the
proposed method gives nearly the same estimates as the rigorous method.

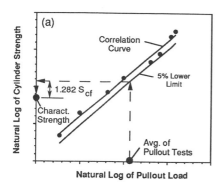 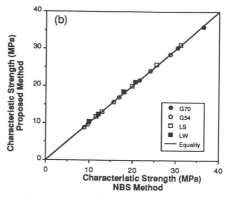

Fig. 4 (a) Schematic of the new procedure to estimate characteristic strength; (b) comparison of estimates using NBS method and new procedure.

Table 4 Comparison of in-place characteristic strengths using proposed method and NBS method [9]

Test Series† [1]	In-Place Pullout Results Avg. Load (kN) [2]	C.V. (%) [3]	Pred. Avg.Str. (MPa) [4]	Characteristic Strength NBS Method 5% Risk (MPa) [5]	Proposed Method (MPa) [6]	Difference (6-5)/5 (%) [7]
G70	13.39	10.9	14.8	13.0	13.0	0.0
G70	13.48	20.7	14.7	11.7	11.7	0.0
G70	22.12	11.9	24.6	21.6	21.5	-0.3
G70	22.52	20.3	24.8	19.9	19.9	0.0
G70	36.13	9.9	40.8	36.3	36.0	-0.8
G70	36.09	21.9	39.8	30.3	30.3	0.0
G54	13.39	10.9	11.6	9.8	9.7	-1.4
G54	13.48	20.7	11.5	9.0	8.9	-0.8
G54	22.12	11.9	19.6	16.9	16.8	-0.8
G54	22.52	20.3	19.7	15.7	15.6	-0.4
G54	36.13	9.9	32.9	28.7	28.6	-0.5
G54	36.09	21.9	32.1	24.1	24.1	0.0
LS	13.39	10.9	14.9	13.0	13.0	0.0
LS	13.48	20.7	14.9	11.6	11.7	1.2
LS	22.12	11.9	23.0	19.9	20.0	0.7
LS	22.52	20.3	23.1	18.1	18.3	1.5
LS	36.13	9.9	35.3	31.1	31.2	0.4
LS	36.09	21.9	34.6	25.4	25.9	1.9
LW	13.39	10.9	15.0	12.1	12.3	1.1
LW	13.48	20.7	14.9	10.1	10.5	3.4
LW	22.12	11.9	26.1	20.6	21.0	1.7
LW	22.52	20.3	26.2	17.9	18.4	3.1

†G70 = river gravel aggregate, apex angle = 70°; G54 = river gravel aggregate, apex angle = 54°. LS = limestone aggregate and LW = lightweight aggregate, both with apex angle = 70°.

5 Summary

This paper has discussed statistical methods to estimate the characteristic strength of concrete based on the results of in-place tests. Mandel's regression analysis procedure which accounts for error in the X-value has been reviewed. The procedure should be used to obtain the correlation relationship for the particular construction project. Regression analysis should be performed using the natural logarithms of the results obtained from the correlation testing program; this accounts for the increasing within-test variability with increasing concrete strength. While ordinary-least-squares analysis results in approximately the same correlation relationship as with Mandel's procedure, the standard deviation of the predicted value of Y for a new value of X is underestimated using OLS-analysis.

A simplified version of the rigorous NBS method to estimate the in-place characteristic strength is proposed. Based on simulated in-place test data, the simplified approach yields estimates which are very close to those obtained by the NBS method. The calculations were performed by modifying a spreadsheet template previously developed for OLS-analysis [12]. The proposed method includes the key aspects of the rigorous procedure but is easier to implement. Thus it is felt that a balance has been attained between statistical rigor and practicality.

6 References

1. Lew. H.S., et al., "Investigation of Construction Failure of Reinforced Cooling Tower at Willow Island, WV," Nat. Bur. Stand. (U.S.), Bldg. Sci. Ser. 148, Sept. 1982, 158 pp.
2. "Building Code Requirements for Reinforced Concrete," ACI 318-83, American Concrete Institute, Detroit, MI.
3. Annual Book of ASTM Standards, V. 04.02, Concrete Aggregates, ASTM, Philadelphia, 1989.
4. "In-Place Methods for Determination of Strength of Concrete," Report ACI 228.1R-89, American Concrete Institute, Detroit, 1989, 26 pp.
5. "Recommended Practice for Evaluation of Strength Test Results of Concrete," ACI 214-77 (1983), American Concrete Institute, Detroit, MI.
6. Ku, H.H., "Notes on the Use of Propagation of Error Formulas," in Precision Measurement and Calibration - Statistical Concepts and Procedures, National Bureau of Standards SP 300, V. 1, 1969, pp 331-341.
7. Mandel, J., "Fitting Straight Lines When Both Variables are Subject to Error," Journal of Quality Technology, V. 16, No. 1, Jan. 1984, pp.1-14.
8. Stone, W.C. and Reeve, C.P., "A New Statistical Method for Prediction of Concrete Strength from In-Place Tests," ASTM Journal of Cement, Concrete, and Aggregates, Summer 1986, V. 8, No. 1, pp 3-12.
9. Stone, W.C., Carino, N.J. and Reeve, C.P., "Statistical Methods for In-Place Strength Predictions by the Pullout Test," ACI Journal, V. 83, No. 5, Sept./Oct. 1986, pp. 745-756.

10. Hindo, K. R. and Bergstrom, W. R., "Statistical Evaluation of the In-Place Strength of Concrete," Concrete International, V. 7, No. 2, February 1985, pp 44-48.
11. Natrella, M.G., _Experimental Statistics_, National Bureau of Standards Handbook 91, 1963.
12. Carino, N.J. and Stone, W.C., "Analysis of In-Place Test Data with Spreadsheet Software," _Computer Use for Statistical Analysis of Concrete Test Data_, ACI SP-101, P. Balaguru and V. Ramakrishnan, Eds., American Concrete Institute, Detroit, 1987, pp. 1-26.

PART EIGHT
CONCLUSION

35 RESULTS OF THE WORKSHOP AND SUGGESTIONS FOR FUTURE WORK

H.W. REINHARDT
Darmstadt University of Technology, Darmstadt, Germany

Abstract
The workshop has been evaluated with respect to four aspects: Control of the building process, quality control, prediction of future performance, and fundamental questions of in-place testing. Future activities are suggested which may be tackled by active RILEM Technical Committees and/or new committees.

1 Relevance of testing during construction

The aim of testing is generally to improve the quality of a product. This holds also for testing during construction. There are four main items which have been dealt with during the workshop and which have been discussed. First, testing enables an active building control from setting of concrete until prestressing of tendons. Second, it allows to reject poor quality material and workmanship. Third, testing can lead to an objective prediction of future performance. And fourth, finally, in-situ testing has to have a reliability scheme. All four aspects will receive due attention in the following chapters.

2 Control of the building process

An important property of fresh concrete is workability since this determines for a great deal the correct placing and compaction of concrete. Although workability is not a clearly defined property, there is a certain time when the end of workability is reached for a given compaction procedure. Since experience tells us that the knowledge of beginning and end of setting according to Vicat as measured on cement does not correlate with the end of workability, i.e. with that time at which the mix cannot compacted appropriately any more. Especially during slipform construction where small quantities of concrete are casted the concrete has to wait a long time before it can be placed completely. Two papers have dealt with this subject and made objective proposals for measuring the end of workability. One applies ultrasonic waves in the fresh concrete and shows that the propagation velocity correlated strongly with setting. The other measured the temperature rise due to heat of hydration and receives good correlation between rise of 1 K and setting of concrete on site. Both types of measurements are easy to perform and allow to control the appropriate slipping rate on the job site.

Thick walled structures suffer often from excessive temperature rise due to heat of hydration and from eigenstresses during cooling or from restraint of connected structural parts also during cooling. It is the state of young concrete between 2 and 4 days when the mechanical properties and the eigenstresses and restraint stresses develop. Research results have been presented on appropriate maturity functions which allow the prediction of strength as function of time and temperature. Furtheron, advanced theory and experimental results have been shown with respect to strength, stiffness, and creep of young concrete which enable the engineer to predict the probability of cracking in a thick walled structure during the cooling period. Interactive use of site measurements and software on the hydration process has been demonstrated. This hardening control system has been successfully applied at various building sites, i.e. artificial cooling has been controlled, stripping times and partial prestressing times have been determined and cracks have been reliably prevented. The workshop has shown that a large progress has been achieved in the last ten years with respect to young concrete and the control of cracks due to heat of hydration.

The same goal is envisaged by surface pullout tests. They give a proper indication of the strength present in-situ and allow to determine early loading operations and stripping times. The methods have been used for many years and are now accepted in a large number of countries. In Denmark, they may be applied instead of the standard cube test even as an acceptance test.

Curing of concrete is essential for the quality of the skin of a structure and hence for many properties related to durability. Although this subject will be treated in Chapter 4 of this resumé, it would belong partly to the items of control of the building process. For instance, if it were demonstrated by appropriate measurements that the required skin quality is not reached due to low temperature, slow type of cement, or improper curing methods, curing could be resumed until sufficient curing is achieved. This aspect is rather new and needs more consideration with respect to testing techniques, scatter of results, and importance as quality indicator.

3 Quality control

Quality refers to materials to be used, to the concrete mix, to cover of concrete, to filling of prestressing ducts, i.e. to materials and workmanship. Testing during construction should not substitute production control since production control is defined very accurately and if carried out correctly can reliably guarantee good quality of materials such as cement, aggregate, reinforcing steel and of concrete mix. Production control can be improved if all steps of checking and all results are formalized and collected on data sheets. However, it does not guarantee appropriate compaction and curing of concrete, placing of reinforcing steel and prestressing of tendons. Therefore, testing during construction is an additional means for providing enhanced quality by rejecting poor quality material and by improving workmanship after first occurance of deficiencies. It is essential that testing takes place as early as possible in the construction process.

During the workshop, a complex survey has been given on the available methods to measure concrete content, water content or water/cement ratio. Furthermore, the determination of flyash in a concrete mix by flotation has been presented which works with little preinformation about the properties of the constituents in a mix. Two newly developed methods were discussed which allow the determination of entrained air either by volume or by air void size distribution in fresh concrete. Compared with the usual method applied to hardened concrete which is time consuming and rather late, the new methods allow the immediate test before concreting and thus a correction of the subsequent mixes.

Bad compaction of concrete can be detected by radiographic methods which are well developed and by ultrasonic and impact echo methods. The first method can be applied on green and young concrete while the two latter methods are suitable on young and hardened concrete only.

There are many testing methods and commercial devices for measuring concrete cover. Some of them allow determination of cover thickness and size of bar, however, with varying accuracy. The detection of voids in partly filled prestressing ducts by means of the impact echo method was convincingly presented. The methods on steel and ducts are nondestructive; besides the production of objective data, the methods are important psychologic tools on a building site for enhanced quality consciousness.

4 Prediction of future performance

Discussing durability can be devided into potential durability and attainable durability. The appropriate selection of materials and mix design determine the potential durability which can be achieved if optimally manufactured, compacted and cured. Opposite to that, the attainable durability is less due to deficiencies in manufacture and curing. The design aims at the potential durability but, usually, it is not checked at the realized structure.

Methods were presented and discussed which allow the measurement of air and/or water permeability through the skin of a concrete structure. Since permeability is correlated to diffusion, degradation processes which are diffusion based can be related to these surface measurements. An example was given with respect to carbonation of Portland cement concrete in a hot climate.

Of course, cover of reinforcement is also durability related. However, the testing methods have already been tackled in the preceeding chapter.

It turned out during the workshop that a need for durability related testing exists but that data are lacking which correlate measuring results with expected or experienced durability.

5 Fundamental questions of in-place testing

Extensive data of the strength distribution in buildings have been presented. They show systematic influences of blending, segregation and overburden of concrete on strength. Furtheron, there are errors inherent to the testing method and to the calibration of the devices. Finally, scatter occurs due to nonuniform mix and compaction distribution. All these effects are neglected in the structural design because strength is related to the standard cube strength as measured on separate specimens.

On the other hand if strength is assessed by in-place testing, a reliability scheme should be available which links measuring results to design requirements. Two approaches were presented: in Denmark, 80 % of the specified strength has to be reached in-place as an acceptance criterion while, in the United States of America, a rigorous statistical procedure has been developed which takes the scatter of standard cubes and in-place measurements into account.

It was agreed that in-place testing deserves most consideration with respect to the number of measurements, the testing devices with their inherent accuracy and error, and the systematic difference between building parts.

6 Topics related to active RILEM Technical Committees

The preceeding chapters contain several topics which fit into the framework of active Technical Committees of RILEM. The following committees are addressed:

116 Permeability of concrete as a criterion of its durability

119 Avoidance of thermal cracking in concrete at early ages

122 Microcracking and lifetime performance of concrete

126 In-place testing of hardened concrete

The workshop results will be made available to those committees. Furtheron, the Technical Advisory Committee and the Coordinating Committee of RILEM may use the results for their steering activities.

7 Future activities proposed

There are several topics which may not be covered by active TCs which deserve more attention. These topics are the following.

7.1 End of workability of fresh concrete

There is no clear definition of end of workability and, furthermore, the test on cement - beginning and end of setting according to Vicat - does not correlate with relevant engineering properties of concrete. The method of ultrasonic measurement of end of workability should be promoted. Probably, other methods should be evaluated as well. It is proposed to set up a new committee on this subject. If it is true as has been shown by experiments and an estimating theoretical approach that there is a constant value of ultrasonic pulse velocity for the end of workability, this would be a simple and attractive engineering tool for the construction site.

7.2 Impact-echo method

It has been demonstrated that a short impact produces a pulse of a few micro seconds which is reflected from places with lower acoustic impedance such as voids. The echo is received and analyzed by fast Fourier transform. The shape of the reflected pulse contains information about the depth and the size of the void. This promising method can be extended to three-dimensional monitoring and crack detection. Further evaluation of this method seems promising.

7.3 Cover of reinforcement

There are numerous devices available for the detection of cover thickness and reinforcing bar diameter. It would be useful to collect the knowledge and to write a state-of-the-art report on this subject.

7.4 Reliability scheme of in-place measurements

Although not confined to early age testing, in-place testing can be considered as a method of testing during construction. If those measuring results are used as acceptance control an appropriate scheme should be developed taking account of the scatter of properties and calibration errors of the instruments. Since this topic touches material and structures, a joint committee may be proposed consisting of members of RILEM and CEB. This topic should include statistical methods for the use of in-place tests for strength determination during critical phases of construction.

7.5 Interrelationship of early age characteristics of concrete measured during construction to its long term engineering properties and durability (Dr. R. Swamy, The University of Sheffield)

The objectives of the proposed committee are:

1. To evaluate and identify those early age characteristics of concrete that are relevant and significant to its subsequent development of engineering properties and durability.

2. To examine and evaluate nondestructive test methods capable of evaluating properties prior to the end of workability stage and beyond ie, during the plastic and hardened states (up to 24 hrs say).

3. To identify those early age characteristics that will have distinct influence and can form an early age index of long term stability; To establish test methods for their statistically significant measurement,
and
To establish the relationship between these early age properties and engineering performance and durability.

8 Acknowledgement

The author is grateful to all participants of the workshop for their valuable suggestions. Special thanks are due to the following persons who provided written comments which have been used in this part of the proceedings: Dr. Carino, Prof. Naaman, Dr. Petersen, Dr. Pomeroy, Dr. Swamy, Prof. Tassios.

INDEX

This index has been compiled using keywords allocated to the papers by the Concrete Information Service of the British Cement Association. The assistance of the Association in this is gratefully acknowledged. The numbers refer to the opening page of the papers.

Site testing 49, 82, 138, 175, 253, 280, 354, 432
Slipforms 162
Standard deviation 97, 417
Standards 97
Statistics 97, 150, 432
Strain 207, 384
Strength 253
Strength estimation 58
Stress 207
Surface area 111
Surface zone 311, 333, 354

Temperature 14, 122, 162, 192, 207, 224, 311, 343
Temperature difference 243
Tensile strength 207, 354, 389
Tensile strength 207, 354, 389
Tensile stress 389
Testing 3, 19, 40, 58, 92, 162, 311, 384, 451
Testing equipment 253, 333, 369, 384, 389, 396
Test specimens 253, 417
Thermal analysis 243
Thermal cracking 207
Time 192

Ultrasonic pulse velocity testing 122, 175, 280, 354, 396

Vibration 384
Viscoelasticity 207
Void detection 369

Water 343
Water absorption 311
Water/cement ratio 19, 58, 111, 122, 150, 280
Water content 19, 58, 138, 150
Water demand 97, 111
Water pressure 354
Weight batching 35

X-radiation 396